Lecture Notes in Computer Science   8840

Commenced Publication in 1973
Founding and Former Series Editors:
Gerhard Goos, Juris Hartmanis, and Jan van Leeuwen

T0212923

Radha Poovendran   Walid Saad (Eds.)

Decision and Game Theory for Security

5th International Conference, GameSec 2014
Los Angeles, CA, USA, November 6-7, 2014
Proceedings

 Springer

Volume Editors

Radha Poovendran
University of Washington
Electrical Engineering Department, Network Security Lab
Box 352500, Seattle, WA 98195-2055, USA
E-mail: rp3@uw.edu

Walid Saad
Virginia Tech, Department of Electrical and Computer Engineering
Whittemore Hall 302, 1185 Perry Street, Blacksburg, VA 24061, USA
E-mail: walids@vt.edu

ISSN 0302-9743   e-ISSN 1611-3349
ISBN 978-3-319-12600-5   e-ISBN 978-3-319-12601-2
DOI 10.1007/978-3-319-12601-2
Springer Cham Heidelberg New York Dordrecht London

Library of Congress Control Number: 2014951399

LNCS Sublibrary: SL 4 – Security and Cryptology

Typesetting: Camera-ready by author, data conversion by Scientific Publishing Services, Chennai, India

Printed on acid-free paper

Springer is part of Springer Science+Business Media (www.springer.com)

Preface

Recent advances in networking, communications, computation, software, and hardware technologies have revolutionized the information technology landscape. Indeed, this cyberspace has become an integral part of every person's daily life and the way we conduct business. Protecting the sensitive content of every nation's cyberspace infrastructure has thus become critical to ensure economic growth, prosperity, and advancement. However, the heterogeneous, dynamic, and large-scale nature of modern-day networked and information technology infrastructure warrants novel analytical and practical approaches for securing its assets and maintaining its trustworthiness.

Owing to its powerful analytical and modeling frameworks, game theory has recently emerged as a key tool for building resilient, secure, and dependable networked systems. Coupled with synergistic techniques such as dynamic control, mechanism design, and economics, game theory, along with incentives and mechanisms design, is expected to provide one of the pillars of the "science of security." The proceedings of the GameSec 2014 conference contain original contributions presenting theoretical and practical contributions that will build the knowledge base in the science of security, in general, and game-theoretic security, in particular.

The topics cover multiple facets of cybersecurity that include: rationality of adversary, game-theoretic cryptographic techniques, vulnerability discovery and assessment, multi-goal security analysis, secure computation, economic-oriented security, and surveillance for security. These aspects are covered in a multitude of domains that include networked systems, wireless communications, border patrol security, and control systems. The GameSec conference aims to provide a forum for engineers, game and control theorists, and computer scientists from research and industry to contribute and develop the foundation of the next generation of security science. We are pleased to bring these proceedings to your hands. Enjoy!

September 2014
Walid Saad
Radha Poovendran

Organization

Steering Board

Tansu Alpcan	University of Melbourne, Australia
Nick Bambos	Stanford University, USA
John S. Baras	University of Maryland, USA
Tamer Başar	University of Illinois at Urbana-Champaign, USA
Anthony Ephremides	University of Maryland, USA
Jean-Pierre Hubaux	EPFL, Switzerland

Organizers

General Chair

Radha Poovendran   University of Washington, USA

TPC Chair

Walid Saad   Virginia Tech, USA

Publicity Chair

Quanyan Zhu   NYU, USA

Publication Chair

Rahul Jain   USC, USA

Local Arrangements Chair

Terry Benzel   USC, USA

Web Chair

Hao Shi   USC, USA

Finance Chair

Albert Jiang   USC, USA

Technical Program Committee

TPC Chair

Walid Saad   Virginia Tech, USA

Table of Contents

Full Papers

Short Papers

Full Papers

Full Papers

Defending Against Opportunistic Criminals:
New Game-Theoretic Frameworks and Algorithms

Chao Zhang[1], Albert Xin Jiang[1] Martin B. Short[2],
P. Jeffrey Brantingham[3], and Milind Tambe[1]

[1] University of Southern California, Los Angeles, CA 90089, USA
{zhan661,jiangx,tambe}@usc.edu
[2] Georgia Institute of Technology, Atlanta, GA 30332, USA
mbshort@math.gatech.edu
[3] University of California, Los Angeles, CA 90095, USA
pjb@anthro.ucla.edu

Abstract. This paper introduces a new game-theoretic framework and algorithms for addressing opportunistic crime. The Stackelberg Security Game (SSG), which models highly strategic and resourceful adversaries, has become an important computational framework within multiagent systems. Unfortunately, SSG is ill-suited as a framework for handling opportunistic crimes, which are committed by criminals who are less strategic in planning attacks and more flexible in executing them than SSG assumes. Yet, opportunistic crime is what is commonly seen in most urban settings. We therefore introduce the Opportunistic Security Game (OSG), a computational framework to recommend deployment strategies for defenders to control opportunistic crimes. Our first contribution in OSG is a novel model for opportunistic adversaries, who (i) opportunistically and repeatedly seek targets; (ii) react to real-time information at execution time rather than planning attacks in advance; and (iii) have limited observation of defender strategies. Our second contribution to OSG is a new exact algorithm EOSG to optimize defender strategies given our opportunistic adversaries. Our third contribution is the development of a fast heuristic algorithm to solve large-scale OSG problems, exploiting a compact representation. We use urban transportation systems as a critical motivating domain, and provide detailed experimental results based on a real-world system.

1 Introduction

Security is a critical societal challenge. We focus on urban security: the problem of preventing urban crimes. The Stackelberg Security Game (SSG) was proposed to model highly strategic and capable adversaries who conduct careful surveillance and plan attacks [1, 2], and has become an important computational framework for allocating security resources against such adversaries. While there are such highly capable adversaries in the urban security domain, they likely comprise only a small portion of the overall set of adversaries. Instead, the majority of adversaries in urban security are criminals who conduct little planning or surveillance before "attacking" [3]. These adversaries capitalize on local opportunities and react to real-time information. Unfortunately, SSG

R. Poovendran and W. Saad (Eds.): GameSec 2014, LNCS 8840, pp. 3–22, 2014.

is ill-suited to model such criminals, as it attributes significant planning and little execution flexibility to adversaries.

Inspired by modern criminological theory [3], this paper introduces the Opportunistic Security Game (OSG), a new computational framework for generating defender strategies to mitigate opportunistic criminals. This paper provides three key contributions. First, we define the OSG model of opportunistic criminals, which has three major novelties compared to SSG adversaries: (i) criminals exhibit Quantal Biased Random Movement, a stochastic pattern of movement to search for crime opportunities that contrasts with SSG adversaries, who are modeled as committed to a single fixed plan or target; (ii) criminals react to real-time information about defenders, flexibly altering plans during execution, a behavior that is supported by findings in criminology literature [4]; (iii) criminals display anchoring bias [5], modeling their limited surveillance of the defender's strategy. Second, we introduce a new exact algorithm, Exact Opportunistic Security Game (EOSG), to optimize the defender's strategy in OSG based on use of a markov chain. The third contribution of this work is a fast algorithm, Compact OPportunistic Security game states (COPS), to solve large scale OSG problems. The number of states in the Markov chain for the OSG grows exponentially with the number of potential targets in the system, as well as with the number of defender resources. COPS compactly represents such states, dramatically reducing computation time with small sacrifice in solution quality; we provided a bound for this error.

Thus, while OSG does share one similarity with SSG — the defender must commit to her strategy first, after which the criminals will choose crime targets — the OSG model of opportunistic adversaries is fundamentally different. This leads us to derive completely new algorithms for OSG. OSG also differs fundamentally from another important class of games, pursuit-evasion games (PEG) [6]; these differences will be discussed in more depth in the related work section.

While OSG is a general framework for handling opportunistic crime, our paper will use as a concrete example crime in urban transportation systems, an important challenge across the world. Transportation systems are at a unique risk of crime because they concentrate large numbers of people in time and space [7]. The challenge in controlling crime can be modeled as an OSG: police conduct patrols within the transportation system to control crime. Criminals travel within the transportation system for such opportunities [8], usually committing crimes such as thefts at stations, where it is easy to escape if necessary [9]. These opportunistic criminals avoid committing crime if they observe police presence at the crime location.

In introducing OSG, this paper proposes to add to the class of important security related game-theoretic frameworks that are widely studied in the literature, including the Stackelberg Security Games and Pursuit Evasion Games frameworks. We use an urban transportation system as an important concrete domain, but OSG's focus is on opportunistic crime in general; the security problems posed by such crime are relevant not only to urban crime, but to other domains including crimes against the environment [10], and potentially to cyber crime [11, 12]. By introducing a new model and new algorithms for this model, we open the door to a new set of research challenges.

2   Related Work

In terms of related work, there are three main areas to consider. First are Pursuit-Evasion Games (PEG), which model a pursuer(s) attempting to capture an evader, often where their movement is based on a graph [6]. However, PEG fail to model criminals who opportunistically and repeatedly strike targets as modeled using QBRM in OSG. Furthermore, in PEG, a pursuer's goal is to capture an evader while in OSG, the defender's goal is to minimize crime; additionally in PEG, the evader's goal is to avoid the pursuer and not seek crime opportunities as in OSG. These critical differences in behaviors of defenders and adversaries lead to new algorithms, i.e., EOGS and COPS, for OSG, that are fundamentally different from algorithms for PEG.

Second are SSG [13–15], which use a model of highly strategic adversaries to generate randomized patrol strategies. The SSG framework has been successfully applied in security domains to generate randomized patrol strategies, e.g., to protect flights [2], for security in the cyber realm [11, 12], and for counter-terrorism and fare evasion checks on trains [16, 17]. Recent work in SSG has begun to consider bounded rationality of adversaries [18] and incorporate some limited flexibility in adversary execution [15]. However, SSG [13–15], again, fails to model criminals who use real-time information to adjust their behavior in consecutive multiple attacks. In SSG, attackers cannot use real-time observation to decide whether to attack at the current time, nor can they use it to update beliefs and plan for their next consecutive attacks. Furthermore, SSG does not investigate efficient algorithms of deriving defender strategies against such opportunistic criminals. The Adversarial Patrolling Game (APG) [19], which is a variant of SSG, does consider the attacker's current observation. However, this game does not consider multiple consecutive attacks. It fails to model attacker's movement during multiple attacks and therefore the influence of current observation on future movement. Recent research has focused on applying game theory in network security [20], especially in communication and computer networks [21, 22]. However, these works again do not consider the flexibility and real-time adjustment of attackers under Stackelberg settings. Besides, the physical constraints (e.g., travel time between targets) in OSG do not exist in communication networks.

A third thread of recent research has made inroads in the modeling of opportunistic criminal behavior, and in how security forces might defend against such adversaries. In [23] burglars are modeled as biased random walkers seeking "attractive" targets, and [24] follows up on this work with a method for generating effective police allocations to combat such criminals. However, these works make the extreme assumption that criminals have no knowledge of the overall strategy of the police, and their behavior is only affected by their observation of the current police allocation in their immediate neighborhood. Also, in [24] police behave in a similarly reactionary way, allocating their resources in an instantaneously optimal way in response to the current crime risk distribution rather than optimizing over an extended time horizon. Furthermore, in [24] there is no notion of the "movement" of police - rather, the distribution of police officers are chosen instantaneously, with no regard for the mechanics of exactly how the allocation may transform from one time step to the next. Our current approach is an attempt to generalize these threads of research.

3 OSG Framework

OSG unfolds on a connected graph that can be seen to model a metro rail system (though many other domains are also possible), where stations are nodes and trains connecting two stations are edges. Fig. 1 shows a simple scenario with three fully connected stations. Stations and trains are collectively referred to as locations. Let the stations be labeled $1, \ldots, N$, with N denoting the number of stations. The train from station i to its neighboring station j is denoted as $i \rightarrow j$. The number of locations is $N_l > N$, e.g., in Fig. 1, $N_l = 9$.

We divide time equally into time steps so that trains arrive at stations at the beginning of each time step. There are two phases in any time step. First is the *decision phase*, the period when trains are at stations for boarding and unboarding. In this phase, each passenger at each location decides where in the system to move next. There are two types of choices available. *Go* $i \rightarrow j$ means that (i) if a passenger is at station i, he gets on the train $i \rightarrow j$; (ii) if he is on a train arriving at station i, he now gets (or stays) on the train $i \rightarrow j$. *Stay* means that the passenger stays at the station, so that if the passenger was on a train, he gets off. After the brief decision phase is the *action phase*, in which trains depart from all stations to all directly connected stations. This model matches the metro systems in Los Angeles, where trains leave stations at regular intervals to all directly connected stations. Without losing generality, we assume that the time it takes to travel between any two adjacent stations is identical; this assumption can be relaxed by including dummy stations. In OSG, the defender ("she") – assisted by our algorithms – is modeled to be perfectly rational. The criminal ("he") is modeled with cognitive biases. Fig. 2 illustrates the OSG flowchart, with relevant equation numbers near variables – these variables and equations are described in the following.

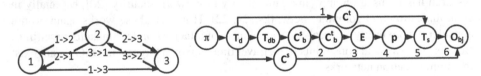

Fig. 1. The metro network   Fig. 2. Flow chart of OSG

3.1 Modeling Defenders

A defender is a team of police officers using trains for patrolling to mitigate crime. We start with a single defender and deal with multiple defenders later. The defender conducts randomized patrols using a *Markov Strategy* π, which specifies for each location a probability distribution over all available actions. At location l, the probabilities of *Go* $i \rightarrow j$ and *Stay* are denoted by $g_l^{i \rightarrow j}$ and s_l respectively.

Example 1: Markov Strategy *In Figure 1, a possible distribution for location* $3 \rightarrow 2$ *in a Markov strategy* π *is,*

$$s_{3 \rightarrow 2} = 0.1, \ g_{3 \rightarrow 2}^{2 \rightarrow 1} = 0.8, \ g_{3 \rightarrow 2}^{2 \rightarrow 3} = 0.1$$

Table 1. Notation used throughout this paper

π	Defender's Markov strategy	c_b^s	Criminal's belief of $\mathbf{c}^s$
T_d	Defender transition matrix	c_b^t	Criminal's belief of $\mathbf{c}^t$
$\mathbf{c}^s$	Defender stationary coverage	T_d	Criminal's belief of T_d
$\mathbf{c}^t$	Defender coverage vector at time step t	E	Target expected value for criminals
T_s	Transition matrix for the OSG Markov chain	p	Criminal's next strike probability

that is, if the defender is on the train from station 3 to 2, then at the next decision phase: she has probability 0.1 to choose Stay, thereby exiting the train and remaining at station 2; 0.8 to Go 2 → 1, meaning she remains on her current train as it travels to station 1; and 0.1 to Go 2 → 3, meaning she exits her current train and boards the train heading the opposite direction toward station 3.

Given π, the defender's movement is a Markov chain over the locations with *defender transition matrix* T_d, whose entry at column k, row l specifies the probability of a defender currently at location k being at location l during the next time step. In T_d, index i ($i \in 1, \ldots, N$) represents station i; indexes larger than N represent trains.

Example 2: *For Example 1, T_d is as follows:*

$$
\begin{array}{c c}
 & \begin{array}{ccccc} 1 & 2 & \cdots 2 \to 3 & 3 \to 1 & 3 \to 2 \end{array} \\
\begin{array}{c} 1 \\ 2 \\ 3 \\ 1 \to 2 \\ 1 \to 3 \\ \cdots \end{array} &
\left(\begin{array}{ccccc}
s_1 & 0 & 0 & s_{3\to1} & 0 \\
0 & s_2 & 0 & 0 & \partial_{3\to2} \\
0 & 0 & s_{2\to3} & 0 & 0 \\
g_1^{1\to2} & 0 & 0 & g_{3\to1}^{1\to2} & 0 \\
g_1^{1\to3} & 0 & 0 & g_{3\to1}^{1\to0} & 0 \\
\cdots & \cdots & \cdots & \cdots & \cdots
\end{array} \right)
\end{array}
$$

Using T_d and $\mathbf{c}^t = (c_1, c_2, \cdots, c_N, c_{1\to2}, \cdots)^T$, defined as the probability distribution of a defender's location at time t, we can calculate the coverage vector at time step $t_1 > t$ through the formula

$$
\mathbf{c}^{t_1} = (T_d)^{t_1 - t} \cdot \mathbf{c}^t \tag{1}
$$

We restrict each element in π to be strictly positive so that T_d is ergodic, meaning it is possible to eventually get from every location to every other location in finite time. For an ergodic T_d, based on Lemma 1, there is a unique *stationary coverage* $\mathbf{c}^s$, such that $T_d \cdot \mathbf{c}^s = \mathbf{c}^s$. The dependence of $\mathbf{c}^s$ on T_d and hence on π is shown in Fig. 2. The defender's initial coverage, $\mathbf{c}^1$, is set to $\mathbf{c}^s$ so that the criminal will face an invariant distribution whenever he enters the system. This invariant initial distribution is analogous to assuming that the defender patrols for a long time and becomes stable, but under our model, criminals can enter the system at any time.

Lemma 1. *(Fundamental Theorem of Markov Chains) For an ergodic Markov chain P, there is a unique probability vector $\mathbf{c}$ such that $P \cdot \mathbf{c} = \mathbf{c}$ and $\mathbf{c}$ is strictly positive.*

Proof. This is a very simple restatement of the property of ergodic Markov chain. [25] provides detailed proof. □

3.2 Modeling Opportunistic Criminals

Our model of the criminal consists of three components.

**Criminal's probability to
commit a crime at the current
time step:** We assume the criminal will only commit crimes at
stations, as discussed earlier [9],
and only during action phases,
since decision phases are considered instantaneous. The prob-

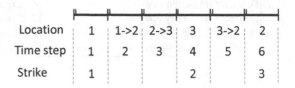

Location	1	1->2	2->3	3	3->2	2
Time step	1	2	3	4	5	6
Strike	1			2		3

Fig. 3. Example of strikes

ability of such a crime is determined by two factors. The first is the *attractivenes* of each
target station [23], which measures the availability of crime opportunities at a station.
Attractiveness measures how likely a criminal located at that station during an action
phase is to commit a crime *in the absence of defenders*; $\mathbf{Att} = (Att_1, Att_2, \cdots, Att_N)$
is the N vector composed of station attractiveness. The second factor is the defender's
presence; i.e., if a criminal is at the same station as a defender, he will not commit a
crime. Thus, his probability of committing a crime at station i will be influenced by
$c^t(i)$. Using this strategy, the criminal will never be caught red handed by the defender,
but may be forced toward a less attractive target. Thus, the probability of the criminal committing a crime if located at station i during the action phase of time step t, is
denoted as $q_c(i, t) = (1 - c^t(i))Att(i)$.

Criminal's Belief State of the Defender: During the decision phase, the criminal decides the next target station; he then moves directly to that station at the next action
phase(s). Hence, the criminal's motion within the metro system can be distilled down
to a sequence of stations where he chooses to locate; we refer to these instances of at-
tempted crime as *Strikes*. Figure 3 is a toy example showing the relationship between
the time steps and strikes for a criminal. As shown in the figure, only the time steps
when the criminal is at stations are counted as strikes.

When making these target decisions, the criminal tends to choose stations with high
expected utilities. He uses his knowledge of π and his real-time observations to make
such decisions. Let T_{db}, $\mathbf{c}_b^t$, and $\mathbf{c}_b^s$ be his *belief* of T_d, $\mathbf{c}^t$, and $\mathbf{c}^s$, respectively. As the
criminals have limited surveillance capability, these beliefs may not be the same as T_d,
$\mathbf{c}^t$, and $\mathbf{c}^s$. To model the criminal's surveillance imperfection we use *anchoring bias*
– a cognitive bias, with extensive experimental backing, which reveals the human bias
toward choosing a uniform distribution when assigning probabilities to events under
imperfect information [5, 18]. We denote the level of the criminal's anchoring bias with
the parameter b, where $b = 0$ indicates no anchoring bias, and $b = 1$ indicates complete
reliance on such bias. We set $T_{db} = (1 - b) \cdot T_d + b \cdot T_u$, with corresponding stationary
coverage $\mathbf{c}_b^s$, where T_u corresponds to the uniform distribution.

At any given time step t when the criminal is at a station, i.e., a strike, he may be
modeled as using his belief and observations to estimate $\mathbf{c}_b^t$. We assume the opportunistic criminal only uses his current observation, $\mathbf{c}_b^s$ and T_{db} to estimate $\mathbf{c}_b^t$ (criminal's
belief of defender's location distribution). Specifically, if the criminal is at station i and
the defender is also there, then $\mathbf{c}_b^t$ is $(0, 0, ..., 1, 0, ..., 0)^T$, where row i is 1 and all others

are 0. Otherwise the defender is not at i, and

$$\mathbf{c}_b^t = \frac{(c_b^s(1), c_b^s(2), ..., 0, c_b^s(i+1), ..., c_b^s(N_l))^T}{[1 - c_b^s(i)]}, \tag{2}$$

where row i is 0 and other rows are proportional to the corresponding rows in $\mathbf{c}_b^s$. Our approach to compute $\mathbf{c}_b^t$ is justified on two grounds. First, it is computationally cheap. Second, as we show in experimental results, even perfect knowledge provides very limited improvement in the criminal's performance given our modeling of the criminal's bounded rationality and anchoring bias; thus a more complex procedure is unnecessary. Given $\mathbf{c}_b^t$ and T_{db}, the belief coverage vector at time step t_1 $(t_1 > t)$, $\mathbf{c}_b^{t_1}$, is calculated via Eq. 1.

Input: i: the criminal's station; π: defender's Markov strategy; m: the defender's location;
       b: parameter of criminal's anchoring bias
Output: $p(\cdot|i, \mathbf{c}_b^{to})$: The criminal's probability distribution for next target
1   Initial N with the number of stations ;
2   Initial T_d by π;
3   Initial $\mathbf{c}^s$ with stationary coverage of T_d;
4   Initial $\mathbf{c}_b^{to}$ with a $1 \times (3N - 2)$ zero vector ;
5   $T_{db} = (1 - b) \cdot T_d + b \cdot T_u$;
6   $\mathbf{c}_b^s = (1 - b) \cdot \mathbf{c}^s + b \cdot \mathbf{c}_u^s$;
7   **if** $i == m$ **then**
8      $\mathbf{c}_b^{to}(i) = 1$;
9   **end**
10 **If** $i \neq m$ **then**
11   **for** $j \in Location$ **do**
12        $\mathbf{c}_b^{to}(j) = \dfrac{\mathbf{c}_b^s(j)}{1 - \mathbf{c}_b^s(i)}$;
13   **end**
14      $\mathbf{c}_b^{to}(i) = 0$;
15 **end**
16 **for** $j \in Station$ **do**
17      $t = |i - j| + 1$;
18      $\mathbf{c}_b^{to+t} = (T_{db})^t \cdot \mathbf{c}_b^{to}$;
19      $E(j|i, \mathbf{c}_b^{to}) = \dfrac{\left(1 - \mathbf{c}_b^{to+t}(j)\right) Att(j)}{t}$;
20 **end**
21 **for** $j \in Station$ **do**
22      $p(j|i, \mathbf{c}_b^{to}) = \dfrac{E(j|i, \mathbf{c}_b^{to})^\lambda}{\sum_{h=1}^N E(h|i, \mathbf{c}_b^{to})^\lambda}$;
23 **end**
24 **return** $p(\cdot|i, \mathbf{c}_b^{to})$;

Algorithm 1. BIASED RANDOM WALK ALGORITHM

We set the actual payoff for a crime to 1, but this can be generalized. The expected payoff for the criminal when choosing station j as the next strike, given that the current strike is at station i at time step t, is $q_{cb}(j, t + \delta_{ij})$, where $\delta_{ij} \geq 1$ is the minimum time needed to arrive at j from i. But, criminals are known to discount more distant locations when choosing targets. Therefore, the *utility* that the criminal places on a given payoff is discounted over time. We implement this by dividing the payoff by the time taken. Finally, the criminal must rely on his belief of the defender's coverage when evaluating $q_{cb}(j, t + \delta_{ij})$. Altogether, station j has the expected utility $E(j|i, \mathbf{c}_b^t) = \frac{q_{cb}(j, t+\delta_{ij})}{\delta_{ij}}$, which is

$$E(j|i, \mathbf{c}_b^t) = \frac{\left(1 - \left[(T_{db})^{\delta_{ij}} \cdot \mathbf{c}_b^t\right](j)\right) Att(j)}{\delta_{ij}}. \tag{3}$$

The Criminal's Quantal Biased Random Movement (QBRM): Finally, we propose QBRM to model the criminal's bounded rationality based on other such models of criminal movements in urban domains [23]. Instead of always picking the station with highest expected utility, his movement is modeled as a random process biased toward stations of high expected utility. Given the expected value for each station $\mathbf{E}(\cdot|i, \mathbf{c}_b^t)$, the probability distribution for each being chosen as the next strike, $\mathbf{p}(\cdot|i, \mathbf{c}_b^t)$ is:

$$p(j|i, \mathbf{c}_b^t) = \frac{E(j|i, \mathbf{c}_b^t)^\lambda}{\sum_{h=1}^N E(h|i, \mathbf{c}_b^t)^\lambda}, \tag{4}$$

where $\lambda \geq 0$ is a parameter that describes the criminal's level of rationality. This is an instance of the *quantal response* model of boundedly rational behavior [26]. The criminal may, as an alternative to choosing a further strike, leave the system at *exit rate* α. Therefore, the criminal eventually leaves the system with probability 1, and in expectation receives a finite utility; he cannot indefinitely increase his utility.

Given the criminal's QBRM, the Opportunistic Security Game can be simplified to a Stackelberg game for specific value of the parameters describing criminal's behaviour (Theorem 2).

Lemma 2. *When the criminal's rantionality level parameter $\lambda = 0$, the defender's optimal strategy is a stationary strategy, meaning that the defender picks a station and does not move in the patrol.*

Proof. According to Eqn. 4, when $\lambda = 0$, $p(j|i, \mathbf{c}_b^t) = \frac{1}{N}$ for all targets, which is independent of defender's Markov strategy π. Therefore, the OSG is equivalent to a Stackelberg Game where the leader (the criminal) makes his choice first, which is independent of the follower's (defender's) choice. Then the follower can decide her action given the leader's action. Therefore, as in a Stackelberg game, the follower's (defender's) optimal strategy is a pure strategy. Furthermore, we know that in this Stackelberg game, the leader (the criminal) is making a uniform random choice, meaning that he chooses each target with the same probability. Therefore, the defender's optimal strategy is staying at the station with highest attractiveness.   □

To summarize, as shown in Figure 2, the opportunistic criminal is modeled as follows: First, he decides whether to commit a crime or not based on the defender's presence at his station at each strike. Next, he uses his imperfect belief T_{db} of the defender's strategy, which is affected by anchoring bias, and his real-time observation to update his belief c_b^t using a simple scheme (Eq. 2). Finally, we use QBRM to model his next attack (Eq. 4) based on the expected utility of different targets (Eq. 3). Algorithm 1 is a full mathematical description of the criminal's movement. In Algorithm 1, steps 1-4 are initialization; steps 5-6 model how the criminal generates his imperfect belief; steps 7-15 model how the criminal updates his belief given his real-time observation; steps 16-20 model how the criminal evaluates each station based on his updated belief; and steps 21-24 use QBRM to model his probability distribution of visiting each station in his next strike.

4   Exact OSG (EOSG) Algorithm

Given the defender and criminal models, the EOSG algorithm computes the optimal defender strategy by modeling the OSG as a finite state Markov chain. As all the criminals behave identically, we can focus on the interaction between the defender and one criminal without loss of generality.

Each state of the EOSG Markov chain is a combination of the criminal's station and the defender's location. Here we only consider situations where the criminal is at a station as states because he only makes decisions at stations. Since there are N stations and N_l locations, the number of states is $N \cdot N_l$ in the EOSG markov chain. State transitions in this EOSG markov chain are based on *strikes* rather than *time steps*. The transition matrix for this Markov chain, denoted as T_s, can be calculated by combining the defender and criminal models. For further analysis, we pick the element $p_{S1 \to S2}$ in T_s that represents the transition probability from state $S1$ to $S2$. Suppose in $S1$ the criminal is at station i while the defender is at location m at time step t, and in $S2$, the criminal is at station j while the defender is at location n at time step $t + \delta_{ij}$. We need two steps to calculate the transition probability $p_{S1 \to S2}$. First, we find the transition probability of the criminal from i to j, $p(j|i, c_b^t)$. Then, we find the defender's transition probability from m to n, which is $c^{t+\delta_{ij}}(n) = \left((T_d)^{\delta_{ij}} \cdot e_m \right)(n)$, where e_m is a basis vector for the current location m. The transition probability $p_{S1 \to S2}$ is therefore given by

$$p_{S1 \to S2} = p(j|i, c_b^t) \cdot c^{t+\delta_{ij}}(n). \tag{5}$$

Since $p(j|i, c_b^t)$ and $c^{t+\delta_{ij}}(n)$ are determined by π, $p_{S1 \to S2}$ is also in terms of π (see Fig. 2), and hence so is T_s.

Given this EOSG model, we can calculate the defender's expected utility at each strike. For each successful crime, the defender receives utility $u_d < 0$, while if there is no crime, she receives utility 0. We do not consider the time discount factor in the defender's expected utility, as the goal of the defender shall be to simply minimize the total expected number of crimes that any criminal will commit. Formally, we define a vector $r_d \in \mathbf{R}^{N \cdot N_l}$ such that entries representing states with both criminal and defender at the same station are 0 while those representing states with criminal at station i and defender not present are $Att(i) \cdot u_d$. Then, the defender's expected utility $V_d(t)$ during strike number t is $V_d(t) = r_d \cdot x_t$, where x_t is the state distribution at strike number t. x_t can be calculated from the initial state distribution x_1, via $x_t = ((1 - \alpha) \cdot T_s)^{t-1} x_1$.

The initial state distribution $\mathbf{x}_1$ can be calculated from the initial criminal distribution and $\mathbf{c}^s$. The defender's total expected utility over all strikes is thus

$$Obj = \lim_{\ell \to \infty} \sum_{t=1}^{\ell} V_d(t)$$
$$= \mathbf{r}_d \cdot (I - (1 - \alpha)T_s)^{-1}\mathbf{x}_1 , \tag{6}$$

where I is an identity matrix and α is the criminal's *exit rate*. In this equation we use the geometric sum formula and the fact that the largest eigenvalue of T_s is 1, so that $I - (1 - \alpha)T_s$ is nonsingular for $0 < \alpha < 1$.

The objective is a function of the transition matrix T_s and $\mathbf{x}_1$, which can be expressed in terms of π via Eqs. (1), (3), (4), and (5). We have thus formulated the defender's problem of finding the optimal Markov strategy to commit to as a nonlinear optimization problem, specifically to choose π to maximize Obj (that is, minimize the total amount of crime).

5   OSG for Multiple Defenders

If K multiple defenders all patrol the entire metro, using the same π, which is denoted as *full length patrolling*, then they will often be at the same station simultaneously, which carries no benefit. On the other hand if we allow arbitrary defenders' strategies that are correlated, we will need to reason about complex *real-time* communication and coordination among defenders. Instead, we divide the metro into K contiguous segments, and designate one defender per segment, as in typical real-world patrolling of a metro system. Each defender will have a strategy specialized to her segment.

Defenders: In the k-th segment, the number of locations is n_l^k. Defender k patrols with the Markov strategy π_k. Her transition matrix is $T_{dk} \in \mathbf{R}^{n_l^k \times n_l^k}$. Her coverage vector at time t is $\mathbf{c}_k^t$, and $\mathbf{c}_k^s$ is her stationary coverage. Hence, defender k's behavior is the same as that in a single-defender OSG, while the collective defender behavior is described by the Markov strategy $\pi = (\pi_1, \pi_2, ..., \pi_K)$. The transition matrix T_d is as follows, where we have dropped the trains between segments from the basis for T_d and ensured that station numbering is continuous within segments:

$$T_d = \begin{pmatrix} T_{d1} & \cdots & 0 \\ \vdots & \ddots & \vdots \\ 0 & \cdots & T_{dK} \end{pmatrix}. \tag{7}$$

The coverage of all units at time step t is $\mathbf{c}^t$, and is defined as the concatenation of coverage vectors $(\mathbf{c}_1^t; \mathbf{c}_2^t; ...; \mathbf{c}_K^t)$. $\mathbf{c}^t$ sums to K since each $\mathbf{c}_k^t$ sums to 1. The vector $\mathbf{c}^t$ evolves to future time steps t_1 in the same way as before, via Eq. 1. The overall stationary coverage is $\mathbf{c}^s = (\mathbf{c}_1^s; \mathbf{c}_2^s; ...; \mathbf{c}_K^s)$.

Opportunistic Criminals: The previous model for criminals still applies. However, any variables related to defenders (T_d, $\mathbf{c}^t$, $\mathbf{c}^s$) are replaced by their counterparts for the multiple defenders. Furthermore, the criminal in segment k at time t cannot observe

Input: i: the criminal's station; π: vector of defender Markov strategies; $\mathbf{m}$: vector of defender locations; b: parameter of criminal's anchoring bias

Output: $p(\cdot|i, \mathbf{c}_b^{to})$: The criminal's probability distribution for next target

1   Initial N with the number of stations ;

2   Initial K with the number of defenders ;

3   Initial k_i with the segment that station i is in ;

4   **for** $k \leq K$ **do**

5   Initial T_{dk} by π_k;

6   Initial $\mathbf{c}_k^s$ by stationary coverage of T_{dk};

7      $T_{dbk} = (1 - b) \cdot T_{dk} + b \cdot T_{uk}$;

8      $\mathbf{c}_{bk}^s = (1 - b) \cdot \mathbf{c}_k^s + b \cdot \mathbf{c}_{uk}^s$;

9      $\mathbf{c}_{bk}^{to} = \mathbf{c}_{bk}^s$

10   **if** $k == k_i$ **then**

11   Initial $\mathbf{c}_{bk}^{to}$ with a $1 \times n_l^k$ zero vector ;

12   **if** $i == \mathbf{m}(k)$ **then**

13            $\mathbf{c}_{bk}^{to}(i) = 1$;

14   **end**

15   **if** $i \neq \mathbf{m}(k)$ **then**

16   **for** $j \in Location\ in\ segment\ k$ **do**

17                $\mathbf{c}_{bk}^{to}(j) = \dfrac{\mathbf{c}_{bk}^s(j)}{1 - \mathbf{c}_{bk}^s(i)}$;

18   **end**

19            $\mathbf{c}_{bk}^{to}(i) = 0$;

20   **end**

21   **end**

22 **end**

23   $T_{db} = \begin{pmatrix} T_{db1} & 0 & \cdots & 0 \\ 0 & T_{db2} & \cdots & 0 \\ \vdots & \vdots & \ddots & \vdots \\ 0 & 0 & \cdots & T_{dbK} \end{pmatrix}$

24   $\mathbf{c}_b^{to} = (\mathbf{c}_{b1}^{to}; \mathbf{c}_{b2}^{to}; ...; \mathbf{c}_{bK}^{to})$.

25   **for** $j \in Station$ **do**

26      $t = |i - j| + 1$;

27      $\mathbf{c}_b^{to+t} = (T_{db})^t \cdot \mathbf{c}_b^{to}$;

28      $E(j|i, \mathbf{c}_b^{to}) = \dfrac{\left(1 - \mathbf{c}_b^{to+t}(j)\right) Att(j)}{t}$;

29   **end**

30   **for** $j \in Station$ **do**

31      $p(j|i, \mathbf{c}_b^{to}) = \dfrac{E(j|i, \mathbf{c}_b^{to})^\lambda}{\sum_{h=1}^{N} E(h|i, \mathbf{c}_b^{to})^\lambda}$;

32   **end**

33   **return** $p(\cdot|i, \mathbf{c}_b^{to})$;

Algorithm 2. BIASED RANDOM WALK ALGORITHM WITH MULITPLE DEFENDERS

defenders other than k. As a result, his belief of defender coverage is updated only for segment k, i.e., $\mathbf{c}_b^t = (\mathbf{c}_{b1}^s; \mathbf{c}_{b2}^s; ...; \mathbf{c}_{b(k-1)}^s; \mathbf{c}_{bk}^t; \mathbf{c}_{b(k+1)}^s; ...; \mathbf{c}_{bK}^s)$. Algorithm 2 describes a criminal's behavior in the multiple defenders settings. Similar to Algorithm 1, in Algorithm 2, steps 1-3 are initialization; steps 4-22 model how the criminal generates and updates his imperfect belief for each defender, such that for defender $k(k \leq K)$, the process of calculating the criminal's belief is exactly the same as the single defender scenario; steps 23-24 combine the criminal's belief for each defender as his belief for all the defenders; steps 25-29 model how the criminal evaluates each station based on his belief; and steps 30-34 use QBRM to model his probability distribution of visiting each station in his next strike.

EOSG: In optimizing defender strategies via a Markov chain, each state records the station of the criminal and the location of *each* defender. As a result, each state is denoted as $S = (i, m_1, ..., m_K)$, where the criminal is at station i and defender k is at location m_k. Since defender k can be at n_l^k different locations, the total number of states is $N \cdot n_l^1 \cdots n_l^K$. To apply EOSG for multiple defenders, T_s is still calculated using the defender and criminal models. The transition probability $p_{S_1 \to S_2}$ from $S_1 = (i, m_1, ..., m_K)$ at time t to $S_2 = (j, n_1, ..., n_K)$ at time $t + \delta_{ij}$ is

$$p_{S_1 \to S_2} = p(j|i, \mathbf{c}_b^t) \prod_k c^{t+\delta_{ij}}(n_k),$$

where $c^{t+\delta_{ij}}(n_k) = ((T_d)^{\delta_{ij}} \cdot \mathbf{e}_{m_1, m_2, ..., m_K})(n_k)$ and $\mathbf{e}_{m_1, m_2, ..., m_K}$ is an indicator vector with 1 at entries representing locations $m_1, m_2, ..., m_K$ and 0 at all others. The state distribution $\mathbf{x}$ and revenue $\mathbf{r}_d$ are both $N \cdot n_l^1 \cdots n_l^K$ vectors. The defenders' total expected utility is given by Eq. (6); our goal remains to find a π to maximize Obj.

6   The COPS Algorithm

The objective of EOSG can be formulated as a non-linear optimization. Unfortunately, as we will show in our experiments, the EOSG algorithm fails to scale-up to real-world sized problem instances due to the size of T_s in Eq. (6), which is exponential ($N \cdot n_l^1 \cdots n_l^K$ by $N \cdot n_l^1 \cdots n_l^K$) for K defenders. We propose the Compact OPportunistic Security game state (COPS) algorithm to accelerate the computation. COPS simplifies the model by compactly representing the states. The size of the transition matrix in COPS is $2N \times 2N$, *regardless of the number of defenders*, which is dramatically smaller than in the exact algorithm. The COPS algorithm is inspired by the Boyen-Koller(BK) algorithm for approximate inference on Dynamic Bayesian Networks [27]. COPS improves upon a direct application of BK's factored representation by maintaining strong correlations between locations of players in OSG.

In OSG with a single defender, there are two components in a Markov chain state for strike t: the station of the criminal S_c^t and the location of the defender θ_d^t. These two components are correlated when they evolve. We introduce an intermediate component, the criminal's observation O_c^t, which is determined by both S_c^t and θ_d^t. Given the criminal's current station and his observation, we can compute his distribution over the next strike station. At the same time, the evolution of θ_d^t is independent of S_c^t. Such evolution

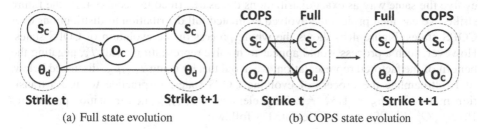

(a) Full state evolution   (b) COPS state evolution

Fig. 4. COPS algorithm

is shown in Figure 4(a). This is an instance of a Dynamic Bayesian Network: S_c^t, O_c^t, and θ_d^t are the random variables, while edges represent probabilistic dependence.

A direct application of the Boyen-Koller algorithm compactly represents the states by using the marginal distribution of these two components, S_c^t and θ_d^t, as approximate states. The marginal distributions of S_c^t and θ_d^t are denoted as $\Pr(S_c^t)$ and $\Pr(\theta_d^t)$ respectively, and it is assumed that these two components are independent, meaning we can restore the Markov Chain states by multiplying these marginal distributions. Note that in Section 4.2, we set $\Pr(\theta_d^t) = \mathbf{c}^s$ for all strikes. Thus, we do not need to store θ_d^t in the state representation. Therefore, the total number of the approximate states in this case is just N. However, such an approximation throws away the strong correlation between the criminal's station and defender unit's location through the criminal's real-time observation. Our preliminary experiments showed that this approximate algorithm leads to low defender expected utility.

To design a better algorithm, we should add more information about the correlation between the criminal and defenders. To that end, our COPS algorithm compactly represents our Markov Chain states with less information lost. Instead of just considering the marginal distributions of each component $\Pr(\theta_d^t)$ and $\Pr(S_c^t)$, we also include the observation of the criminal O_c^t while constructing the approximate states. The criminal's observation is binary: 1 if the defender is at the same station with him, 0 otherwise. The new approximate states, named *COPS states*, only keep the marginal probability distribution of $\Pr(S_c^t, O_c^t)$. So, the new state space is the Cartesian product of the sets of S_c^t and O_c^t, which has size $2N$.

One subtask of COPS is to recover the distributions over the full state space (S_c^t, θ_d^t), given our state representation $\Pr(S_c^t, O_c^t)$. We cannot restore such distribution by multiplying $\Pr(\theta_d^t)$ and $\Pr(S_c^t)$ in COPS. This is because S_c^t, O_c^t, and θ_d^t are not independent. For example, in COPS state $S_c^t = 1$, $O_c^t = 1$, θ_d^t cannot be any value except 1. In other words, the defender's location distribution $\Pr(\theta_d^t|S_c^t, O_c^t)$ is no longer $\mathbf{c}^s$. Instead, we approximate $\Pr(\theta_d^t|S_c^t, O_c^t)$ as follows. In each COPS state (S_c^t, O_c^t), the *estimated marginal distribution* for the defender, $\widehat{\Pr}(\theta_d^t|S_c^t, O_c^t)$, is found in a manner similar to that used to find the criminal's belief distribution $\mathbf{c}_b^t$. Specifically, if $O_c^t = 1$, $\widehat{\Pr}(\theta_d^t|S_c^t, O_c^t) = (0, 0, ..., 1, 0, ..., 0)^T$, where the row representing station S_c^t is 1 and all others are 0; if $O_c^t = 0$, then $\widehat{\Pr}(\theta_d^t|S_c^t, O_c^t)$ is found through Equation 2, but with the $c_b^s(j)$ replaced by the true stationary coverage value $c^s(j)$. We can then recover the estimated distribution over full states $\widehat{\Pr}(S_c^t = i, \theta_d^t|S_c^t = i, O_c^t) = \widehat{\Pr}(\theta_d^t|S_c^t = i, O_c^t)$ for all i and $\widehat{\Pr}(S_c^t = j, \theta_d^t|S_c^t = i, O_c^t) = 0$ for all $j \neq i$. Estimated full distributions evolve the same way as exact distributions do, as described in Section 4. At the future strike, we can then project the evolved estimated full distribution to distributions over COPS states. Figure 4(b) shows the whole process of the evolution of COPS states. However, such a process would appear to involve representing a full T_s, negating the benefit of the factored representation; we avoid that by using T_{COPS}, discussed below.

To streamline the process of evolving COPS states, in practice we use a transition matrix $T_{COPS} \in \mathbb{R}^{2N \times 2N}$. Each element of T_{COPS}, i.e., transition probability $\Pr(S_c^{t'}, O_c^{t'} | S_c^t, O_c^t)$, can be calculated as follows:

$$
\begin{aligned}
&\Pr(S_c^{t'}, O_c^{t'} | S_c^t, O_c^t) \\
&= \sum_{\theta_d^{t'}} \sum_{\theta_d^t} \Pr(S_c^{t'}, O_c^{t'} | S_c^{t'}, \theta_d^{t'}) \cdot \Pr(S_c^{t'}, \theta_d^{t'} | S_c^t, \theta_d^t) \cdot \widehat{\Pr}(S_c^t, \theta_d^t | S_c^t, O_c^t) \\
&= \Pr(S_c^{t'} | S_c^t, O_c^t) \sum_{\theta_d^{t'}} \Pr(O_c^{t'} | S_c^{t'}, \theta_d^{t'}) \cdot \sum_{\theta_d^t} \Pr(\theta_d^{t'} | S_c^{t'}, S_c^t, \theta_d^t) \cdot \widehat{\Pr}(\theta_d^t | S_c^t, O_c^t),
\end{aligned}
\tag{8}
$$

where $\Pr(S_c^{t'} | S_c^t, O_c^t)$ and $\Pr(\theta_d^{t'} | S_c^{t'}, S_c^t, \theta_d^t)$ correspond to $p(j|i, c_b^{to})$ and $c^{t_o+|i-j|+1}(n)$, respectively, in Section 4.

The defenders' total expected utility in COPS is calculated in a similar way as the exact algorithm, which is

$$
Obj_{COPS} = \mathbf{r}_{d,COPS} \cdot (I - (1-\alpha)T_{COPS})^{-1} \mathbf{x}_{1,COPS},
\tag{9}
$$

where $\mathbf{r}_{d,COPS}$, $\mathbf{x}_{1,COPS}$ are the expected utility vector and the initial distribution for COPS states. Similar to $\mathbf{r}_d$, $\mathbf{r}_{d,COPS}(S)$ is 0 if in state S the defender is at the same station with the criminal, else $\mathbf{r}_{d,COPS}(S) = u_d$. COPS is faster than the exact algorithm because the number of states is reduced dramatically. Meanwhile, the approximation error of COPS algorithm is bounded according to Theorem 1.

Definition 1. *Let m_i be the location corresponding to station i. For a distribution over OSG full states $\mathbf{x}$, the corresponding distribution over COPS states $\mathbf{x}_{COPS}$ is:*

$$
\mathbf{x}_{COPS}(i, o) = \begin{cases} \mathbf{x}(i, m_i) & \text{if } o = 1 \\ \sum_{m \neq m_i} \mathbf{x}(i, m) & \text{if } o = 0 \end{cases}
$$

For a distribution over COPS states $\mathbf{x}_{COPS}$, the corresponding approximate distribution over OSG full states $\mathbf{x}'$ is:

$$
\mathbf{x}'(i, m) = \begin{cases} \mathbf{x}_{COPS}(i, 1) & \text{if } m = m_i \\ \mathbf{x}_{COPS}(i, 0) \cdot \frac{c^s(m)}{1 - c^s(i)} & \text{otherwise} \end{cases}
$$

This conversion can be summarized through a single matrix multiplication, such that $\mathbf{x}' = A\mathbf{x}$.

Lemma 3. *Let μ_2 be the magnitude of the second largest eigenvalue of transition matrix T_s. Let δ be the largest possible L_2 approximation error introduced when full state distribution $\mathbf{x}$ is transformed into the COPS representation vector $\mathbf{x}_{COPS}$ and back into the approximate distribution $\mathbf{x}'$ over full states: $\|\mathbf{x} - A\mathbf{x}\| \leq \delta$. At strike number t, the L_2 norm between the EOSG distribution $\mathbf{y}_t$ and the distribution found through COPS algorithm $\mathbf{x}_t$ is bounded, such that $\|\mathbf{y}_t - \mathbf{x}_t\|_2 \leq (1-\alpha)^{t-1} \frac{\delta(1-\mu_2^t)}{1-\mu_2}$.*

Proof. Let x_t be the state vector as found through the COPS algorithm at time t. The time evolution for x proceeds then as follows: $x_t = (1 - \alpha)^{t-1}(AT_s)^{t-1}x_1$, where $x_1 = Ay_1$, and y_1 is the initial state vector for the EOSG algorithm. So, consider the L_2 error introduced at iteration t by the COPS approximation alone

$$||T_s x_t - AT_s x_t||_2 = (1 - \alpha)^{t-1}||T_s(AT_s)^{t-1}x_1 - AT_s(AT_s)^{t-1}x_1||_2.$$

Since the vector $T_s(AT_s)^{t-1}x_1$ is a full state vector, the error bound here is simply

$$||T_s x_t - AT_s x_t||_2 \leq \delta(1 - \alpha)^{t-1}. \tag{10}$$

Now, assume that the error between the state vectors x_t and y_t at some time t is bound by ϵ: $||y_t - x_t||_2 \leq \epsilon$. Since in the EOSG Markov chain it is possible to travel from any state to any other state in a finite amount of time, this Markov chain is ergodic. Let the stationary distribution of T_s be x^s, which is normalized such that $\vec{1} \cdot x^s = 1$. $\mu_1 = 1 > \mu_2 \geq ... \geq \mu_{N \cdot N_l}$ are the magnitudes of the eigenvalues of T_s corresponding to eigenvectors $v_1(= x^s), v_2, ..., v_{N \cdot N_l}$. Since T_s is the transition matrix of an ergodic Markov chain, $\mu_k < 1$ for $k \geq 2$. For eigenvectors $v_k, k \geq 2$, we have $|T_s \cdot v_k| = |\mu_k \cdot v_k|$. Multiplying by $\vec{1}$ and noting that $\vec{1} \cdot T_s = \vec{1}$, we get $|\vec{1} \cdot v_k| = |\mu_k \cdot \vec{1} \cdot v_k|$. Since $\mu_k \neq 1$, $\vec{1} \cdot v_k = 0$.

Write x_t and y_t in terms of $v_1, v_2, ..., v_{N \cdot N_l}$ as:

$$y_t = \beta_1 x^s + \sum_{i=2}^{N \cdot N_l} \beta_i v_i$$

$$x_t = \beta_1' x^s + \sum_{i=2}^{N \cdot N_l} \beta_i' v_i$$

Since $y_t = (1-\alpha)^{t-1}T_s^{t-1}y_1$, then $\vec{1} \cdot y_t = (1-\alpha)^{t-1}$; similarly, $\vec{1} \cdot x_t = (1-\alpha)^{t-1}$. Multiplying both equations above by $\vec{1}$, we get $\beta_1 = \beta_1' = (1 - \alpha)^{t-1}$. Therefore,

$$||T_s \cdot y_t - T_s \cdot x_t||_2 \leq ||\sum_{i=2}^{N \cdot N_l} (\beta_i - \beta_i')\mu_i v_i||_2$$

$$\leq |\mu_2|\sqrt{(\beta_2 - \beta_2')^2 + (\beta_3 - \beta_3')^2 + \cdots + (\beta_{N \cdot N_i} - \beta_{N \cdot N_i}')^2}$$

$$\leq \mu_2||x_t - y_t||_2$$

$$\leq \mu_2\epsilon$$

Accordingly, at $t = 1$, we have

$$||y_1 - x_1||_2 = ||y_1 - Ay_1||_2 \leq \delta.$$

At $t = 2$, we have

$$||y_2 - x_2||_2 = (1-\alpha)||T_s y_1 - AT_s x_1|| = (1-\alpha)||T_s y_1 - AT_s x_1 + T_s x_1 - T_s x_1|| \leq$$
$$(1 - \alpha)||T_s y_1 - T_s x_1||_2 + (1 - \alpha)||T_s x_1 - AT_s x_1||_2.$$

From above, the bound for the first term is $\mu_2\delta$, given the error bound at $t = 1$. The bound for the second term is directly given by (10), and is simply δ. Hence

$$\|\mathbf{y}_2 - \mathbf{x}_2\|_2 \leq (1-\alpha)\delta(\mu_2 + 1) .$$

At $t = 3$, we have

$$\|\mathbf{y}_3 - \mathbf{x}_3\|_2 = (1-\alpha)\|T_s\mathbf{y}_2 - AT_s\mathbf{x}_2\| = (1-\alpha)\|T_s\mathbf{y}_2 - AT_s\mathbf{x}_2 + T_s\mathbf{x}_2 - T_s\mathbf{x}_2\| \leq$$
$$(1-\alpha)\|T_s\mathbf{y}_2 - T_s\mathbf{x}_2\|_2 + (1-\alpha)\|T_s\mathbf{x}_2 - AT_s\mathbf{x}_2\|_2 .$$

From above, the bound for the first term is $\mu_2(1-\alpha)\delta(\mu_2 + 1)$, given the error bound at $t = 2$. The bound for the second term is taken from (10), and is $\delta(1-\alpha)$. Hence

$$\|\mathbf{y}_3 - \mathbf{x}_3\|_2 \leq (1-\alpha)^2\delta(\mu_2^2 + \mu_2 + 1) .$$

By extension, then, the error bound at time step t between EOSG and COPS states is:

$$\|\mathbf{y}_t - \mathbf{x}_t\|_2 \leq (1-\alpha)^{t-1}\delta\sum_{i=0}^{t-1}\mu_2^i = (1-\alpha)^{t-1}\delta\frac{1-\mu_2^t}{1-\mu_2} .$$

$\square$

Theorem 1. *The difference between the EOSG objective and the COPS approximate objective $|Obj - Obj_{COPS}|$ is bounded by $\frac{\sqrt{N \cdot N_l}\delta|u_d|}{[1-(1-\alpha)\mu_2]\,\alpha}$*

Proof. Since Lemma 3 gives the bound of L_2 distance while $|Obj - Obj_{COPS}|$ is L_1 distance, we use the fact that for any two vectors v_1, v_2, the relationship between the L_1 distance and L_2 distance is: $\|v_1 - v_2\|_2 \leq \|v_1 - v_2\|_1 \leq \sqrt{n}\|v_1 - v_2\|_2$, where n is the dimension of the vectors. Therefore, $\|\mathbf{y}_t - \mathbf{x}_t\|_1 \leq \frac{\sqrt{N \cdot N_l}(1-\alpha)^{t-1}(1-\mu_2^t)\delta}{1-\mu_2}$. Hence we have:

$$|Obj - Obj_{COPS}| = \sum_{t=1}^{\infty}|\mathbf{r}_d \cdot \mathbf{y}_t - \mathbf{r}_d \cdot \mathbf{x}_t|$$

$$= \sum_{t=1}^{\infty}|\mathbf{r}_d \cdot (\mathbf{y}_t - \mathbf{x}_t)|$$

$$\leq \sum_{t=1}^{\infty}|r_{max}|\|\mathbf{y}_t - \mathbf{x}_t\|_1$$

$$\leq |r_{max}|\sum_{t=1}^{\infty}\frac{\sqrt{N \cdot N_l}(1-\alpha)^{t-1}(1-\mu_2^t)\delta}{1-\mu_2}$$

$$= |r_{max}|\frac{\sqrt{N \cdot N_l}\,\delta}{[1-(1-\alpha)\mu_2]\,\alpha}$$

where r_{max} is the element in $\mathbf{r}_d$ with largest magnitude, which is $min(Att(i) \cdot u_d)$ because $\mathbf{r}_d$ is a non-positive vector by definition. Given $Att(i) \leq 1$, we have $|Obj - Obj_{COPS}| \leq \frac{\sqrt{N \cdot N_l}\delta|u_d|}{[1-(1-\alpha)\mu_2]\,\alpha}$

$\square$

7   Experimental Results

Settings: We use the graphs in Figure 5 – metro structures commonly observed in the world's mega cities – in our experiments. We also tested our algorithm on line structure systems, and the results are similar (online appendix: http://osgcops.webs.com/). We solve the non-linear optimization in OSG using the `FindMaximum` function in Mathematica, which computes a locally optimal solution using an Interior Point algorithm. Each data point we report is *an average of 30 different instances,* each based on a different attractiveness setting; these instances were generated through a uniform random distribution from 0 to 1 for the attractiveness of each station. For multiple patrol unit scenarios, we use segment patrolling (except for Fig. 6(d)), and divide the graph so that the longest distances in each segments are minimized; the dashed boxes in Fig. 5 show the segments used. Results for other segmentations are similar (online appendix). The defender's utility of a successful crime is $u_d = -1$. The criminal's initial distribution is set to a uniform distribution over stations. The criminal exit rate is $\alpha = 0.1$. Strategies generated by all algorithms are evaluated using Equation 6. All key results are *statistically significant* ($p < 0.01$).

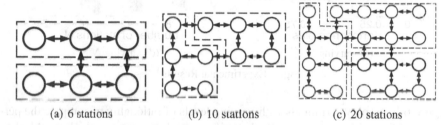

(a) 6 stations   (b) 10 stations   (c) 20 stations

Fig. 5. Part of metro systems in mega cities

Results: Fig. 6(a) shows the performance of the COPS algorithm and the EOSG algorithm using the settings from Fig. 5(a) and Fig. 5(b). In both, we set $\lambda = 1$. The Interior Point algorithm used by Mathematica is a locally optimal solver and there is always a current best feasible solution available, although the quality of the solution keeps improving through iterations. Therefore, one practical way to compare solutions is to check the solution quality after a fixed run-time. The x-axis in this figure shows runtime in seconds on a log scale, while the y-axis maps the defenders' average expected utility against one criminal, achieved by the currently-best solution at a given run time. Focusing first on results of 6 stations, where we have one defender, COPS outperforms EOSG for any runtime within 100 s, even though COPS is an approximate algorithm. This is because COPS reaches a local optimum faster than EOSG. Further, even for runtime long enough for EOSG to reach its local optimum (3160 s), where it outperforms COPS, the difference in solution quality is less than 1%. Focusing next on results of 10 stations with 2 defenders (using segment patrolling), the conclusions are similar to 6 stations, but the advantage of COPS is more obvious in this larger scale problem. In most instances, COPS reaches a local optimum in 1000 s while the output of EOSG are the same as initial values in 3160 s.

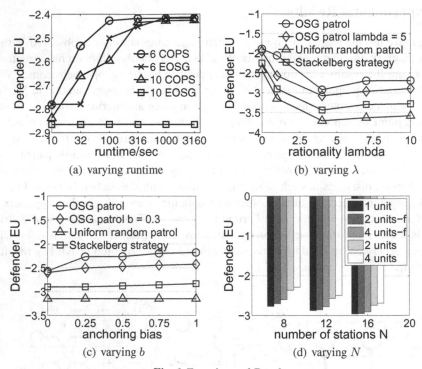

Fig. 6. Experimental Results

Figure 6(b) employs criminals with varying levels of rationality to compare the performance of three different strategies: the uniform random strategy, which is a Markov strategy with equal probability for all available actions at each location; an SSG strategy, which is the optimal strategy against a strategic attacker that attacks a single target; and a COPS OSG strategy (given 1800 s so it reached a local optimum). In Fig. 6(b), we set $b = 0$; results with other b are similar, which are shown in online appendix. The system consists of 10 stations and 2 defenders. The COPS OSG strategy outperforms the random and SSG strategies significantly for any λ. Next, two more settings are tested: the first is the OSG strategy against criminals who have perfect knowledge of defenders' current location. This is a purely hypothetical setting, and created only to check if a more complex criminal belief model than the one in Eq. 2 would have led to significantly different defender performance. The degradation in performance against perfect criminals is less than 6%, indicating that a more complex belief update for defenders' current location would have insignificant impact on the results. The second is also an OSG strategy, but the defenders set a fixed λ during computation to test performance when the defender has an inaccurate estimate of λ. We picked $\lambda = 1$ from a set of sampled λ, since the OSG strategy with $\lambda = 1$ performs best against criminals with various levels of rationality. Even though the OSG strategy assuming $\lambda = 1$ performs slightly worse than that using the correct λ, it is still better than SSG and uniform strategies. We conclude that OSG is a better model against opportunistic criminals even with an inaccurate estimation of λ.

The COPS strategy, the SSG, and the uniform random strategy are compared again in Fig. 6(c), this time against criminals with different levels of anchoring bias b. In order to evaluate the performance of COPS when the defender has an inaccurate estimate of the anchoring bias b, we plotted both the expected utility of COPS where the defender has an accurate estimate of the criminal's anchoring bias and that using a fixed anchoring bias $b = 0.5$. $b = 0.5$ was picked from a set of sampled b since the OSG strategy with this b performs best. In Fig. 6(c), λ is fixed to 1, but experiments with other λ generate similar results, which are shown in the online appendix. Again, COPS outperforms uniform random and SSG strategies.

To show COPS's scalability, we compare its performance with different numbers of defenders in metro systems with a varying number of stations; Five different settings are compared in Fig. 6(d): one defender, two defenders with full length patrolling, three defenders with full length patrolling, two defenders with segment patrolling, and three defenders with segment patrolling. The max runtime is 1800 s. With the same patrol techniques, more defenders provide higher expected utility. But, with the same amount of resources, segment patrolling outperforms full length patrolling.

8   Summary

This paper introduces OSG, a new computational framework to address opportunistic crime, opening the door for further research on this topic. Furthermore, we propose a new exact algorithm, EOSG, to compute defender resource allocation strategies, and an approximate algorithm, COPS, to speed up defender allocation to real-world scale scenarios. Our experimental results show that the OSG strategy outperforms baseline strategies with different types of criminals. We also show that COPS is more efficient than EOSG in solving real-world scale problems. Given our experimental results, COPS is being evaluated in the Los Angeles Metro system. Finally, in introducing OSG, this paper has added to the class of important security-focused game-theoretic frameworks in the literature, opening the door to a new set of research challenges for the community of researchers focused on game theory for security.

Acknowledgement. This research is supported by MURI grant W911NF-11-1- 0332.

References

1. Letchford, J., MacDermed, L., Conitzer, V., Parr, R., Isbell, C.L.: Computing optimal strategies to commit to in stochastic games. In: AAAI (2012)
2. Tambe, M.: Security and Game Theory: Algorithms, Deployed Systems, Lessons Learned. Cambridge University Press (2011)
3. Brantingham, P.J., Tita, G.: Offender mobility and crime pattern formation from first principles. In: Artificial Crime Analysis Systems (2008)
4. Ratcliffe, J.H.: A temporal constraint theory to explain opportunity-based spatial offending patterns. Journal of Research in Crime and Delinquency (2006)
5. See, K.E., Fox, C.R., Rottenstreich, Y.S.: Between ignorance and truth: Partition dependence and learning in judgment under uncertainty. Journal of Experimental Psychology (2006)
6. Hespanha, J.P., Prandini, M., Sastry, S.: Probabilistic pursuit-evasion games: A one-step nash approach (2000)

7. Brantingham, P., Brantingham, P.: Criminality of place. European Journal on Criminal Policy and Research 3(3), 5–26 (1995)
8. Cornish, D.B., Clarke, R.V.: Understanding crime displacement: An application of rational choice theory. Criminology (1987)
9. Loukaitou-Sideris, A., Liggett, R., Iseki, H.: The geography of transit crime documentation and evaluation of crime incidence on and around the green line stations in Los Angeles. JPER (2002)
10. Qian, Y., Jiang, X., Haskell, W., Tambe, M.: Online planning for optimal protector strategies in resource conservation games. In: AAMAS (2014)
11. Zhu, Q., Clark, A., Poovendran, R., Basar, T.: Deceptive routing games. In: 2012 IEEE 51st Annual Conference on Decision and Control (CDC), pp. 2704–2711. IEEE (2012)
12. Clark, A., Zhu, Q., Poovendran, R., Başar, T.: Deceptive routing in relay networks. In: Grossklags, J., Walrand, J. (eds.) GameSec 2012. LNCS, vol. 7638, pp. 171–185. Springer, Heidelberg (2012)
13. Basilico, N., Gatti, N., Amigoni, F.: Leader-follower strategies for robotic patrolling in environments with arbitrary topologies. In: AAMAS (2009)
14. Basilico, N., Gatti, N.: Automated abstractions for patrolling security games. In: AAAI (2011)
15. Basilico, N., Gatti, N., Rossi, T., Ceppi, S., Amigoni, F.: Extending algorithms for mobile robot patrolling in the presence of adversaries to more realistic settings. In: Proceedings of the 2009 IEEE/WIC/ACM International Joint Conference on Web Intelligence and Intelligent Agent Technology, vol. 02 (2009)
16. Jiang, A.X., Yin, Z., Zhang, C., Tambe, M., Kraus, S.: Game-theoretic randomization for security patrolling with dynamic execution uncertainty. In: AAMAS (2013)
17. Varakantham, P., Lau, H.C., Yuan, Z.: Scalable randomized patrolling for securing rapid transit networks. In: IAAI (2013)
18. Pita, J., Jain, M., Ordóñez, F., Tambe, M., Kraus, S., Magori-Cohen, R.: Effective solutions for real-world stackelberg games: When agents must deal with human uncertainties. In: AAMAS (2009)
19. Vorobeychik, Y., An, B., Tambe, M.: Adversarial patrolling games. In. In: AAAI Spring Symposium on Security, Sustainability, and Health (2012)
20. Manshaei, M.H., Zhu, Q., Alpcan, T., Bacşar, T., Hubaux, J.P.: Game theory meets network security and privacy. ACM Computing Surveys, CSUR (2013)
21. Zhu, Q., Basar, T.: Dynamic policy-based ids configuration. In: Proceedings of the 48th IEEE Conference on CDC/CCC 2009. IEEE (2009)
22. Nguyen, K.C., Alpcan, T., Basar, T.: Stochastic games for security in networks with interdependent nodes. In: GameNets 2009. IEEE (2009)
23. Short, M.B., D'Orsogna, M.R., Pasour, V.B., Tita, G.E., Brantingham, P.J., Bertozzi, A.L., Chayes, L.B.: A statistical model of criminal behavior. Mathematical Models and Methods in Applied Sciences 18(suppl. 01), 1249–1267 (2008)
24. Zipkin, J.R., Short, M.B., Bertozzi, A.L.: Cops on the dots in a mathematical model of urban crime and police response (2013)
25. Plavnick, A.: The fundamental theorem of markov chains. University of Chicago VIGRE REU (2008)
26. McKelvey, R.D., Palfrey, T.R.: Quantal Response Equilibria for Normal Form Games. Games and Economic Behavior 10(1), 6–38 (1995)
27. Boyen, X., Koller, D.: Tractable inference for complex stochastic processes. In: UAI, pp. 33–42. Morgan Kaufmann Publishers Inc. (1998)

Addressing Scalability and Robustness in Security Games with Multiple Boundedly Rational Adversaries

Matthew Brown, William B. Haskell, and Milind Tambe

University of Southern California, Los Angeles, CA, USA

Abstract. Boundedly rational human adversaries pose a serious challenge to security because they deviate from the classical assumption of perfect rationality. An emerging trend in security game research addresses this challenge by using behavioral models such as quantal response (QR) and subjective utility quantal response (SUQR). These models improve the quality of the defender's strategy by more accurately modeling the decisions made by real human adversaries. Work on incorporating human behavioral models into security games has typically followed two threads. The first thread, scalability, seeks to develop efficient algorithms to design patrols for large-scale domains that protect against a single adversary. However, this thread cannot handle the common situation of multiple adversary types with heterogeneous behavioral models. Having multiple adversary types introduces considerable uncertainty into the defender's planning problem. The second thread, robustness, uses either Bayesian or maximin approaches to handle this uncertainty caused by multiple adversary types. However, the robust approach has so far not been able to scale up to complex, large-scale security games. Thus, each of these two threads alone fails to work in key real world security games. Our present work addresses this shortcoming and merges these two research threads to yield a scalable and robust algorithm, MIDAS (MaxImin Defense Against SUQR), for generating game-theoretic patrols to defend against multiple boundedly rational human adversaries. Given the size of the defender's optimization problem, the key component of MIDAS is incremental cut and strategy generation using a master/slave optimization approach. Innovations in MIDAS include (i) a maximin mixed-integer linear programming formulation in the master and (ii) a compact transition graph formulation in the slave. Additionally, we provide a theoretical analysis of our new model and report its performance in simulations. In collaboration with the United States Coast Guard (USCG), we consider the problem of defending fishery stocks from illegal fishing in the Gulf of Mexico and use MIDAS to handle heterogeneity in adversary types (i.e., illegal fishermen) in order to construct robust patrol strategies for USCG assets.

1 Introduction

Incorporating human behavioral models [11,3] into security games represents an important progression that has been demonstrated to improve the performance

R. Poovendran and W. Saad (Eds.): GameSec 2014, LNCS 8840, pp. 23–42, 2014.

of defender patrol strategies in both simulations and human subject experiments [15,19,18,13]. Behavioral models allow for the relaxation of the one of the strongest assumptions in classical game theory: namely, that the adversary is a perfectly rational utility maximizer. Instead, behavioral models, such as the quantal response (QR) model [11] and the subjective utility quantal response (SUQR) model [13], feature stochasticity in human decision making. These models are able to better predict the actions of real human adversaries and thus lead the defender to choose strategies that perform better in practice. Boundedly rational human behavioral models raise two fundamental research challenges that previous work has tried to address separately: scalability and robustness.

While perhaps counter-intuitive, modeling adversaries which behave suboptimally actually makes the defender's optimization problem computationally more difficult. Both QR and SUQR are non-linear models and are difficult to use directly in large-scale security domains. This issue of scalability for large-scale security games with boundedly rational adversaries has received attention in the literature. [19] presented a mixed-integer linear programming (MILP) approximation for QR and SUQR models which improves tractability. Additionally, [18] introduces a cutting planes approach which can handle general patrol schedules and uses a master-slave formulation to iteratively generate deep cuts. We emphasize that the work [19,18] only allows for a single boundedly rational adversary.

However, in many domains the defender could encounter multiple different types of boundedly rational human adversaries. Thus, a separate line of security games research has focused on achieving robustness against uncertainty in the true adversary model. [17] proposed a Bayesian approach which learns a Gaussian distribution over adversary types. This approach has two potential drawbacks. First, the assumption that the adversary types are normally distributed is difficult to justify in practice. Second, even if the adversaries are normally distributed, a large amount of data is needed to learn the Gaussian distribution. Alternatively, [5] introduced a *maximin* approach which does not use a distribution over the adversary types. Instead, the defender chooses a patrol that maximizes the worst-case expected defender reward over a set of adversary types. In an effort to scale up, [17,5] focused on security games with a simplified defender strategy space that do not have complicated patrol schedules.

In this paper, we merge these two research threads for the first time by addressing scalability and robustness simultaneously. Each thread alone is impractical for important real-world security domains, such as environmental crime. Security games with complicated patrol schedules *and* multiple boundedly rational adversary types present a number of modeling and computational challenges. However, overcoming these challenges is critical as they are precisely the characteristics that define real-world security games. Our main contribution here is MIDAS (MaxImin Defense Against SUQR) which computes robust defender patrols for large-scale security games with a heterogeneous adversary population. Building off the insights of [19,18,17,5], we offer two key innovations: (i) a *robust* model that generates patrols that hedge against uncertainty about a heterogeneous population of adversaries and (ii) a *tractable* MILP approximation of our robust problem. We develop key theoretical properties of MIDAS and also compare MIDAS against previous approaches in simulation.

In collaboration with the United States Coast Guard (USCG), we have applied MIDAS to protect fisheries in the Gulf of Mexico, where illegal, unreported, and unregulated (IUU) fishing seriously threatens the health of local fish stocks. The USCG has both surface and air assets with which to deter IUU fishing. We frame the interaction between the USCG and illegal fisherman from Mexico (henceforth called Lanchas) as a Stackelberg security game. By using historical data on Lancha sightings, we learn and construct a set of SUQR adversary types. However, there is not sufficient data to accurately construct a probability distribution over Lancha types. Generation of robust defender strategies for this domain has previously been explored in [5]. However, that work was more of a hot spot prediction model and it did not account for actual USCG schedules. In contrast, MIDAS constructs patrol schedules directly, resulting in higher quality patrol schedules for the USCG. The USCG began live testing of patrol schedules generated using MIDAS in July 2014.

2 Related Work

Game theory has been successfully applied to security problems such as the protection of networks [9,12,14] and physical infrastructure [16]. In particular, the Stackelberg game model with its leader-follower paradigm has been used extensively in security domains. Stackelberg games capture the fact that, in the real world, the defender (i.e., the security agency) must commit first to a strategy that may be observed and then exploited by adversaries. Given this first mover advantage, it is critical to understand and predict how adversaries will respond to a given strategy in order to find the best strategy. Classical game theory assumes that the adversary is perfectly rational and will always select the best available action in response to the defender's strategy. In some domains, such as network security [4,8], this assumption is reasonable as the game is played by software agents. For other domains, particularly those with human adversaries, a theoretically optimal defender strategy under standard rationality assumptions can perform poorly in practice. Under the assumption of perfect rationality, the adversary will always select just one action (the utility maximizing action). This assumption can lead to non-robust strategies for the defender.

As such, human behavioral models are becoming an increasingly important aspect of security games research. [19] was the first to address human adversaries in security games by incorporating the quantal response (QR) model [10] from the social psychology literature. QR predicts a probability distribution over adversary actions where actions with higher utility have a greater chance of being chosen. By anticipating possible adversary deviation from the optimal action, strategies computed with QR are more robust to uncertainty in human decision making. [7] generalized the QR model to be robust against all adversary models satisfying monotonicity (i.e., higher utility actions are selected more frequently than lower utility actions), but this approach struggles to scale up to larger security games. [13] extended the QR model by proposing that humans use "subjective utility", a weighted linear combination of factors (such as defender coverage, adversary reward, and adversary penalty), to make decisions. [13] proposes the subjective utility quantal response (SUQR) model which was shown to outperform QR in predicting the actions of participants of human subject experiments, thus leading to better defender strategies.

Building off that foundation, [18] presented an efficient cutting planes approach for solving security games with a large defender strategy space and a single adversary following a QR model. Meanwhile, two approaches have emerged for handling security games with multiple human adversary types. [17] utilized a Bayesian approach which learns a distribution over a set of SUQR types from available data. This distribution was assumed to be normal so as to minimize the number of parameters that need to be learned. Alternatively, [5] developed a robust version of [17] and applies it to the fishery protection domain where only limited data about the adversaries is available. Borrowing from the robust optimization literature [1,2], a *maximin* approach is used to optimize defender expected utility against the worst-case type from the set of possible adversary types. However, [18] handles only one adversary type, while [17] and [5] both fail to scale up. Neither of these two threads of research is individually able to handle the needs of security game applications in real-world domains such as environmental crime.

Most security problems do not feature static deployments, but rather have dynamic deployments that evolve in time and space. Thus, it is imperative to consider the capabilities of and restrictions on security resources such as personnel, cars, boats, and aircraft. Additionally, the adversaries in most physical security domains are likely to be humans, who have biases and limitations in their decision making process. This bounded rationality makes it difficult to predict the actions of the adversary and in turn for the defender to optimize their strategy. As a further complication, rather than a single adversary type there is usually a set of potential adversary types that may be encountered and it is critical to be robust against uncertainty in adversary type. Prior work on boundedly rational adversaries in security games has addressed only one of the challenges of scalability and robustness.

In this paper, we propose MIDAS which improves upon prior work by providing a holistic model that better captures the practicalities of large-scale, real-world security domains. More specifically, MIDAS enhances the incremental cut generation technique for solving large-scale security games with a single boundedly rational adversary type from [18] by using a robust *maximin* formulation for handling the uncertainty posed by multiple potential boundedly rational adversary types. Additionally, the QR model used in [18] for modeling boundedly rational adversary types is replaced with the SUQR model. Thus, MIDAS addresses the challenges of both scalability and robustness simultaneously, representing the first and only approach for solving security games with patrols schedules *and* multiple boundedly rational adversary types.

3   Background

We consider a Stackelberg security game (SSG) where the defender uses M available resources to protect a set of targets $T = \{1, \ldots, |T|\}$ from a set of boundedly rational adversaries Ω. For the remainder of this paper we will focus on the SUQR behavioral model and treat $\omega \in \Omega$ as an SUQR adversary type. SUQR outperforms QR and other human behavioral models in human subject experiments. As a result, SUQR is widely considered to be the state of the art for modeling boundedly rational adversaries in security games.

Each target $t \in T$ is assigned a set of payoffs $\{R_t^a, P_t^a, R_t^d, P_t^d\}$: R_t^a is the reward earned by an adversary if they successfully attack target t, while P_t^a is the penalty received by an adversary for an unsuccessful attack on target t. Conversely, if the defender assigns a resource to protect target t and an adversary attacks target t, the defender receives a reward R_t^d. If an adversary attacks target t and the defender has not assigned a resource to protect target t, the defender receives a penalty P_t^d. It should be noted that the payoffs for all adversary types in Ω are identical, it is the parameters of the SUQR behavioral model that distinguish between types in Ω.

The defender commits to a mixed strategy that the adversaries are able to observe and then respond to by choosing a target to attack (Korzhyk, Conitzer, and Parr 2010; Basilico, Gatti, and Amigoni 2009). We denote the jth defender pure strategy as A_j, which is an assignment of all the security resources. A_j is represented as a column vector $A_j = \langle A_{tj} \rangle^T$, where A_{tj} indicates whether target t is covered by A_j. For example, in an SSC with 4 targets and 2 resources, $A_j = \langle 1, 1, 0, 0 \rangle$ represents the pure strategy of assigning one resource to target 1 and another to target 2. Let $\mathcal{A} = \{A_j\}$ be the collection of feasible assignments of resources, i.e., the set of defender pure strategies. The defender's mixed strategy can then be represented as a vector $\mathbf{a} = \langle a_j \rangle$, where $a_j \in [0, 1]$ is the probability of choosing A_j. For large-scale security games, the number of pure strategies can grow so large that $\mathcal{A}$ cannot be represented explicitly in practice making it impossible to optimize $\mathbf{a}$ directly. However, there is a more compact "marginal" representation for defender strategies. Let $\mathbf{x}$ be the marginal strategy, where $x_t = \sum_{A_j \in \mathcal{A}} a_j A_{tj}$ is the probability that target t is covered. The set of all feasible marginal distributions is

$$\mathcal{X}_f = \left\{ \mathbf{x} : x_t = \sum_{A_j \in \mathcal{A}} a_j A_{tj}, \, t \in T, \, \sum_{A_j \in \mathcal{A}} a_j = 1, \, \mathbf{a} \geq 0 \right\}.$$

We treat $\omega \in \Omega$ as an SUQR adversary type with the weight vector $\omega = \{\omega_1, \omega_2, \omega_3\}$ which encodes the relative importance of x_t, R_t^a, and P_t^a, respectively, in the decision making process of the adversary. Recall that the SUQR model selects a probability distribution over adversary actions rather than deterministically selecting the utility maximizing adversary action. Given defender strategy $\mathbf{x}$, the probability that adversary ω will attack target t is

$$q_t\left(\omega \mid \boldsymbol{x}\right) = \frac{e^{\omega_1 x_t + \omega_2 R_t^a + \omega_3 P_t^a}}{\sum_{t'} e^{\omega_1 x_{t'} + \omega_2 R_{t'}^a + \omega_3 P_{t'}^a}}.$$

If an adversary chooses to attack target t, then for a given defender strategy $\boldsymbol{x}$, the defender's expected utility is

$$U_t\left(\boldsymbol{x}\right) = x_t R_t^d + \left(1 - x_t\right) P_t^d.$$

Against a known adversary type $\omega \in \Omega$, the defender's optimization problem is then

$$\max_{\boldsymbol{x} \in \mathcal{X}} F\left(\boldsymbol{x} \mid \omega\right) \triangleq \sum_t U_t\left(\boldsymbol{x}\right) q_t\left(\omega \mid \boldsymbol{x}\right), \tag{1}$$

which can be solved for a defender mixed strategy $\boldsymbol{a}$. However, in this paper we consider an entire population of heterogeneous adversaries in Ω. Thus, the optimization problem above is inadequate.

4   Adversary Uncertainty

4.1   Bayesian Estimation

If we have a distribution $\mathbb{P}$ over the set Ω of all possible types, then the expected utility maximizing problem is

$$\max_{x \in \mathcal{X}_f} \int_\Omega F\left(\boldsymbol{x} \mid \omega\right) \mathbb{P}\left(d\omega\right). \tag{2}$$

Problem (2) maximizes the expected defender utility, where the expectation is over the adversary types. In practice Problem (2) requires $\mathbb{P}$ to be estimated from sample data. Estimation of $\mathbb{P}$ presents two potential issues: first, it assumes that the types in Ω are normally distributed in order to use convenient update rules; second, large amounts of data are required. This method is referred to as Bayesian SUQR [17].

4.2   Maximin

Robust optimization offers up remedies for the shortcomings of Bayesian SUQR. *Maximin* does not require large amounts of data, but it can still utilize data when it is available even if only in small quantities. It is also less sensitive to assumptions about the nature of the underlying data, for instance the assumption that $\mathbb{P}$ is a normal distribution.

We treat Ω as an uncertainty set in line with robust optimization. For convenience, we assume that Ω is finite. This assumption is reasonable in practice since we will only ever have finitely many observations of the adversary. Then we solve the robust optimization problem

$$\max_{x \in \mathcal{X}_f} \min_{\omega \in \Omega} F\left(\boldsymbol{x} \mid \omega\right) \tag{3}$$

to get a patrol for the defender, where again $F(\boldsymbol{x}\,|\,\omega)$ is the expected utility corresponding to type ω. Problem (3) is a nonlinear, nonconvex, nonsmooth optimization problem. For easier implementation, we transform Problem (3) into the constrained problem

$$\max_{x\in\mathcal{X}_f,\,s\in\mathbb{R}} \{s : s \le F(\boldsymbol{x}\,|\,\omega),\, \forall\omega\in\Omega\}, \tag{4}$$

by introducing a dummy variable $s\in\mathbb{R}$ to replace the nonsmooth objective with a collection of smooth constraints.

5   Mixed-Integer Linear Programming

By considering a human behavior model such as SUQR, Problem (4) becomes a nonlinear nonconvex optimization problem. In the general case, this problem class has been shown to be NP-hard to solve to optimality. Our idea in this section is to introduce a tractable MILP approximation scheme.

An approximate approach for solving Problem (1) with a single boundedly rational adversary was presented in [19,18]. This approach is based on a piece-wise linear approximation that leads naturally to an MILP. In this section, we generalize this approach to create MIDAS, an algorithm for solving the robust Problem (4) with a set of boundedly rational adversaries.

First notice that, $F(\boldsymbol{x}\,|\,\omega)$, the defender's payoff against a single adversary type $\omega\in\Omega$ can be written out as

$$F(\boldsymbol{x}\,|\,\omega) = \sum_t U_t(\boldsymbol{x})\, q_t(\omega\,|\,\boldsymbol{x}) = \frac{\sum_t \left(\left(R_t^d - P_t^d\right)x_t + P_t^d\right) e^{\omega_1 x_t + \omega_2 R_t^a + \omega_3 P_t^a}}{\sum_t e^{\omega_1 x_t + \omega_2 R_t^a + \omega_3 P_t^a}}$$

which is a fractional function $N(\boldsymbol{x}\,|\,\omega)\,/\,D(\boldsymbol{x}\,|\,\omega)$ where

$$N(\boldsymbol{x}\,|\,\omega) = \sum_t \left(\left(R_t^d - P_t^d\right)x_t + P_t^d\right) e^{\omega_1 x_t + \omega_2 R_t^a + \omega_3 P_t^a}$$

and $D(\boldsymbol{x}\,|\,\omega) = \sum_t e^{\omega_1 x_t + \omega_2 R_t^a + \omega_3 P_t^a}$. The goal in this section is to estimate the optimal value, which we will denote s^*, of Problem (4), i.e., the defender receives a payoff of at least s^* against every adversary type $\omega\in\Omega$. We use a binary search to compute s^* by updating a parameter r. We know that $r\le s^*$ if there exists some $\boldsymbol{x}\in\mathcal{X}_f$ such that

$$r \le \frac{N(\boldsymbol{x}\,|\,\omega)}{D(\boldsymbol{x}\,|\,\omega)},\ \forall\omega\in\Omega.$$

Equivalently, we can rearrange the terms to require

$$r\,D(\boldsymbol{x}\,|\,\omega) - N(\boldsymbol{x}\,|\,\omega) \le 0,\ \forall\omega\in\Omega.$$

Therefore, to check if $r\le s^*$, we solve

$$\min_{x\in\mathcal{X}_f,\,\xi\in\mathbb{R}} \{\xi : \xi \ge r\,D(\boldsymbol{x}\,|\,\omega) - N(\boldsymbol{x}\,|\,\omega),\, \forall\omega\in\Omega\}. \tag{5}$$

If the optimal value of the above problem is less than or equal to zero, then $r \leq s^*$; otherwise, $r > s^*$; then r is adjusted appropriately. However, Problem (5) is still nonlinear and nonconvex. Thus, we need to find a tractable approximation to implement this scheme.

5.1   Linear Approximation

The nonlinearity and nonconvexity of Problem (5), whose objective function is a summation of nonlinear functions in x, can be overcome by approximating each nonlinear function with a piecewise linear function with K pieces. The functions $r\,D\,(x\,|\,\omega) - N\,(x\,|\,\omega)$ in the constraints of Problem (5) can be approximated with piecewise linear functions $L\,(x\,|\,\omega)$ of the form:

$$L\,(x\,|\,\omega) = \sum_{t\in T} \left(r - P_t^d\right)\left(f_t(0|\omega) + \sum_{k=1}^{K} \gamma_{\omega tk} x_{tk}\right) - \sum_{t\in T}\left(R_t^d - P_i^d\right)\sum_{k=1}^{K}\mu_{\omega tk} x_{tk}$$

where $\gamma_{\omega tk}$ is the slope of the function $f_t(x_t|w)$ in the kth segment while $\mu_{\omega tk}$ is the corresponding slope of $x_t f_t(x_t|\omega)$. With this approximation, we then solve the feasibility check problem

$$\min_{x,\xi} \xi \tag{6}$$

$$\text{s.t.}\, \xi \geq L\,(x\,|\,\omega), \quad \forall \omega \in \Omega, \tag{7}$$

$$0 \leq x_{tk} \leq 1/K, \quad \forall t, \quad k = 1\dots K, \tag{8}$$

$$z_{tk}/K \leq x_{tk}, \quad \forall t, \quad k = 1\dots K - 1, \tag{9}$$

$$x_{t(k+1)} \leq z_{tk}, \quad \forall t, \quad k = 1\dots K - 1, \tag{10}$$

$$z_{tk} \in \{0,1\}, \quad \forall t, \quad k = 1\dots K - 1, \tag{11}$$

$$x_t = \sum_{A_j \in \mathcal{A}} a_j A_{tj}, \quad \forall t, \tag{12}$$

$$\sum_{A_j \in \mathcal{A}} a_j = 1, \tag{13}$$

$$x,\, a \geq 0. \tag{14}$$

5.2   Column Generation

In this subsection we produce a tractable scheme for solving Problem (6) - (14). First, we derive a relaxation of Problem (6) - (14). Second, we show how to iteratively improve this approximation via a network flow problem: to that end Problem (6) - (14) is used to add new constraints to the relaxed version of the problem, and column generation is used in service of solving Problem (6) - (14) which then uses the network flow representation. Our network flow problem differs substantially from earlier work, which focused on aviation security and environmental crime, because of the generality of our formulation.

To begin, we approximate the constraint $x \in \mathcal{X}_f$ with a linear relaxation

$$\left\{ x : \hat{H} x \leq \hat{h} \right\},$$

which represents a subset of linear boundaries of $\mathcal{X}_f$. Then we solve the relaxation

$$\max_{x, s \in \mathbb{R}} \left\{ s : s \leq F\left(x \mid \omega\right), \forall \omega \in \Omega, \hat{H} x \leq \hat{h} \right\} \tag{15}$$

using the binary search method, i.e. Problem (6) - (14).

Given a candidate $\tilde{x}$, we check if $\tilde{x} \in \mathcal{X}_f$ by solving the projection problem

$$\min_{z \in \mathbb{R}^{|T|}, a \in \mathbb{R}^J} \sum_{t \in T} z_t \tag{16}$$

$$\text{s.t. } A a - \tilde{x} \leq z, \tag{17}$$

$$- z \leq A a - \tilde{x}, \tag{18}$$

$$\sum_{A_j \in \mathcal{A}} a_j = 1, \tag{19}$$

$$a \geq 0. \tag{20}$$

Problem (16) - (20) finds the best 1-norm approximation of x in $\mathcal{X}_f$, and returns the optimal value zero if $x \in \mathcal{X}_f$. Otherwise, we find a violated constraint which we add to the approximation $\hat{H} x \leq \hat{h}$.

Problem (16) - (20) has a large number of variables since $\mathcal{A}$ is exponentially large. We solve (16) - (20) using a column generation method similar to the one introduced in [6]. We solve a restriction of Problem (16) - (20) with a subset of columns $\hat{A} \subset A$ where a is now understood as a vector in $a \in \mathbb{R}^{|\hat{A}|}$, with $a_j = 0$ for all j with $A_j \notin \hat{A}$. Then we check for columns A_j to add to $\hat{A}$ by computing the reduced costs of variables a_j with $A_j \notin \hat{A}$ via the dual problem.

The dual to Problem (16) - (20) is

$$\max_{y, u} \tilde{x}^T y + u \tag{21}$$

$$\text{s.t. } A^T y + u \leq 0, \tag{22}$$

$$- 1 \leq y \leq 1, \tag{23}$$

which has a large number of constraints due to the presence of the matrix A. For a subset of columns $\hat{A} \subset A$ (abusing notation since these are matrices), we have the relaxation of the dual

$$\max_{y, u} \tilde{x}^T y + u \tag{24}$$

$$\text{s.t. } \hat{A}^T y + u \leq 0, \tag{25}$$

$$- 1 \leq y \leq 1, \tag{26}$$

$$g \geq 0. \tag{27}$$

We are looking for a column A_j such that

$$A_j^T y + u \leq 0$$

is violated. So, we solve the slave problem

$$\max_{A_j \in \mathcal{A}} \{y^T A_j\} + u \tag{28}$$

and identify a violated constraint if the optimal value of this problem is positive. Specifically, we solve Problem (28) using the technique in [6], i.e. we use a maximum reward network flow problem (since Problem (28) is a maximization problem).

To setup this network flow problem, we create a source node with supply 1, and a sink node with demand 1. We have a fixed time horizon, $n = 0, 1, \ldots, N$ stages, so we create a node (n, t) for every target and every time. The variables in this problem are the flow between nodes,

$$\mu_{(t,n),\,(t',n+1)}$$

which indicate a transition in the asset from target t at time n to target t' at time $n + 1$ in the next period. Effectively, we are taking a transition graph representation on the state space T^{N+1}. This formulation has the advantage of allowing us to express constraints on feasible patrols. The maximum reward network flow problem is then of the form

$$\max_{\mu} \left\{ \sum_{n \in \mathbb{N}} y_t \sum_{n,t} \mu_{(t,n),\,(t',n+1)} : \text{network flow constraints on } \mu \right\}.$$

The preceding network flow problem is a linear programming problem. This problem class is well studied and many efficient solution algorithms (such as the Simplex algorithm) exist that can obtain an exact optimal solution. We also point out that the preceding network flow problem can be solved efficiently for any underlying network topology.

6   Problem Properties

This section summarizes some key problem properties. The main points are to better understand our approximation scheme, to confirm that our cut generation scheme produces deep cuts, and to see how the standard Bayesian estimation approach relates to our robust approach.

6.1   MILP Approximation Error

Our underlying approach is a piecewise linear approximation to a nonconvex problem. We want to better understand the error bound for this approximation and the resulting solution quality of the corresponding MILP. We will show that all of the nonconvex functions we are approximating have bounded Lipschitz constants. Thus, since their variability is bounded, we have an upper bound on the piecewise linear approximation error as a function of the fineness of the discretization.

Recall that we are approximating the feasibility check problem, which solves

$$\min_{x \in \mathcal{X}_f} \max_{\omega \in \Omega} \{r\, D\,(x\,|\,\omega) - N\,(x\,|\,\omega)\},$$

by linearly interpolating the functions $r\, D\,(x\,|\,\omega) - N\,(x\,|\,\omega)$ for all $\omega \in \Omega$. The first step in our approximation analysis is to estimate the Lipschitz constant of $r\, D\,(x\,|\,\omega) - N\,(x\,|\,\omega)$ for fixed $\omega \in \Omega$.

Lemma 1. *The Lipschitz constant of $r\, D\,(x\,|\,\omega) - N\,(x\,|\,\omega)$ for any $\omega \in \Omega$ is bounded above by*

$$\sum_t e^{1+\max_t R_t^a + \max_t P_t^a} + \sum_t \left(R_t^d - P_t^d\right) e^{1+\max_t R_t^a + \max_t P_t^a}.$$

Proof. By direct computation, $r\, D\,(x\,|\,\omega) - N\,(x\,|\,\omega)$ is equal to

$$r \sum_t e^{\omega_1 x_t + \omega_2 R_t^a + \omega_3 P_t^a} - \sum_t \left(\left(R_t^d - P_t^d\right) x_t + P_t^d\right) e^{\omega_1 x_t + \omega_2 R_t^a + \omega_3 P_t^a}.$$

So

$$|r\, D\,(x\,|\,\omega) - N\,(x\,|\,\omega) - r\, D\,(x'\,|\,\omega) + N\,(x'\,|\,\omega)|$$
$$\leq \sum_t |e^{\omega_1 x_t + \omega_2 R_t^a + \omega_3 P_t^a} - \sum_t \left(\left(R_t^d - P_t^d\right) x_t + P_t^d\right) e^{\omega_1 x_t + \omega_2 R_t^a + \omega_3 P_t^a}$$
$$- e^{\omega_1 x_t' + \omega_2 R_t^a + \omega_3 P_t^a} - \sum_t \left(\left(R_t^d - P_t^d\right) x_t' + P_t^d\right) e^{\omega_1 x_t' + \omega_2 R_t^a + \omega_3 P_t^a}|$$
$$\leq \sum_t |e^{\omega_1 x_t + \omega_2 R_t^a + \omega_3 P_t^a} - e^{\omega_1 x_t' + \omega_2 R_t^a + \omega_3 P_t^a}|$$
$$+ \sum_t |\left(\left(R_t^d - P_t^d\right) x_t + P_t^d\right) e^{\omega_1 x_t + \omega_2 R_t^a + \omega_3 P_t^a} - \left(\left(R_t^d - P_t^d\right) x_t' + P_t^d\right) e^{\omega_1 x_t' + \omega_2 R_t^a + \omega_3 P_t^a}|.$$

We have

$$|e^{\omega_1 x_t + \omega_2 R_t^a + \omega_3 P_t^a} - e^{\omega_1 x_t' + \omega_2 R_t^a + \omega_3 P_t^a}| \leq e^{\omega_2 R_t^a + \omega_3 P_t^a} e^{\omega_1} |x_t - x_t'|.$$

Additionally,

$$|x_t e^{\omega_1 x_t} - x_t' e^{\omega_1 x_t'}| \leq |x_t e^{\omega_1 x_t} - x_t e^{\omega_1 x_t'}| + |x_t e^{\omega_1 x_t'} - x_t' e^{\omega_1 x_t'}|$$
$$\leq x_t e^{\omega_1} |x_t - x_t'| + e^{\omega_1} |x_t - x_t'|$$
$$\leq 2 e^{\omega_1} |x_t - x_t'|.$$

Now use the fact that $e^{\omega_2 R_t^a + \omega_3 P_t^a} e^{\omega_1}$ is bounded above by

$$e^{1+\max_t P_t^a + \max_t R_t^a},$$

and $2e^{\omega_1}$ is bounded above by $2\,e$. Using Lemma 2 and the triangle inequality, for any x, $x' \in \mathcal{X}_f$ we compute

$$\left| \max_{\omega \in \Omega} \{r\, D\,(x \,|\, \omega) - N\,(x \,|\, \omega)\} - \max_{\omega \in \Omega} \{r\, D\,(x' \,|\, \omega) - N\,(x' \,|\, \omega)\} \right|$$
$$\leq r \max_{\omega \in \Omega} |D\,(x \,|\, \omega) - D\,(x' \,|\, \omega)| + \max_{\omega \in \Omega} |N\,(x \,|\, \omega) - N\,(x' \,|\, \omega)|.$$

We can expand on the previous Lipschitz computation to produce an error estimate for the overall piecewise linear approximation, by using the following fact to bound the Lipschitz constant of

$$\max_{\omega \in \Omega} \{r\, D\,(x \,|\, \omega) - N\,(x \,|\, \omega)\}.$$

Lemma 2. *Let X be a given set, and $f_1 : X \to \mathbb{R}$ and $f_2 : X \to \mathbb{R}$ be two real-valued functions on X. Then,*

(i) $|\inf_{x \in X} f_1\,(x) - \inf_{x \in X} f_2\,(x)| \leq \sup_{x \in X} |f_1\,(x) - f_2\,(x)|$, and

(ii) $|\sup_{x \in X} f_1\,(x) - \sup_{x \in X} f_2\,(x)| \leq \sup_{x \in X} |f_1\,(x) - f_2\,(x)|$.

Proof. To verify part (i), note

$$\inf_{x \in X} f_1\,(x) = \inf_{x \in X} \{f_1\,(x) + f_2\,(x) - f_2\,(x)\}$$
$$\leq \inf_{x \in X} \{f_2\,(x) + |f_1\,(x) - f_2\,(x)|\}$$
$$\leq \inf_{x \in X} \left\{ f_2\,(x) + \sup_{y \in Y} |f_1\,(y) - f_2\,(y)| \right\}$$
$$\leq \inf_{x \in X} f_2\,(x) + \sup_{y \in Y} |f_1\,(y) - f_2\,(y)|,$$

giving

$$\inf_{x \in X} f_1\,(x) - \inf_{x \in X} f_2\,(x) \leq \sup_{x \in X} |f_1\,(x) - f_2\,(x)|.$$

By the same reasoning,

$$\inf_{x \in X} f_2\,(x) - \inf_{x \in X} f_1\,(x) \leq \sup_{x \in X} |f_1\,(x) - f_2\,(x)|,$$

and the preceding two inequalities yield the desired result. Part (ii) follows similarly.

6.2   Projection

The feasible region of our problem, $\mathcal{X}_f$, is exactly the same as the one found in [18]. Thus, the results of the cut generation algorithm are unchanged and we obtain deep cuts. The results are repeated here for completeness.

Lemma 3. *(i) If $\tilde{x} \notin \mathcal{X}_f$, let (y^*, g^*, u^*) be the dual variables at the optimal solution of Problem ((16)) - ((20)). Then the hyperplane $(y^*)^T x - (g^*)^T b + u^* = 0$ separates $\tilde{x}$ and $\mathcal{X}_f$.*

(ii) Furthermore, $(y^*)^T x - (g^*)^T b + u^* = 0$ is a deep cut.

As in [18], we now consider a modified norm minimization problem. The idea is that we weight the norm towards an optimal solution using local rate of change information about the objective. In our case, the objective $G(x) = \min_{\omega \in \Omega} F(x \mid \omega)$ is a nondifferentiable function, so we use the subgradient instead of the gradient. The subgradient is

$$\partial G(x) = \text{conv} \{\nabla_x F(x \mid \omega) : F(x \mid \omega) = G(x)\}.$$

For a subgradient $s \in \partial G(x)$, we use the objective $\sum_t (s_t + \xi) z_t$ where $\xi > 0$ is chosen so that $s_t + \xi > 0$ for all t.

6.3 Duality

Here we comment on the relationship of our approach to Bayesian estimation. Bayesian estimation is a classical and widespread tool for incorporating information under uncertainty. To reveal this relationship, we compute the dual of the constrained variant of Problem (3) which we reprint here for convenience:

$$\max_{x \in \mathcal{X}_f, s \in \mathbb{R}} \{s : s \le F(x \mid \omega), \forall \omega \in \Omega\}.$$

The constraints above cause Lagrange multipliers to appear; so we can compute the standard Lagrangian dual. To proceed we first introduce the Lagrange multipliers which lie in $\mathbb{R}^{|\Omega|}$ (since there are only finitely many adversary types). We let $\mathbb{R}_+^{|\Omega|}$ denote the set of nonnegative vectors in $\mathbb{R}^{|\Omega|}$.

Let

$$\mathcal{P}(\Omega) \triangleq \left\{\Lambda \in \mathbb{R}_+^{|\Omega|} : \sum_{\omega \in \Omega} \Lambda(\omega) = 1\right\}$$

be the space of probability measures on Ω, it is a subset of $\mathbb{R}^{|\Omega|}$. We will see shortly that these probability measures are the decision variables in the dual to Problem (4).

Theorem 1. *The dual to Problem (4) is*

$$\min_{\Lambda \in \mathcal{P}(\Omega)} \left\{ d(\Lambda) \triangleq \max_{x \in \mathcal{X}_f} \sum_{\omega \in \Omega} F(x \mid \omega) \Lambda(\omega) \right\}. \tag{29}$$

Proof. Let $\Lambda \in \mathbb{R}_+^{|\Omega|}$ be the Lagrange multiplier for the constraint $s \le F(x \mid \omega)$ for all $\omega \in \Omega$. We obtain the Lagrangian

$$L(x, s, \Lambda) = s + \sum_{\omega \in \Omega} [F(x \mid \omega) - s] \Lambda(\omega).$$

The Lagrangian dual problem is then

$$\min_{\Lambda \in \mathbb{R}_+^{|\Omega|}} \max_{x \in \mathcal{X}_f, s \in \mathbb{R}} \{L(x, s, \Lambda)\}.$$

We see that the inner maximization problem $d(\Lambda)$ yields the implied constraint $\int_\Omega \Lambda(d\omega) = 1$ via

$$\max_{s \in \mathbb{R}} s \left(1 - \sum_{\omega \in \Omega} \Lambda(\omega)\right),$$

which is equal to infinity unless the equality $\sum_{\omega \in \Omega} \Lambda(\omega) = 1$ holds. Thus, we have the dual problem

$$\min_{\Lambda \in \mathbb{R}_+^{|\Omega|}} \left\{\max_{x \in \mathcal{X}_f} \sum_{\omega \in \Omega} F(x \,|\, \omega) \Lambda(\omega) \;:\; \sum_{\omega \in \Omega} \Lambda(\omega) = 1\right\}.$$

We emphasize that the dual decision variables are prior distributions on the set of types. Notice that for any fixed $\Lambda \in \mathcal{P}(\Omega)$, we see that we have a Bayesian problem since we can treat Λ as a prior distribution. For Λ, we can then perform Bayesian estimation as usual. Thus, we see that the dual problem is a search for the "best" prior distribution. As a corollary, we reason that standard Bayesian estimation gives us an upper bound on the optimal value to Problem (3).

Corollary 1. *(i)* $\max_{x \in \mathcal{X}_f} \min_{\omega \in \Omega} F(x \,|\, \omega) \leq \min_{\Lambda \in \mathcal{P}(\Omega)} d(\Lambda)$.
(ii) Let $\Lambda \in \mathcal{P}(\Omega)$ be any prior distribution, then $\max_{x \in \mathcal{X}_f} \min_{\omega \in \Omega} F(x \,|\, \omega) \leq d(\Lambda)$.

Proof. Follows from weak duality for Problem (4),

$$\max_{x \in \mathcal{X}_f, s \in \mathbb{R}} \{s : s \leq F(x \,|\, \omega), \forall \omega \in \Omega\} \leq \min_{\Lambda \in \mathcal{P}(\Omega)} \max_{x \in \mathcal{X}_f} \sum_{\omega \in \Omega} F(x \,|\, \omega) \Lambda(\omega)$$

which gives

$$\max_{x \in \mathcal{X}_f} \min_{\omega \in \Omega} F(x \,|\, \omega) \leq \min_{\Lambda \in \mathcal{P}(\Omega)} \max_{x \in \mathcal{X}_f} \sum_{\omega \in \Omega} F(x \,|\, \omega) \Lambda(\omega)$$

since

$$\max_{x \in \mathcal{X}_f, s \in \mathbb{R}} \{s : s \leq F(x \,|\, \omega), \forall \omega \in \Omega\} = \max_{x \in \mathcal{X}_f} \min_{\omega \in \Omega} F(x \,|\, \omega).$$

7   Evaluation

In this section, we evaluate MIDAS in the fishery protection domain, where the USCG must patrol the Gulf of Mexico to prevent Mexican fishermen (Lanchas) from entering the United States Exclusive Economic Zone (EEZ) and fishing illegally. The zero-sum Stackelberg game we consider is played on a square grid, where each grid cell is a potential target. The defender (USCG) commits to a mixed strategy over fixed length patrols, where each target can be visited at most once. Additionally, all patrols must start and end in the first row of the grid. Meanwhile, the Lanchas select their mixed strategies over targets based on the SUQR behavioral model where each adversary has a unique weight vector ω.

For our experiments, the game payoffs are randomly generated with R_t^a uniformly distributed in [1,10] and P_t^d uniformly distributed in [-10,-1]. The remaining game payoffs, R_t^d and P_t^a, are fixed at 10 and -10, respectively. Note that R_t^a and P_t^a are the same for all adversaries. All the adversary types $\omega \in \Omega$ used in the experiments were learned from USCG data. The default settings for each experiment are: five piecewise linear segments, a set of ten adversary types (i.e., $|\Omega| = 10$), and a patrol length equal to half the number of targets rounded down (i.e. $\lfloor \frac{|T|}{2} \rfloor$). We varied the dimensions of the square grid from 5×5 to 8×8 and created thirty randomly generated game instances for each grid size.

7.1   Linear Approximation

In MIDAS, we use a linear approximation to estimate the nonlinear SUQR behavioral model. The classic tradeoff when using approximation techniques is between solution quality and runtime. Thus, it is important to understand how the granularity of the approximation affects the performance of MIDAS. Figure 1(a) shows how varying the number of segments (5, 10, and 20) used in the linear approximations impacts the defender's utility. The x-axis indicates the size of the grid, while the y-axis is the *maximin* utility obtained by the defender mixed strategy computed by MIDAS. For all grid sizes, we observe that increasing the number of segments results in higher utility for the defender as we would expect. In particular, going from 5 to 10 segments has a significant impact on the defender utility, whereas going from 10 to 20 segments produces diminishing returns and a much smaller improvement.

The other half of the tradeoff is how the number of segments impacts the runtime of MIDAS. Increasing the number of segments increases the number of variables and constraints in MIDAS, leading to a larger optimization problem which presumably would take longer to solve. The results from varying the number of segments used in the linear approximation are shown in Figure 1(b). The x-axis again indicates the size of the grid, while the y-axis is now the runtime of MIDAS in seconds. For grid sizes 5×5 through 7×7, we see that the runtime increases as the number of segments is increased. However, for the 8×8 grid, MIDAS actually runs faster for 10 and 20 segments than it does with 5 segments. One possible explanation is that while each iteration of MIDAS algorithm takes longer to compute with more segments, the quality of the cuts generated by the separation oracle improves as the feasible marginal space is represented with higher granularity. Closer examination of the data for the 8×8 grid suggests that this is indeed the case as MIDAS with 5 segments averages with 125 calls to the separation oracle and patrol generation slave, while 10 and 20 segments average 82 and 70, respectively.

In practice, it is up to the end user to determine the right tradeoff between approximation quality and runtime. Our numerical experiments here offer guidance in this regard.

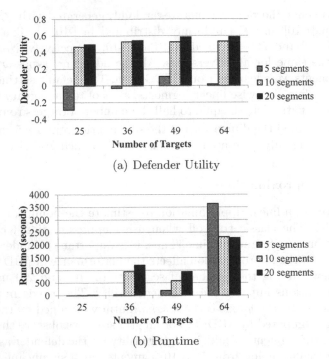

(a) Defender Utility

(b) Runtime

Fig. 1. Effect of the number of piecewise linear segments on MIDAS

7.2   Adversary Types

The primary purpose of MIDAS is to provide a scalable approach for generating game-theoretic patrols protecting against a set of adversaries with complex human behavior models such as SUQR. Therefore, we want to evaluate the effect of the number of adversary types on MIDAS to ensure that it serves its intended function. In Figure 2(a), we present the results for the defender *maximin* utility obtained by varying the number of adversary types on different grid sizes. Given that MIDAS computes a robust *maximin* strategy, we would expect that the defender utility monotonically decreases as the set of adversary types expands, as each additional type could present a new possible worst case for the defender. While overall this trend holds, we occasionally observe that the defender utility increases as the size of Ω is increased. One possible explanation may be the interaction between the linear approximation and the robust maximin formulation. Using 5 piecewise segments may be leading to a coarse approximation in which the monotincity properties no longer hold. As with the number of piecewise linear segments, we would expect that increasing the number adversary types would also lead to an increase in the runtime. In Figure 2(b), we present the runtime results for MIDAS as the size of Ω is increased, which fall in line with our expectations. In particular, for the 8×8 grid we see a significant runtime increase as Ω is expanded. However, we also see that the runtimes are relatively constant for a small number of targets.

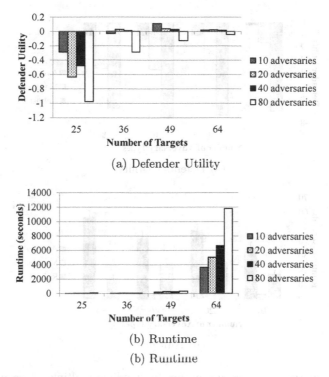

(a) Defender Utility

(b) Runtime

(b) Runtime

Fig. 2. Effect of the number of adversary types on MIDAS

7.3   Approach Comparison

Thus far, we have evaluated the performance of MIDAS as the scale of security games is increased with respect to size of the grid or the size of Ω. Now we want to compare how well MIDAS performs against other approaches that have introduced for solving security games with multiple boundedly rational adversaries. The first approach we will compare against is *Average*, in which a single adversary type ω_{avg} is constructed by averaging the weight vectors of the adversary types in Ω. After obtaining ω_{avg}, we can use MIDAS to solve the security game for $\Omega = \{\omega_{avg}\}$. The second approach we will compare against is *Marginal*, which is the robust *maximin* formulation from [5] that ignores resource assignment constraints to produce a marginal coverage distribution over the targets. To compute the *Marginal* strategy, we run MIDAS for a single iteration which produces a marginal defender strategy without considering resource assignment constraints that is then mapped into a probability distribution over patrols using the one-norm projection. The third approach is *Robust* which involves running the MIDAS algorithm to completion.

In Figure 3(a), we compare the worst case defender utility of the three approaches against sets of varying numbers of boundedly rational adversaries. The x-axis shows the number of adversary types in Ω, while the y-axis indicates the worst case defender utility of the strategies computed by the different approaches

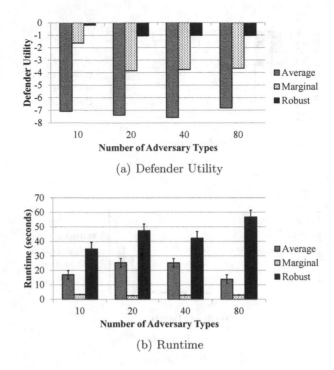

(a) Defender Utility

(b) Runtime

Fig. 3. Comparison of three approaches for handling multiple adversary types

against Ω. Perhaps unsurprisingly, the *Average* approach performs the worst out of the three across all sizes of Ω. The defender is optimizing against an artificially constructed adversary type ω_{avg} that is not in the set Ω. By not considering the extreme points in Ω, the resulting defender's strategy is highly susceptible to being exploited by at least one adversary type which would define the worst case defender utility. The *Marginal* approach shows improvement by being robust against all the types in Ω, even while it initially ignores the resource assignment constraints. Finally, *Robust* uses MIDAS to its full potential and shows additional benefit of considering resource assignment constraints by outperforming *Marginal* for all sizes of $|\Omega|$.

In addition to defender utility, runtime can provide another point of comparison between the three approaches, which we analyze in Figure 3(b). Here the x-axis again indicates the number of adversary types in Ω, while the y-axis is now the runtime needed to generate the defender's strategy using each approach. One would expect that *Average*, considering one adversary type, would run faster than *Robust*, considering $|\Omega|$ adversary types. By considering more types, the defender's optimization becomes larger with more variables and constraints. Indeed, we observe that *Robust* takes longer than *Average* for all sizes of Ω. The gap between the two approaches seems to grow as the number of adversaries is increased, particularly for $|\Omega| = 80$. However, the runtime improvement of *Average* is likely not enough to make up for the poor solution quality in real-world domains. Meanwhile, *Marginal* produces an essentially fixed runtime by solving only a single iteration of MIDAS and thus requires the least amount of runtime between the three approaches. Given the high stakes of real-world security domains, it is easy to imagine scenarios where security agencies would prefer the improved solution quality of *Robust* over the improved runtime of *Marginal*.

8   Conclusion

The use of bounded rationality models like QR and SUQR in security games is becoming increasingly popular in order to generate strategies that perform better against real human adversaries. These models raise two main research challenges: (i) scalability when handling resource assignment constraints and (ii) robustness when handling multiple boundedly rational adversaries. Up to this point, previous work has addressed these challenges individually. This paper addresses both scalability and robustness simultanesouly by introducing a new algorithm, MIDAS. The key feature of MIDAS is the combination of incremental cut generation with a robust *minimax* formulation. Our experiments demonstrate that MIDAS can scale up to security games with complex resource allocation constraints in the form of spatio-temporal patrols. Additionally, MIDAS outperforms previous approaches for protecting against multiple adversaries by providing better solution quality guarantees in terms of worst-case performance. The overall performance of MIDAS suggests that it represents the state of the art for complex security game with boundedly rational adversaries.

Acknowledgments: This research was supported by the United States Department of Homeland Security through the National Center for Risk and Economic Analysis of Terrorism Events (CREATE) under award number 2010-ST-061-RE0001 and MURI grant W911NF-11-1-0332.

References

1. Ben-Tal, A., Nemirovski, A.: Robust optimization–methodology and applications. Mathematical Programming 92(3), 453–480 (2002)
2. Bertsimas, D., Brown, D.B., Caramanis, C.: Theory and applications of robust optimization. SIAM Review 53(3), 464–501 (2011)
3. Camerer, C.: Behavioral game theory: Experiments in strategic interaction. Princeton University Press (2003)
4. Clark, A., Zhu, Q., Poovendran, R., Başar, T.: Deceptive routing in relay networks. In: Grossklags, J., Walrand, J. (eds.) GameSec 2012. LNCS, vol. 7638, pp. 171–185. Springer, Heidelberg (2012)
5. Haskell, W., Kar, D., Fang, F., Cheung, S., Denicola, E., Tambe, M.: Robust protection of fisheries with compass. In: Innovative Application of Artificial Intelligence, IAAI (2014)
6. Jain, M., Kardes, E., Kiekintveld, C., Ordóñez, F., Tambe, M.: Security games with arbitrary schedules: A branch and price approach. In: AAAI (2010)

7. Jiang, A.X., Nguyen, T.H., Tambe, M., Procaccia, A.D.: Monotonic maximin: A robust stackelberg solution against boundedly rational followers. In: Das, S.K., Nita-Rotaru, C., Kantarcioglu, M. (eds.) GameSec 2013. LNCS, vol. 8252, pp. 119–139. Springer, Heidelberg (2013)

8. Lu, W., Xu, S., Yi, X.: Optimizing active cyber defense. In: Das, S.K., Nita-Rotaru, C., Kantarcioglu, M. (eds.) GameSec 2013. LNCS, vol. 8252, pp. 206–225. Springer, Heidelberg (2013)

9. Manshaei, M.H., Zhu, Q., Alpcan, T., Bacşar, T., Hubaux, J.-P.: Game theory meets network security and privacy. ACM Computing Surveys (CSUR) 45(3), 25 (2013)

10. McKelvey, R.D., Palfrey, T.R.: Quantal response equilibria for normal form games. Games and Economic Behavior 2, 6–38 (1995)

11. McKelvey, R.D., Palfrey, T.R.: Quantal response equilibria for normal form games. Games and Economic Behavior 10(1), 6–38 (1995)

12. Nguyen, K.C., Alpcan, T., Basar, T.: Stochastic games for security in networks with interdependent nodes. In: International Conference on Game Theory for Networks, GameNets 2009, pp. 697–703. IEEE (2009)

13. Nguyen, T.H., Yang, R., Azaria, A., Kraus, S., Tambe, M.: Analyzing the effectiveness of adversary modeling in security games. In: AAAI (2013)

14. Píbil, R., Lisý, V., Kiekintveld, C., Bošanský, B., Pěchouček, M.: Game theoretic model of strategic honeypot selection in computer networks. In: Grossklags, J., Walrand, J. (eds.) GameSec 2012. LNCS, vol. 7638, pp. 201–220. Springer, Heidelberg (2012)

15. Pita, J., Jain, M., Ordonez, F., Tambe, M., Kraus, S.: Robust solutions to stackelberg games: Addressing bounded rationality and limited observations in human cognition. Artificial Intelligence Journal 174(15), 1142–1171 (2010)

16. Tambe, M.: Security and Game Theory: Algorithms, Deployed Systems, Lessons Learned. Cambridge University Press, New York (2011)

17. Yang, R., Ford, B., Tambe, M., Lemieux, A.: Adaptive resource allocation for wildlife protection against illegal poachers. In: International Conference on Autonomous Agents and Multiagent Systems, AAMAS (2014)

18. Yang, R., Jiang, A.X., Tambe, M., Ordóñez, F.: Scaling-up security games with boundedly rational adversaries: A cutting-plane approach. In: Proceedings of the Twenty-Third International Joint Conference on Artificial Intelligence, pp. 404–410. AAAI Press (2013)

19. Yang, R., Ordonez, F., Tambe, M.: Computing optimal strategy against quantal response in security games. In: Proceedings of the 11th International Conference on Autonomous Agents and Multiagent Systems, vol. 2, pp. 847–854. International Foundation for Autonomous Agents and Multiagent Systems (2012)

Empirical Game-Theoretic Analysis of an Adaptive Cyber-Defense Scenario (Preliminary Report)

Michael P. Wellman and Achintya Prakash

University of Michigan, Ann Arbor, MI, USA

Abstract. We investigate an adaptive cyber-defense scenario, where an attacker's ability to compromise a targeted server increases progressively with probing, and the defender can erase attacker progress through a moving-target technique. The environment includes multiple resources, interdependent preferences, and asymmetric stealth. By combining systematic simulation over a strategy space with game-theoretic analysis, we identify equilibria for six versions of this environment. The results show how strategic outcomes vary qualitatively with environment conditions, and demonstrate the value of reliable probe detection in setting up an effective deterrent to attack.

1 Introduction

Game-theoretic analyses of cyber-security domains typically start with stylized models of environments and agent strategies, and seek analytical characterization of solutions (e.g., equilibria) in terms of qualitative strategy properties. Such an approach often yields valuable insight, which may apply generally for broad classes of scenarios. An alternative, less frequently employed, is to start with a detailed environment model and specific dynamic strategies, and solve games based on these. We take the latter approach because it complements the former, and allows us to explore a rich set of questions without premature simplification, such as isolating all the key strategic variables in advance. This flexibility is particularly valuable for the study of *adaptive* cyber-defense, due to the complexity of analyzing strategies that interact over time.

The defining characteristic of adaptive cyber-defense, for our purposes, is that the defender policy takes into account the attack state of the system, in consideration of how successful attacks require a succession of actions to gain knowledge about and eventually compromise targeted resources. The present study incorporates only simple forms of adaptive behavior, but these are sufficient to exhibit strategically interesting decisions by both attacker and defender.

The approach we adopt is *empirical game-theoretic analysis* (EGTA) (Wellman, 2006), in which game-theoretic models are estimated from simulation data. The advantage of simulation is its ability to handle complex, stochastic, and temporally extended scenarios. Its main disadvantage is that conclusions may be difficult to generalize beyond the specific environments and strategy instances

R. Poovendran and W. Saad (Eds.): GameSec 2014, LNCS 8840, pp. 43–58, 2014.

studied. This complements traditional game-theoretic treatments, which sacrifice complexity for generality (within the simplified model). We have employed EGTA for strategic reasoning in a variety of domains, including for example collusion in privacy attacks (Duong et al., 2010) and incentives for compliance with a network security protocol (Wellman et al., 2013).

The EGTA exercise presented here demonstrates some of the interesting strategic behavior emerging from a simple adaptive cyber-defense scenario, and shows how solutions vary qualitatively depending on environmental conditions. Our results shed light on important ingredients of attacker and defender strategies, and pivotal features of environment models. We label the report "preliminary" at this stage, however, as the investigation has as yet not sufficiently explored the space of strategies and variations in environment settings that would lead us to consider the findings conclusive about strategic behavior in this domain.

2   Scenario: General Description and Related Work

Our adaptive cyber-defense scenario comprises two players: a *defender* who operates an array of networked computational assets, and an *attacker* who seeks to control or compromise these assets. For concreteness, we refer to the assets as *servers*. Servers are initially under the control of the defender, but the attacker may gain control through targeted attacks. A key feature of our scenario is that attack effort is cumulative, in that the more time an attacker has spent probing a server, the greater its prospect for successfully taking control. The main defense action is a *moving-target* technique (Jajodia et al., 2011), where the defender effectively resets the state of the server such that the attacker must restart its effort from scratch. For example, the defender could reimage a server, dynamically randomizing the layout of its address space. Our scenario model abstracts from the implementation details of this defense operation, but it falls in the category of *dynamic runtime environment* techniques, within the taxonomy of moving-target defenses presented by Okhravi et al. (2014). The point of dynamically modifying the runtime environment is that specific knowledge that the attacker has accumulated from probes to that point (e.g., based on specific memory locations of attack surfaces) is rendered obsolete by the runtime modification.

Figure 1 illustrates how a sequence of attack and defense actions can play out over time in our scenario. Attacker probes are indicated by demon heads, and defense reimage operations by reset icons. Each row represents a server, which may be under control of the defender (light blue) or attacker (dark red). Attacker probes succeed in changing control probabilistically, whereas defender reimages always work. Although the figure presents a sequence of discrete time periods, in our actual model time is continuous and actions are asynchronous.

The setup we investigate shares several features with the FlipIt abstract cybersecurity game (van Dijk et al., 2013). As in FlipIt, a server is at any time under the control of one of the players, and gaining this control is the players' main interest. Also like FlipIt, our scenario exhibits *stealth*, in that defenders do not know when the attacker has taken control. Our scenario also incorporates some

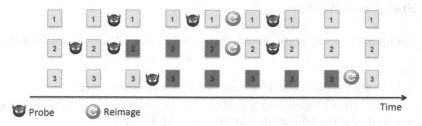

Fig. 1. Illustrative timeline of our adaptive cyber-defense scenario

major extensions of the FlipIt model, along with other important differences. First, we consider multiple servers, which is not simply the same as playing multiple FlipIts simultaneously because the utility of controlling or compromising one server depends on the state of others. The second major extension is a finer-grained model of attack actions, which compromise a server through cumulative acquisition of knowledge rather than in a one-shot takeover. Each probe action succeeds in taking over a server with some probability, which is increasing in the number of probes since the last defender reset. This is an essential feature, since as pointed out by Evans et al. (2011), moving target defenses are effective only when the attack process is incremental or progressive in some way. Finally, the stealth in our scenario is asymmetric. Attackers know when they have compromised a server, and when the defender has retaken control. And though defenders cannot tell whether an attack has succeeded, they can detect the attacker probe actions.

Extensions related to these were also studied in a series of recent papers, most written by Laszka and colleagues. In one extension, the authors incorporate multiple servers, and model objectives at two extremes where attacker control of one or all is required to control the system (Laszka et al., 2013a). Pham and Cid (2012) introduce sensing actions that reveal the compromise state of the server, at some cost. In the FlipIt version studied by Laszka et al. (2013b), the effect of attack actions is not immediate, but rather the compromise takes a stochastic amount of time. These same authors also investigate a variation in which defense actions are non-stealthy (Laszka et al., 2013c); that is, as in our scenario, attackers are aware of the state of server control.

Each of these extensions is well motivated by practical realism, but seriously complicates analysis of the FlipIt game, which to date has eluded complete analytic solution, even in its basic version. The works cited provide partial analytic solutions, contributing significant strategic insights on individual issues. In order to combine multiple issues and enable extension to yet richer environments, we adopt a simulation-based approach.

3   Detailed Game Specification

The two players in our game vie for the control of m servers, with $m = 3$ in the environment instances investigated here. The scenario runs for a finite horizon of T time units. We set $T = 10,000$ for the present study.

3.1   States and Actions

At any point of time, the state of each server can be described by a triple $\langle \chi, \upsilon, \rho \rangle$, where:

- $\chi \in \{att, def\}$ represents the player who controls the server;
- $\upsilon \in \{up\} \cup [0, T]$ represents whether the server is up ($\upsilon = up$), or, is still down from a reimage initiated at time $\upsilon \in [0, T]$; and
- ρ represents the number of attacker probes since the last defender reimage action.

The state of the overall system is defined by the joint state of the servers, plus the current clock time $t \in [0, T]$.

Each player has one available action, which it can choose to execute at any time on a specified server. The action is atomic and instantaneous, with effect described in terms of an associated state transition.

The attacker action is called *probe*. Probing a server has the effect of compromising it with some probability, depending on the extent of probing to that point. To describe the action precisely, let $\langle \chi_t, \upsilon_t, \rho_t \rangle$ be the state at time t, when a probe action is executed. We denote the state immediately following the probe by $\langle \chi_{t+}, \upsilon_{t+}, \rho_{t+} \rangle$. We specify the probe action's effects by the following rules:

- If $\upsilon_t \neq up$, the probe has no effect: $\langle \chi_{t+}, \upsilon_{t+}, \rho_{t+} \rangle = \langle \chi_t, \upsilon_t, \rho_t \rangle$.
- If $\upsilon_t = up$, the number of probes is incremented: $\rho_{t+} = \rho_t + 1$.
- If $\upsilon_t = up$ and $\chi_t = att$, the attacker maintains control: $\chi_{t+} = att$.
- If $\upsilon_t = up$ and $\chi_t = def$, the attacker takes control with probability $1 - e^{-\alpha(\rho_t + 1)}$, where $\alpha > 0$ is an environmental factor representing the information value of probes. That is, with aforementioned probability $\chi_{t+} = att$, and with remaining probability $e^{-\alpha(\rho_t + 1)}$, $\chi_{t+} = def$. In our focal environment, we set $\alpha = 0.05$.

The defender action is called *reimage*. The purpose of reimaging a server is to reset its state, so that if compromised it reverts to defender control, and if not compromised the cumulative effect of probes is erased. As for the attacker's action, we define the effect of reimage in terms of state transition rules. Suppose the defender executes a reimage at time t.

- If $\upsilon_t \neq up$, the reimage has no effect: $\langle \chi_{t+}, \upsilon_{t+}, \rho_{t+} \rangle = \langle \chi_t, \upsilon_t, \rho_t \rangle$.
- If $\upsilon_t = up$, the state is reset as follows: $\langle \chi_{t+}, \upsilon_{t+}, \rho_{t+} \rangle = \langle def, t, 0 \rangle$.

We model the reimaging duration by taking the server down for a specified time interval Δ. (In our environment instances, $\Delta = 7$.) If a reimage resets a server's state at time t, then the server comes back up Δ time units later. That is, we have $\upsilon_{t'} = t$ for $t' \in [t+, t + \Delta)$, followed by an update to the state variable $\upsilon_{t+\Delta} = up$. Aside from this one exception, all state changes in our scenario are the immediate effects of player actions.

3.2 Observation Model

As noted above, our observation model is asymmetric with respect to the two players. The defender is aware of every probe that is executed on any server, but is unaware of which probes succeed in compromising their targets. The attacker is aware of which probes succeeded, and when the defender retakes a compromised server through reimaging. To state this more precisely, we specify conditions on action-generated state transitions that the players observe.

Following a probe action:

– The attacker perfectly observes the state at $t+$.
– If $\upsilon_t = up$, the defender detects the probe, and can therefore infer ρ_{t+}.

Following a reimage action:

– The attacker detects the reimage if and only if (iff) it loses control of that server due to the reimage, that is, iff $\chi_t = att$ and $\chi_{t+} = def$. In that case, it observes the full state at $t+$.
– The defender perfectly observes the state at $t+$.

Note that the attacker always knows the control state χ, but can only imperfectly track ρ between actions. The reason is that a defender in control of a server may reset the number of probes with a reimage, and the attacker does not find out about this until its next probe. The defender always knows ρ, but except right after a reimage does not know χ.

3.3 Utility

Each player accrues utility depending on the number of servers in their control per unit time. Let u_k^i denote the rate of utility accrual for player $i \in \{att, def\}$ when i controls $k \in [0, m]$ servers. For example, if i controls m servers for $T/2$ time units, then loses control of one server for the remaining $T/2$ time units, its utility accrued would be $(T/2)u_m^i + (T/2)u_{m-1}^i$. We normalize by setting $u_0^i = 0$ and $u_m^i = 1$. Utility for control is monotonic: $k' > k \Rightarrow u_{k'}^i \geq u_k^i$. In one example instance (with $m = 3$), we set $u_1^{def} = 0.1$, $u_2^{def} = 0.7$, $u_1^{att} = 0.3$, and $u_2^{att} = 0.7$. With these settings, the attacker's utility per server compromised is close to linear, whereas the defender takes a particularly large penalty for losing control of its second server.

We can interpret these utility values in terms of the so-called "CIA triad": Confidentiality, Integrity, Availability (Pfleeger and Pfleeger, 2012). From the defender's perspective, confidentiality is maintained when all servers are under its control, availability when any of them are, and integrity (in a rough sense) when the predominance of servers are controlled by the defender. A low value for u_2^{def} would indicate that confidentiality is paramount, as most utility is lost if even one server goes out of control. Conversely, a high value of u_1^{def} would represent that the defender is concerned primarily with availability.

In addition, our model imposes a cost for executing actions. Invoking a reimage costs the defender $c_D > 0$ per unit of downtime, or a total of $c_D \Delta$ per reimage.

The attacker pays a cost of $c_A > 0$ per probe. In our study we set $c_A = 0.2$, and consider downtime costs $c_D \in \{0.3, 0.6\}$.

4   Heuristic Strategies

A *strategy* for the attacker or defender is a policy by which the player chooses when to execute its actions on what servers, as a function of its observation history and the current time. Even with a single action type, the space of available strategies is vast, owing to the combinatorial explosion of possible histories. Rather than explore the strategy space directly, we therefore focus on parameterized families of heuristic strategies, defined by regular structures and patterns of behavior over time. We define a restricted game over a selected set of such strategies, and systematically refine this set through an iterative process of strategy exploration and empirical game analysis.

Our strategy implementations interact with a discrete-event simulation of the environment. Any time that a player's knowledge state changes (see §3.2), the player strategy is queried for its next action—time and target server—assuming that it gets no further observations in the meantime. Depending on the strategy, the player may choose to retain its pending (previously scheduled) action, or to replace it on the queue with the action selected based on its latest knowledge. The environment simulator is driven by the scheduling queue, continually processing the next scheduled player action or environment event (i.e., server transition to *up*), according to time precedence. Among events scheduled for the same time, ties are broken randomly.

4.1   Attacker Strategies

The heuristic attacker strategies we consider are basically periodic, differing on the period P and the criteria by which they choose the server to target. We have thus far defined three different selection strategies.

- *Uniform-Uncompromised.* The attacker selects uniformly at random among those servers under the defender's control ($\chi_t = def$).
- *MaxProbe-Uncompromised.* The attacker selects the server that has been probed the most since last reimage (that the attacker knows about), among those servers under the defender's control, breaking ties uniformly.
- *Uniform-Uncompromised-or-Threshold.* The attacker considers servers eligible for probe if they are under the defender's control, *or* if they have not been probed within the last τ time units. The rationale for attacking servers already compromised is to prevent the defender from inferring from lack of probes that it has lost control of a server. The attacker selects uniformly among the eligible servers.

We implemented two different policies for employing these selections in a periodic manner. In *Periodic-A* strategies, the attacker schedules a probe at time $t + P$ on the designated server (or null, if no servers meet the eligibility criteria).

If the attacker observes a state change at time $t < t' < t + P$, it withdraws its pending probe, reconsiders according to the specified criteria, and schedules a new probe for $t' + P$. In *Periodic-B* strategies, the attacker selects a server based on its criteria at time t, and executes the probe immediately. It then schedules a dummy action for $t + P$ so that it can evaluate its choice at that time.

In addition, we consider the *No-Op* strategy, in which the attacker never takes any action.

4.2 Defender Strategies

For the defender, we define two selection criteria for periodic strategies, and one heuristic based on probe activity. The periodic strategies are defined by their period P, criteria for selecting which server to reimage, and the periodic management policy. A defender using *Periodic-A* strategies schedules a reimage at time $t + P$ on the designated server. If all servers are down, the defender schedules a dummy action for $t + P$ and checks whether any servers are up at that time. In case a server comes up at a time $t' \in [t, t + P]$, the defender schedules a reimage action at $t' + P$. In contrast to *Periodic-A* attackers, the *Periodic-A* defenders do not reconsider their scheduled reimage based on observations within the period. *Periodic-B* defenders select a server to reimage based on their knowledge state at t, and initiate the reimage immediately.

- *Uniform.* The defender selects uniformly at random among all active servers ($v_t = up$)
- *MaxProbe.* The defender selects the active server that has been probed most since the last reimage, breaking ties uniformly.

The third strategy triggers a reimage operation based on probe activity or inactivity.

- *ProbeCount-or-Period.* The defender reimages a server whenever it detects that it has been probed more than π times since the last reimage, *or* if it has been probed at least once but not within the last P time units. The rationale for reimaging a server that is not being probed is that this could be an indication that the attacker has already compromised it and thus ceased attack.

Finally, defenders may also adopt the null strategy *No-Op*.

5 Empirical Game Analysis

We analyze our adaptive cyber-defense scenario using *empirical game-theoretic analysis* (EGTA), an approach combining simulation with game-theoretic reasoning. Starting with a representative set of heuristic strategies for attackers and defenders, we evaluate their interactions by repeated simulation. Outcomes from these scenarios are used to define payoffs for the respective strategies in the game, and the resulting game model is analyzed to determine strategy profiles that are in game-theoretic equilibrium.

5.1   Simulation Setup and Game Model Generation

Our scenario specification includes several configurable parameters, described in §3. As noted above, the scenario instances studied here take $m = 3$, $T = 10{,}000$, $\alpha = 0.05$, $c_A = 0.2$, $c_D \in \{0.3, 0.6\}$, and $\Delta = 7$. We analyze six environment instances, differing only in defender utility for server control, and downtime cost for reimaging. Specifically, our environments employ the following utility settings, each for both Low ($c_D = 0.3$) and High ($c_D = 0.6$) downtime costs:[1]

Int: $u_1^{def} = 0.1$, $u_2^{def} = 0.7$
Con: $u_1^{def} = 0.1$, $u_2^{def} = 0.2$
Ava: $u_1^{def} = 0.8$, $u_2^{def} = 0.9$

We label these utility settings by the "CIA" features they emphasize, respectively: integrity, confidentiality, and availability. For all environments we take $u_1^{att} = 0.3$ and $u_2^{att} = 0.7$, which means that the attackers value server control in an approximately linear manner.

We implement the scenario using a discrete-event simulator. The simulator maintains state as described in §3.1. It manages a queue of scheduled actions and state transitions, repeatedly processing the next element of the queue. Actions may be directly executable (fully specified probes or reimages), or may be dummy action objects that require a call to the agent strategy to flesh out in detail. Whenever a state transition includes something observable by an agent, that agent is notified, and based on the strategy may also lead to insertion of further actions (or dummy actions) on the queue.

Table 1 lists the strategy instances that we included in our evaluation. For each attacker (Att) or defender (Def) heuristic, we specify the parameter values covered. For example, we included Uniform-Uncompromised attacker strategies, in both *Periodic-A* and *Periodic-B* policy versions, for each of the nine periods (P) listed. For the Uniform-or-Threshold attackers we considered all combinations of the indicated P and τ values (25 total). For the defender, one P value was inadvertently omitted in the *Periodic-A* case. We instantiated ProbeCount-or-Period defender strategies for all combinations of parameters π and P listed, except that for $\pi = 1$ the period is irrelevant so only one instance was included. Altogether, we included 87 attacker and 58 defender strategy instances.

For each environment, we ran simulations of all $87 \times 58 = 5046$ strategy profiles. Each profile was run at least 600 times (often many more), and for each profile we take the sample-average payoff for attacker and defender as the payoff vector in the estimated normal-form game.

5.2   Game-Theoretic Analysis Process

Once we have a normal-form game model for a specified scenario, we proceed to analyze it using standard game-theoretic algorithms. Our analysis followed the

---

[1] For the Ava/Low environment, we also ran a version with $T = 1000$, which produced identical game analysis results.

Table 1. Strategy instances included in our EGTA study

Att/Def	Heuristic	A/B	P	π or τ
Att	Uniform	A,B	0.1, 0.5, 0.7, 4, 7, 11, 15, 17, 20	—
Att	MaxProbe	A,B	0.1, 0.5, 0.7, 4, 7, 11, 15, 17, 20	—
Att	Uniform-or-Threshold	A,B	0.1, 5, 11, 17, 20	1, 5, 13, 27, 35
Def	Uniform	A	3, 15, 23, 31, 46, 57, 67, 75	—
Def	Uniform	B	3, 7, 15, 23, 31, 46, 57, 67, 75	—
Def	MaxProbe	A,B	3, 7, 15, 23, 31, 46, 57, 67, 75	—
Def	ProbeCount-or-Period	—	10, 50, 100, 150, 200, 250, 300	1, 2, 3, 4
Att,Def	No-Op	—	—	—

process displayed in Figure 2. As described above, simulating all combinations of attacker and defender strategies yields estimated payoffs for a normal-form game. We then simplify the game by removing dominated strategies. This produces a game model we can solve with standard algorithms, employing Nash equilibrium as a solution concept. Further analysis yields insight on the qualitative performance of heuristic strategies. The remainder of this section and the next elaborate on the last three steps, and their application to the six environment instances studied here.

Fig. 2. Empirical game-theoretic analysis pipeline

We start by eliminating strategies that are strictly dominated. Such strategies cannot be part of any Nash equilibrium, and removing such strategies simplifies the game. Moreover, eliminating a dominated strategy may render other strategies dominated, hence we iterate the pruning process. One pass of *iterated elimination of strictly dominated strategies* (IESDS) removes a strategy for one player such that there exists another strategy that performs strictly better regardless of the other player's choice among its remaining strategies. We implemented IESDS using the algorithm of Knuth et al. (1988). Starting from games of size 87 × 58 for each of the six environments, IESDS is able to prune 2–8 attacker strategies, and 6–41 defender strategies. Using linear programming to compute domination by mixtures eliminated just a few additional strategies. The residual subgames are still too large for our available game solver, so we require a more aggressive pruning regimen.

Toward this end, we eliminate some strategies that are not strictly dominated, but would be if the dominating strategy (or mixture of strategies) were given a boost by some small payoff increment δ. This concept, called *δ-dominance*

(Cheng and Wellman, 2007), has been found to achieve significant simplification with modest loss of accuracy. Although such aggressive pruning can eliminate some equilibrium strategies, all equilibria of the game after iterated "weaker-than-weak" elimination are guaranteed to be approximate equilibria of the original game. Specifically, they are ϵ-Nash equilibria for any ϵ less then or equal to cumulative δ (i.e., sum over iterations) employed for elimination.

In our study, we employed δ-dominance elimination as necessary to reduce the number of strategies for each player to 42 or fewer: the size we determined our solver could handle. In each round, we identified and removed the strategy requiring minimum δ for elimination, then further pruned by IESDS (i.e., $\delta = 0$). In all cases, we were able reduce the game sufficiently with relatively small cumulative δ. For our six environments, δ-IESDS achieved reduced sizes as follows:

Int/Low: 42×18, with cumulative $\delta = 1.3$
Con/Low: 42×11, with cumulative $\delta = 0.8$
Ava/Low: 22×28, with cumulative $\delta = 0.03$
Int/High: 38×39, with cumulative $\delta = 0.02$
Con/High: 28×35, with cumulative $\delta = 0.02$
Ava/High: 31×38, with cumulative $\delta = 0.03$

We calculate Nash equilibria using Gambit (McKelvey et al., 2014), a general tool for game-theoretic computation. Gambit has some difficulty with games even of this size, so we feed it a series of smaller games, with all combinations of three undominated attacker strategies against the full set of undominated defender strategies. This produces a set of candidate equilibria, which we then filter by testing deviations from the rest of the undominated attacker set. This process will produce all equilibria with attacker support three or fewer in the shrunken game, as long as the subgame solutions are exhaustive.

5.3   Equilibrium Results and Analysis

We found three qualitatively distinct equilibria for this adaptive cyber-defense scenario, which manifest across the six environments in an intuitively sensible pattern (see Table 2). We did not conduct an explicit statistical analysis of the results, as it was quite apparent that the sampling error made no difference to the qualitative equilibrium conclusions.

Maximal Defense. In the *Maximal Defense* (MaxDef) equilibrium, the defender responds to probing activity with aggressive reimaging, to the point that the attacker cannot achieve any worthwhile amount of compromise, and in consequence simply gives up. For example, suppose the defender plays ProbeCount-or-Period with $\pi = 1$ (abbreviated by $PCP(1,x)$, as the period is irrelevant at that setting), which means it reimages as soon as it sees a probe. Even if the attacker's probe were successful, it would not maintain control for more than an instant, so the probe had cost without benefit. The best response for the attacker is therefore No-Op. Against the No-Op attacker, the aggressive PCP defense never actually has to reimage, so it achieves the greatest possible utility (continual control of all servers, no reimaging cost) in this equilibrium.

Table 2. Nash equilibria for the six environments studied

c_D	Defender Utility (u_1^{def}, u_2^{def})		
	Int (0.1, 0.7)	**Con** (0.1, 0.2)	**Ava** (0.8, 0.9)
Low (0.3)	MaxDef	MaxDef	MaxDef, PerΔ
High (0.6)	MaxDef, MaxAtt	MaxDef, MaxAtt	MaxDef, PerΔ

MaxDef is an equilibrium profile for all six of our environments. The attacker utility is constant across these instances, and defender utility is maximal in all as well given full control and no reimaging. Technically, there are a large set of MaxDef equilibria, where the attacker plays No-Op and the defender plays some mixture of PCP strategies that are sufficient to deter probes. Specifically, No-Op is a best response for the attacker against $PCP(1,x)$, $PCP(2,10)$, or $PCP(3,10)$, and any PCP strategy is a best response against No-Op. Any mixture of the strategies listed against No-Op would therefore constitute an equilibrium, as would mixtures of these along with some probability of playing other PCP strategies, as long as the components of the most aggressive PCP strategies are probable enough to deter the attacker from probing.

In fact, our argument that MaxDef is in equilibrium applies under the general assumptions of this scenario, even with respect to the full space of possible attacker and defender strategies.

Proposition 1. *For any environment parameter settings and any strategy sets that include No-Op for the attacker and $PCP(1,x)$ for the defender, the profile comprising these strategies is a Nash equilibrium.*

Proof. Suppose the defender plays $PCP(1,x)$. The only way an attacker can gain positive utility is to compromise servers through probing. However, with $PCP(1,x)$ the defender immediately takes back control on compromise, and so the attacker ends up accruing zero utility, $u_0^{att} = 0$, regardless. The No-Op strategy achieves zero payoff, which is better than any strategy that involves any probing. Therefore No-Op is a best response to $PCP(1,x)$, among all possible attacker strategies.

Suppose the attacker plays No-Op. In that case, the defender keeps control of all servers, and accrues maximum utility $u_m^{def} = 1$ with any strategy. The strategy $PCP(1,x)$ never reimages and thus incurs zero cost, so overall payoff is maximal. Therefore $PCP(1,x)$ is a best response to No-Op, among all possible defender strategies.

We have established the Nash equilibrium with no reference to variable parameters m, T, α, c_A, c_D, Δ, or u_k^i for $0 < k < m$, thus the result holds for any legal settings. $\square$

For two of our environments, MaxDef is the only equilibrium found. The other four have additional equilibria.

Maximal Attack *Maximal Attack* (MaxAtt) is the flip-side to MaxDef, where the attacker probes sufficiently aggressively to deter active defense. Such deterrence applies when the utility the defender can achieve by taking control of servers through reimaging is not worth its cost. In such a case, the defender's best response is No-Op. When the defender plays No-Op, the attacker maximizes payoff by taking control of the servers as quickly as possible, which for our strategy set is achieved by the periodic (Uniform or MaxProbe) strategies, with $P = 0.1$. The differences between Uniform and MaxProbe, and between *Periodic-A* and *Periodic-B* in this situation are statistically indistinguishable.

MaxAtt is an equilibrium for environments Int/High and Con/High, but not the others. To see why this is the case, consider that when the attacker is probing at high frequency, maintaining control of the server requires reimaging it almost as soon as it comes back up from the last reimage. We can gauge whether this is worthwhile by comparing the utility for controlling servers with the downtime cost (c_D), since both parameters are in units of payoff/time. For environment Int/Low, keeping control of one server is not worthwhile ($u_1^{def} = 0.1 < 0.3 = c_D$), but it is worthwhile to keep control of *two* ($u_2^{def} = 0.7 > 0.6 = 2c_D$). When we double c_D to get environment Int/High, however, it is not worthwhile to keep control of any number of servers against a high-frequency attacker. Therefore MaxAtt is an equilibrium for Int/High, and for Con/High as well. It is not an equilibrium for Ava/High, as defense is worthwhile for one server even at the high downtime cost. At the low cost, MaxAtt is not in equilibrium for any values of u_1^{def} and u_2^{def}, as defending all three servers is worthwhile regardless: $u_3^{def} = 1.0 > 0.9 = 3c_D$.

We illustrate the outcomes of responding to an aggressive attacker (Uniform selection, *Periodic-A*, $P = 0.1$), in Figure 3. The top plot shows defender payoffs for a range of periodic strategies. In both Ava environments, the defender accrues the greatest payoff by choosing a period that maintains control of one server. The maximum is achieved at a period coinciding with reimaging downtime. A higher frequency of reimaging incurs more downtime cost, whereas at lower frequency the defender controls no servers for much of the time. For the Int and Con cases, respectively, the defender's utility function particularly rewards controlling two and three servers. With High downtime cost, the periodic defender cannot make reimaging worthwhile, which is reflected in payoffs increasing toward zero with longer periods. For Int with Low downtime cost, the defender can achieve small positive payoff with a short period, which refutes MaxAtt as an equilibrium for this environment.

The bottom plot shows the response of PCP strategies. Since these have two parameters (π and P), they cannot be ordered linearly on the x-axis. For each environment, we can discern a pattern of payoffs as P is increased for each Probe-Count threshold π. Some are increasing, some decreasing, and the Ava environments in particular exhibit interior maxima. The positive payoffs for $PCP(1, x)$ and $PCP(2, 10)$ refute MaxAtt as an equilibrium in Con/Low.

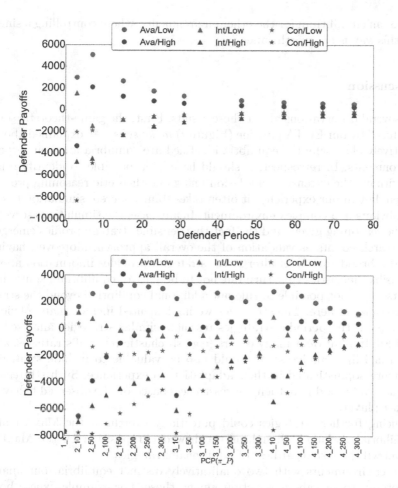

Fig. 3. Defender payoffs versus a maximally aggressive attacker, for all six environments. Marker shapes represent the different utility settings and colors represent different downtime costs. (top) Payoffs for periodic defender strategies, varying P, against $Unif(A, 0.1)$. (bottom) Payoffs for PCP defender strategies, varying π_P, against $Unif(A, 0.1)$.

5.4 Periodic Δ Reimage

Our final equilibrium, *Periodic Δ Reimage* (PerΔ), has the defender playing a periodic (MaxProbe, *Periodic-B*) strategy with $P = 7$, against a high-frequency defender (MaxProbe, *Periodic-B* with $P = 0.1$). As suggested above, that the defender's period equals the downtime interval Δ is not a coincidence. By this strategy, the defender keeps one server under its control (albeit down all the time), leaving the other two to be grabbed by the attacker. These strategies are in equilibrium for the Ava utility environments (both Low and High downtime cost), where the defender gets the lion's share of its possible utility by controlling one server.

PerΔ is not an equilibrium for the other environments, where controlling a single server in this way is not worthwhile.

5.5   Discussion

We offer several observations about these results. First, the game-theoretic solutions produced by our EGTA pipeline (Figure 2) *make sense.* As explained above, it is intuitively clear why the equilibria identified are equilibria for their respective environments. In retrospect, it should have been possible to identify some of these without the extensive simulation and game-theoretic reasoning process undertaken, but in our experience it often takes some concrete simulation to expose the obvious in a complex environment. In any case, the simulations serve to confirm the reasoning given, and the fact that sensible strategy profiles emerged from the search counts as validation of the overall approach. Moreover, having considered a broad variety of alternative strategies provides information about other plausible heuristics that turn out not to be part of equilibrium solutions.

Of course, it is not possible to rule out additional equilibria beyond the strategy sets considered here. The strategies we implemented include many obvious candidates (e.g., the periodic strategies resemble similar strategies analyzed in studies of FlipIt), but omit many others (e.g., stochastic renewal strategies, also considered in FlipIt analyses). It would also be valuable to include strategies that are more sophisticated in their adaptation to experience. Such strategies, for example, could modulate their aggressiveness based on the observed behavior of the other player.

Introducing further strategies could potentially overthrow the MaxAtt and PerΔ equilibria we found, though as we showed (Proposition 1), the MaxDef equilibrium will persist regardless of additional strategies.

For the environments with two qualitatively distinct equilibria, our analysis has nothing to say about selection among these. For example, where both MaxDef and MaxAtt are in equilibrium, which would prevail depends on the relative fortitude of the attacker and defender. More technically, we would ask which player can more credibly threaten its maximalist policy. Such questions could be addressed through a more extensive-form (dynamic) analysis, for example by explicitly considering multiple stages of decision and adopting equilibrium refinement based on perfection. Alternatively, we could consider Stackelberg models, where one player or the other is presumed to have commitment power based on the scenario setup.

It is also important to question whether features of the environment that produce these results are entirely realistic. For instance, it seems strange to give the defender so much credit for controlling a server that is down, particularly if availability is the basis for a particular utility function. We saw that changing the relative cost of downtime compared to server control utility (i.e., the Low versus High environments) could indeed affect equilibria. Moreover, the analysis underscores the necessity of interpreting a particular setting of c_D relative to the utility settings u^{def}.

The fact that MaxDef is always in equilibrium also prompts scrutiny about environment assumptions. The credibility of the defender responding to every probe relies crucially on the power the defender has to perfectly detect such probes. Any inaccuracy in this detection would undermine the maximal defense. If there were a significant prospect of false positives, this policy might be too costly to the defender. Or with false negatives, the attacker could get some traction even against the maximal defense.

6 Conclusions

We studied a simple scenario in adaptive cyber-defense. The model employs abstract models of actions and attacker and defender objectives, yet goes beyond previous models in simultaneously accommodating multiple resources, progressive attack behavior, and asymmetric stealth. Through empirical methods, relying heavily on simulation coupled with game-theoretic reasoning, we identified equilibrium strategy profiles for a variety of environment settings. Though the results must be considered preliminary (sparse coverage of the space of environments, provisional equilibria based on incomplete strategy sets), the pattern of equilibria we found reveal interesting strategic interactions between the attacker and a moving-target defender. In particular, having perfect ability to detect probes gives a defender a powerful deterrent strategy, applicable in a broad range of environment settings.

Our study also illustrates empirical game-theoretic methodology in a salient security domain. The simulation approach allows us to deal with dynamic complexity in the environment, yet still apply standard game-theoretic solution concepts.

Work on this scenario, and modeling adaptive cyber-defense more generally in this framework, is ongoing. In addition to the extensions noted in §5.5, we intend to explore environments with a range of probe efficacy (e.g., settings of α), stochastic downtimes, and alternative attacker utility models. We are also focusing on extending the space of strategies to include those far more adaptive to opponent behavior, including intent inference, and explicit reasoning about threats and counter-threats.

Acknowledgment. This work was supported in part by MURI grant W911NF-13-1-0421 from the US Army Research Office. We thank George Cybenko and other participants in this project for contributions to defining the scenario, and Satinder Singh Baveja for constructive comments on an earlier draft.

References

Cheng, S.F., Wellman, M.P.: Iterated weaker-than-weak dominance. In: 20th International Joint Conference on Artificial Intelligence, Hyderabad, pp. 1233–1238 (2007) van Dijk, M., Juels, A., Oprea, A., Rivest, R.L.: FlipIt: The game of "stealthy takeover". Journal of Cryptology 26, 655–713 (2013)

Duong, Q., LeFevre, K., Wellman, M.P.: Strategic modeling of information sharing among data privacy attackers. Informatica 34, 151–158 (2010)

Evans, D., Nguyen-Tuong, A., Knight, J.: Effectiveness of moving target defenses. In: Jajodia, et al. (2011)

Jajodia, S., Ghosh, A.K., Swarup, V., Wang, C., Wang, X.S. (eds.): Moving Target Defense: Creating Asymmetric Uncertainty for Cyber Threats. Springer (2011)

Knuth, D.E., Papadimitriou, C.H., Tsitsiklis, J.N.: A note on strategy elimination in bimatrix games. Operations Research Letters 7, 103–107 (1988)

Laszka, A., Horvath, G., Felegyhazi, M., Buttyán, L.: FlipThem: Modeling targeted attacks with FlipIt for multiple resources. In: Poovendran, R., Saad, W. (eds.) GameSec 2014. LNCS, vol. 8840, pp. 173–192. Springer, Heidelberg (2014)

Laszka, A., Johnson, B., Grossklags, J.: Mitigating covert compromises: A game-theoretic model of targeted and non-targeted covert attacks. In: Chen, Y., Immorlica, N. (eds.) WINE 2013. LNCS, vol. 8289, pp. 319–332. Springer, Heidelberg (2013b)

Laszka, A., Johnson, B., Grossklags, J.: Mitigation of targeted and non-targeted covert attacks as a timing game. In: Das, S.K., Nita-Rotaru, C., Kantarcioglu, M. (eds.) GameSec 2013. LNCS, vol. 8252, pp. 175–191. Springer, Heidelberg (2013c)

McKelvey, R.D., McLennan, A.M., Turocy, T.L.: Gambit: Software tools for game theory, version 13.1.2 (2014), www.gambit-project.org

Okhravi, H., Hobson, T., Bigelow, D., Streilein, W.: Finding focus in the blur of moving-target techniques. IEEE Security and Privacy 12(2), 16–26 (2014)

Pfleeger, C.P., Pfleeger, S.L.: Analyzing Computer Security: A Threat/Vulnerability/Countermeasure Approach. Prentice Hall (2012)

Pham, V., Cid, C.: Are we compromised? Modelling security assessment games. In: Grossklags, J., Walrand, J. (eds.) GameSec 2012. LNCS, vol. 7638, pp. 234–247. Springer, Heidelberg (2012)

Wellman, M.P.: Methods for empirical game-theoretic analysis (extended abstract). In: 21st National Conference on Artificial Intelligence, Boston, pp. 1552–1555 (2006)

Wellman, M.P., Kim, T.H., Duong, Q.: Analyzing incentives for protocol compliance in complex domains: A case study of introduction-based routing. In: Twelfth Workshop on the Economics of Information Security (2013)

Strategic Discovery and Sharing
of Vulnerabilities in Competitive Environments

M.H.R. Khouzani, Viet Pham, and Carlos Cid

Information Security Group, Royal Holloway, University of London, UK
{arman.khouzani,carlos.cid}@rhul.ac.uk, viet.pham.2010@live.rhul.ac.uk

Abstract. We investigate the incentives behind investments by competing companies in discovery of their security vulnerabilities and sharing of their findings. Specifically, we consider a game between competing firms that utilise a common platform in their systems. The game consists of two stages: firms must decide how much to invest in researching vulnerabilities, and thereafter, how much of their findings to share with their competitors. We fully characterise the Perfect Bayesian Equilibria (PBE) of this game, and translate them into realistic insights about firms' strategies. Further, we develop a monetary-free sharing mechanism that encourages both investment and sharing, a missing feature when sharing is arbitrary or opportunistic. This is achieved via a light-handed mediator: it receives a set of discovered bugs from each firm and moderate the sharing in a way that eliminates firms' concerns on losing competitive advantages. This research provides an understanding of the origins of inefficiency and paves the path towards more efficient sharing of cyber-intelligence among competing entities.

1 Introduction

Businesses across different sectors of the economy, from telecommunication and finance to energy, healthcare and transportation, increasingly rely on cyberspace and IT services. Past incidents of cyber-attacks and consequent damages have left little doubt in the minds of business managers and policy makers about the importance of investment in cybersecurity. Gathering and exchange of security intelligence are identified as key factors in enhancing the effectiveness of cybersecurity measures. Steps have been taken by governments to provide the environments to galvanise and coordinate the exchange of cybersecurity information: UK launched the "Cyber Security Information Sharing Partnership" [1] after a pilot program in 2011/12 as a "joint, collaborative initiative between industry and government to share cyber threat and vulnerability information in order to increase overall situational awareness of the cyber threat". In the US, the "National Coordinating Center for Communications (NCC)" acts as the "Information Sharing and Analysis Center (ISAC)" for telecommunication [2].

While "Information Sharing and Analysis Centers (ISACs)" – such as Information Technology (IT)-ISAC and Financial Services (FS)-ISAC – can provide the platform for exchange of cyber-intelligence, the role of incentives must not be

R. Poovendran and W. Saad (Eds.): GameSec 2014, LNCS 8840, pp. 59–78, 2014.

ignored. Providing the means of communication in the presence of strategic and competing profit-maximizing entities does not necessarily lead to exchange of their cybersecurity information. In order to understand the incentives of firms in creating and sharing information security knowledge, it is important to identify the distinct nature of the security information being shared. Some example categories of the type of cyber-intelligence to be shared are: (a) steps, protocols and measures a firm has taken to improve its security; (b) past incidents of successful or unsuccessful attacks and the resulting privacy, intellectual property and financial losses; and (c) discovered security vulnerabilities. Sharing each of these types of information have specific incentive implications. For instance, "public disclosure" of security breach incidents can harm the consumers and investors' confidence and lead to a statistically significant decreases in the market value of firms [3–5]. In this paper, we particularly focus on the third type of information: sharing discovered security vulnerabilities, or *bugs* for short.

From the societal point of view, sharing knowledge of security vulnerabilities among firms is a positive move: it improves the overall efficiency of bug discovery efforts. It moreover enhances the cyber protection of an entire industry against future attacks by reducing the common exploitable threats. It is often the case that different organizations of an economic sector bear similar vulnerabilities in their information systems [6]. This is partly due to the adoption of common implementations, libraries or operating systems. For instance, the `Heartbleed` bug (formally, CVE-2014-0160), a buffer-over-read vulnerability in the `OpenSSL` cryptographic library exposed in April 2014, affected around half a million certificates issued by trusted certificate authorities [7]. Another reason why different technological companies face common threats is the incorporation of discovered vulnerabilities into hacking toolkits which enables even less sophisticated users to configure the same malware to attack across different organizations [6].

Recognizing the need for cyber-protection, companies may invest in finding their security vulnerabilities. These can be "bugs" for example in their application level software, operating system or implementation of a network protocol, which we will hence generically refer to as the common *platform*. No company knows exactly how many bugs there are in a software they are using. More investment and effort in security research increases the chances of discovering them, but there is always a factor of luck involved. Each company patches and rectifies the vulnerabilities it finds, which is usually the much easier part than finding them in the first place. Each bug that is not discovered by a company, and hence not rectified, is potentially exploitable by cyber-attackers.

When a bug is indeed successfully exploited, the victim suffers direct losses. These can include outage of their services, recovery costs, losses of important data, user compensation, legal fines, etc. However, a company may also be affected by incidents of cyber-attacks on other companies in that economic sector: On one hand, the whole sector of the economy may suffer a blow: as customers may lose confidence in the whole "service" and seek alternative "safer" means. For instance, if one or a few major online banking companies fall victim to a cyber-attack, then some customers may lose confidence in the whole sector and switch to traditional banking altogether. Moreover, investors and stock holders may too lose confidence in the whole industry in favour of alternative options for investment. These two effects translate to a net market value loss of the whole sector, which bites all of the companies upon a successful attack on anyone of them. However, on the other hand, if (and once) a bug is exploited in competitor(s) that a company has discovered before (and has hence taken care of), it can have the opposite effect of boosting the confidence of customers as well as the investors: customers may switch to use and investors redirect their capital to the "safer" company. In other words, discovering a bug in a common software may give a company a "competitive edge" compared to others.

The two effects work in the opposite direction of each other in terms of incentives for sharing the found vulnerabilities. The sharing strategies, in turn, affect the investment decisions to discover the bugs in the first place: On the one hand, sharing information translates to a more effective discovery process and hence encourages investment, as the findings of one company is fortified by another's since the process of finding the bugs is probabilistic in nature. But on the other hand, there can be a tendency of free-riding on the discovery investment of other companies and hence get away with less investment. Further complicating the problem is the presence of uncertainty and information asymmetry: companies ought to make their discovery investment decisions in the face of uncertainties about the total number of bugs, and they need to make decision about sharing of their findings not knowing the number of findings of the other company.

Contributions of this paper are as follows: In Section 2, we model the interdependent vulnerability research investment and information sharing decisions of two strategic and competing firms as a two stage Bayesian game. We fully determine the Perfect Bayesian Equilibria of the game in closed-form in Section 3. Specifically, in Subsection 3.1, we derive the Bayesian equilibrium strategies of the firms about sharing of their finding for a given investment pair, and given their findings. In particular, we establish that the sharing strategies are unique and dominant, and are in the simple forms of "full-sharing" or "no sharing", completely determined by the competitive nature of the security findings. In Subsection 3.2, we derive the investment strategies of the firms knowing their subsequent sharing strategies. We show how "full sharing" leads to free-riding and inefficiently low investments. Also how "no sharing" is socially inefficient by preventing mutual benefit of sharing, double-efforts and potential over-investment. Finally, in Section 4, we provide a light-weight mediation mechanism free of monetary-transfers that enable (partial) sharing of the information when the firms fail to achieve any sharing on their own.

Comparison to Literature: Information sharing in the context of cybersecurity is investigated in papers like [8–15]. These works build on microeconomic models of information sharing in a general oligopoly (e.g. [16–18]) where the effect of information sharing is captured as improvement in the efficiency of production, i.e., reducing the marginal cost, or improving demand, or both. A common feature of the models is that there is no specification of the type of security information to be shared. The decision of how much information to share is modelled as a normalized continuous variable between zero and one, zero corresponding to no sharing and one corresponding to full sharing. In contrast, we specifically model the information as the discovered security vulnerabilities by each player, and hence, the sharing decisions in our model is the "number" of bugs to be shared. In addition, the relation between security investments and information sharing is rather loose in the previous literature. For instance, the effective amount of shared information is heuristically chosen as the product of the investment decision and sharing decision. In contrast, we specifically model the process of investment for "generation" of security information and subsequently, sharing of them. Moreover, we develop a mediation monetary-free mechanism that enables sharing in the face of competition as a novel contribution. More distantly, this work is related to research on R&D rivalries, e.g. [19], with at least one major difference that vulnerability discoveries are inherently not patentable.

2   Model

Our model considers a game between firm i and firm j where each decides how much to invest in security research on a common "platform", and subsequently how many of their found security vulnerabilities to share with the other. The platform has an unknown number of security vulnerabilities, or "bugs", which, if not discovered and rectified, may be exploited with ramifications for both firms. Before the game starts, the nature determines the total number of bugs following some distribution. Let the random variable representing the total number of bugs be B with the sample space of $\mathbb{N}^{+1}$ and known mean value λ. The realisation of B is not observed by any of the firms. The game play consists of two stages: *investment* and *sharing*, as described in the following:

1- Investment: In this stage, the players, while unaware of the total number of bugs in the platform, "simultaneously" decide how much to invest in bug discovery, and make it publicly known. Note that simultaneous move in the context of game theory just implies that neither one of the players can assume pre-commitment to a decision by the other players. A player's investment c determines the probability $p \in [0, 1)$ that each bug is discovered. For simplicity, we assume that the bugs are homogeneous, in that they are equally difficult to discover. Moreover, we assume discovery of each bug is independent across the bugs and across the firms. The research investment c and discovery probability p are related through function π as $p = \pi(c)$, with $\lim_{c \to \infty} \pi(c) = 1$. We naturally assume that $d\pi(c)/dc > 0$, as well as $d^2\pi(c)/dc^2 \leq 0$: The chance of finding bugs should be improved with more investment, and it is increasingly more difficult to improve the success of bug discoveries. In general we assume that the two firms have distinct cost-probability relations, denoted as $\pi_i(c)$ and $\pi_j(c)$. Because we assume both π_i and π_j are strictly incising, there is a one-to-one mapping between investment and discovery probability. Indeed, $c_i = \pi_i^{-1}(p_i)$

---

[1] We adopt the convention that random variables are denoted by capital letters and their realisations by lower case. Also, $\mathbb{N}^+ := \mathbb{N} \cup \{0\}$.

and $c_j = \pi_j^{-1}(p_j)$. Hence, we can equivalently represent each player's strategy in this stage by its choice of discovery probability, i.e., p_i and p_j.

2- Sharing: After investments are made, each player privately and independently "discovers" some bugs in the platform. Subsequently, each decides how many of its findings to share with the other. Note that the discoveries are not part of the strategies of the players and is rather determined probabilistically –by "nature"– once the investments are made. Since the discoveries are private, they cause an "incompleteness of information" of players about each other. We therefore model this sharing decisions as a Bayesian game. Firms i and j respectively discover N_i and N_j bugs in the platform, which are random variables with the common sample space of $\{0, 1, \ldots, B\}$.[2] The set of discovered bugs may have an overlap, i.e., some identical bugs may be discovered by both firms. We denote the number of common bugs by N_{ij}. The sample space of N_{ij} is $\{0, 1, \ldots, \min(N_i, N_j)\}$. Given the total number of bugs B and investment levels c_i and c_j, the nature determines the number of bugs discovered by each firm and the number of commonly discovered bugs N_i, N_j and N_{ij}. The quadruple (B, N_i, N_j, N_{ij}) is the random variable over the set of possible "states of the world" Ω. Note that due to the revelation of investments at the end of the first stage, the probability distribution of (B, N_i, N_j, N_{ij}) over Ω is publicly known. For each nature state $(b, n_i, n_j, n_{ij}) \in \Omega$, firm i (resp. j) observes n_i (resp. n_j), i.e., the number of bugs it has discovered, as its "type". For each realisation of the number of found bugs and announced investments, a firm must decide how many of its found bugs to share with the other. Due to the homogeneity assumption of bugs, the bugs to be shared can be assumed to be picked uniformly randomly. A (pure) strategy of firm i is thus a mapping $s_i(p_j, n_i) : [0, 1] \times \mathbb{N}^+ \to \mathbb{N}^+$ such that $s_i(p_j, n_i) \leq n_i$.[3] Let $\sigma_i = (p_i, s_i)$ denote the pure strategies of player i for the whole game. After both σ_i and σ_j are decided, the the overall utilities of each player is determined as the result of its investment together with the expected losses/gains from security incidents.

In what follows, we describe the expected utility of the two players after two stages of actions. We assume risk-neutral players, that is, the players care equally about their utility of expected outcome and their expected utility. Hence, the utilities are linear sums of the (negative of the) expected costs per each bug minus the investment cost for discovery of the bugs. Note that at the time of taking the decision about sharing the discovered bugs, the investments for discovering the bugs are "sunk" costs, i.e., they are already spent and will not affect the cost to go of different actions to take. Each bug, if not discovered by or informed to a player, will be successfully exploited on that player by attackers with a probability, which without loss of generality, we take to be one. We assume that the exploitation probabilities and the severity of bugs are homogeneously distributed. For each bug there are three types of losses/damages:[4]

---

[2] By $\{0, 1, \ldots, B\}$, it is meant that given the realisation $B = b$, the set is $\{0, 1, \ldots, b\}$.

[3] Since p_i is part of player i's strategy, it needs not be included as an argument to s_i.

[4] For simplicity of exposition, we assume the losses and damages are symmetrical; it is straightforward to generalise the results to non-symmetric cases.

Table 1. List of main notations

Parameter	Definition
B, b	Random variable for the total number of bugs, and a realisation
N_i, n_i	Random variable for the number of bugs discovered by i, and a realisation
N_{ij}	Random variable for the number of common bugs discovered by both
a_i	Action of player i: how many discovered bugs to share
λ	Expected number of the total number of bugs
p_i, p_j	Probability that each bug is discovered by player i,j
u_i, u_j	Expected utilities of player i, j
c_i, c_j	Discovery investment cost of player i,j
l	Direct loss upon exploitation of an (undiscovered) bug by attackers
δ	Loss (gain) in utility of the player who is the only one attacked (not attacked) – capturing the market competition effect
τ	Loss in utility of both players if a bug is exploited in either one of them – capturing the total market section shrinkage effect
$p = \pi(c)$	The relation relating the level of investment c to the discovery probability of a bug p. In this paper, we use $p = \pi(c) = 1 - e^{-\theta c}$.

- **Direct loss** $\boxed{l > 0}$: affecting only the compromised firm (e.g. outage/denial of its services, compromise/corruption of its data, etc.).
- **Market shrinkage** $\boxed{\tau \geq 0}$: the common loss as a result of a successful attack that affects both, even the firm that is not compromised. This is the effect of the market shrinkage after a successful attack as a result of a portion of both demand and investment moving away from (abandoning) the whole service/technology in favour of "safer" alternatives, or simply relinquishing that sector altogether.
- **Competitive loss** $\boxed{\delta \geq 0}$: when *only one* firm is compromised by attackers, the compromised firm loses δ while the other gains δ. This represents the shifting of demand and/or public investment (stocks) upon a successful attack.

Given the notions described above, there are four possibilities of net cost for each bug that a player may incur: (a) The bug is known by both players (either through own discovery or through the information shared by the other firm). In this case, the utility of the players is $(0,0)$, as neither one of the players loses anything.[5] (b) The bug is known by player i, but not player j. In this case, the utility pair is $(\delta - \tau, -\delta - \tau - l)$: the bug will be exploited at firm j, which causes its direct loss l and a competitive advantage δ for firm i, while both of them will lose τ due to market shrinkage. (c) The bug is known by player j, but not player i. This is the mirror situation to case-b: the utility pair is $(-\delta - \tau - l, \delta - \tau)$. (d) The bug is known by neither one of the players. Here, there is no competitive advantage of one over the other, but there is still the market shrinkage effect, besides the direct losses to both. Hence, the utilities are $(-\tau - l, -\tau - l)$.

To facilitate the computation of the expected utilities, we define the following auxiliary random variables (as also depicted by a Venn diagram in Fig. 1): Let $B_{i,j}$, $B_{i,\neg j}$, $B_{\neg i,j}$ and $B_{\neg i,\neg j}$ represent the number of bugs that, respectively, both players, only player i, only only player j, and neither player knows about.

---

[5] The assumption is that once the bug is discovered, its "fix" is immediate and costless.

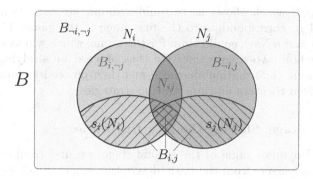

Fig. 1. Venn diagram illustration of the sets of bugs

Let the (expected) utility of players be denoted by u, which is a function from the strategy profile of the players and the state of the world to the set of real numbers. The expectation is taken with respect to the realisation of $B_{i,j}$, $B_{i,\neg j}$, $B_{\neg i,j}$ and $B_{\neg i, \neg j}$ given B, N_i, N_j and N_{ij}, and the sharing strategies. We are now ready to compute the expected utility of player i given a realisation of the state of the world $\omega = (b, n_i, n_j, n_{ij})$, and $\sigma_i = (p_i, s_i)$, $\sigma_j = (p_j, s_j)$:

$$u_i(\omega, \sigma_i, \sigma_j) = -c_i(p_i) + 0 \cdot \mathbb{E}(B_{i,j}) + (\delta - \tau) \cdot \mathbb{E}(B_{i,\neg j})$$
$$+ (-\delta - \tau - l) \cdot \mathbb{E}(B_{\neg i,j}) + (-\tau - l) \cdot \mathbb{E}(B_{\neg i, \neg j}) \quad (1)$$

In what follows we analyse further the structure of this utility function and derive the "outcome" of the game and study its properties.

3 Analysis of the Game

When dealing with strategic entities with inter-dependent utilities, investigating equilibria, most notably Nash Equilibria, is a method of predicting their decisions. Our game contains sequential moves, and thus an ordinary Nash equilibrium concept would potentially cause the problem of "non-credible threats". Also note that our game contains simultaneous actions in each stage, and hence is of "imperfect information". We therefore examine possible perfect Bayesian equilibria (PBE), a solution concept that effectively eliminates non-credible threats in sequential games with incomplete and imperfect information.

Informally, a PBE is a profile of strategies such that, given any belief about the game history that is consistent with that profile, the induced strategy profile must be a Nash equilibrium for the induced subgame (the game from the belief in an information Set onward). To find the set of PBEs, we notice that since the investment decisions are announced before sharing, each Bayesian game in the second stage is a proper subgame of the whole game. This means that we can use backward induction and first construct $((p_i, s_i), (p_j, s_j))$ such that s_i and s_j form a Bayesian Nash equilibrium (BNE) of the Bayesian game of sharing induced by choices of p_i and p_j. This in turn determines the utility of the players for each choice of (p_i, p_j), which allows us to build a simple strategic-form game with actions p_i and p_j corresponding to the first stage of the game. The remaining task will be to find a Nash equilibrium for this game, which will lead to a proper PBE for the whole two-stage game. We thus proceed by studying the second stage of the game (information sharing), and then proceed to analyse players' investments given their equilibrium sharing strategies.

3.1   Second Stage: Sharing the Bug Discoveries

To study the Bayesian game of the second stage, we first compute the utility functions of the players from the basic description in (1). Since $\mathbb{E}(B_{i,j})$ is multiplied by zero, we can safely ignore it. For the rest, we have:

$$\mathbb{E}[B_{i,\neg j}|\omega, \sigma_i, \sigma_j] = (n_i - n_{ij})(1 - \frac{s_i(p_j, n_i)}{n_i}) \tag{2a}$$

$$\mathbb{E}[B_{\neg i,j}|\omega, \sigma_i, \sigma_j] = (n_j - n_{ij})(1 - \frac{s_j(p_i, n_j)}{n_j}) \tag{2b}$$

$$\mathbb{E}[B_{\neg i,\neg j}|\omega, \sigma_i, \sigma_j] = b - n_i - n_j + n_{ij} \tag{2c}$$

In (2a),(2b), we have in part used the fact that the bugs to be shared are chosen uniformly randomly across the discovered bugs. Replacing in (1), we obtain:

$$u_i(\omega, \sigma_i, \sigma_j) = -c_i(p_i) + (\delta - \tau)(n_i - n_{ij})(1 - \frac{s_i(p_j, n_i)}{n_i}) +$$
$$(-\delta - \tau - l)(n_j - n_{ij})(1 - \frac{s_j(p_i, n_j)}{n_j}) + (-\tau - l)(b - n_i - n_j + n_{ij}) \tag{3}$$

We are looking for strategy profiles (strategy pairs (s_i, s_j) in our two-player context) that are simultaneous best responses to each other, given the information that each player has, notably including its number of discovered bugs. In the Bayesian Nash equilibria of the game, each candidate strategy for a player must be a maximizer of its expected utility given the strategy of the other player and given its observed type (number of discovered bugs).[6] Formally, for a given p_i and p_j, we are looking for the strategy pairs (s_i^*, s_j^*), such that:

$$\forall n_i \in \mathbb{N}^+, \ s_i^*(p_j, n_i) \in \arg\max_{s_i(p_j, n_i)} \mathbb{E}[u_i(\omega, (p_i, s_i(p_j, n_i)), (p_j, s_j^*(p_i, n_j)))|n_i] \tag{4}$$

and simultaneously vice versa for j. Such pairs constitute the (pure) Bayesian Nash Equilibria of the second stage of our game. The pair (s_i^*, s_j^*) is further, a Dominant (pure) Bayesian Nash Equilibrium iff:

$$\forall n_i \in \mathbb{N}^+, \forall s_j, \ s_i^*(p_j, n_i) \in \arg\max_{s_i(p_j, n_i)} \mathbb{E}[u_i(\omega, (p_i, s_i(p_j, n_i)), (p_j, s_j(p_i, n_j)))|n_i] \tag{5}$$

---

[6] To analyse the game, each player must specify its actions for all of its possible types, and not just the realised (and observed) type. This is because, the expected utility of each player depends on the possible actions of the other player(s) weighted against their potential types, since the type of other player(s) are not directly observed.

and simultaneously vice versa for j. We are now ready to express the main result of this section:

Proposition 1. *Suppose $p_i, p_j < 1$. If $\delta < \tau$, the unique dominant pure Bayesian Nash Equilibrium of the second stage of the game is $(s_i^*(p_j, n_i), s_j^*(p_i, n_j)) = (n_i, n_j)$, i.e., sharing all the discovered bugs. If $\delta > \tau$, it is $(s_i^*(p_j, n_i), s_j^*(p_i, n_j)) = (0, 0)$, i.e., sharing no information at all. When $\delta = \tau$, any strategy pair becomes a Bayesian Nash Equilibrium. This proposition holds irrespective of the distribution of the total number of bugs.*

Proof. According to (5), a pair (s_i^*, s_j^*) constitutes a Dominant Bayesian Equilibrium if, for each type of a player, its corresponding action is the best (provided the knowledge of its type), irrespective of the strategy of the other player. From (3), the only term in the the expression of $u_i(\omega, \sigma_i, \sigma_j)$ that involves s_i is the second term: $(\delta - \tau)[(n_i - n_{ij})(1 - s_i(p_j, n_i)/n_i)]$. Hence, with the assumption of $p_j < 1$ in mind, the maximization of $\mathbb{E}[u_i(\omega, \sigma_i, \sigma_j)|n_i]$ with respect to $s_i(p_j, n_i)$ reduces to maximizing $(\delta - \tau)(1 - s_i(p_j, n_i))$, which yields the proposition.[7] $\square$

Discussion The proposition makes intuitive sense: when $\delta > \tau$, each bug that is only known by a player wins it a strictly positive (expected) competitive gain of $(\delta - \tau)$, as the competitive shift in the demand and public investment outweighs the overall drop in the demand and fall in the stock market of the whole market section. Hence it rather not share any of its findings, irrespective of what the other player chooses. This is because the players have no means of making their decisions "contingent" on the decision of the other.[8] Similarly, when $\delta < \tau$, the competitive shift in the demand and capital, falls short of the whole market section shrinkage. Therefore, the players prefer to share all their findings to (selfishly) keep themselves from being hurt. Perhaps the surprising result is that the dominant strategy of the players turned out to be completely determined by the relative values of only two parameters δ and τ. This proposition fully determines the sharing strategy of the firms. Notably, aside from the special case of $\delta = \tau$, the equilibrium is unique and hence, there is no ambiguity in selection of the equilibrium. Next, we investigate how each firm invests for discovering the bugs knowing the subsequent sharing strategies.

3.2   First Stage: Investment for Bug Discovery

In the first stage of the game, each player decides about its investment amount for the discovery of bugs, heeding the strategy of the other player in the second stage.

---

[7] Although the proposition leaves out the cases in which the condition $p_i, p_j < 1$ are not satisfied, they are not difficult to analyse: suppose $p_j = 1$, then $\mathbb{E}[(n_i - n_{ij})(1 - s_i(p_j, n_i)/n_i)|n_i] = 0$, and hence the expression for $\mathbb{E}[u_i(\omega, \sigma_i, \sigma_j)|n_i]$ will not depend on s_i at all. Hence, in any Bayesian Nash Equilibria, the choice of s_i becomes arbitrary. Similar situation happens for s_j when $p_i = 1$. Intuitively, if the other player "knows every bug for certain", then a player cannot affect its utility through its action: it cannot gain any competitive advantage if $\delta > \tau$, or help prevent market shrinkage when $\delta > \tau$. Note that realistically, we can safely assume $p_i, p_j < 1$, as no practical amount of investment leads to absolute certainly of finding all bugs.

[8] We will see in §4 how this situation can be altered in the presence of a mediator.

To obtain closed-form results, we need to model the relation between investment decision and the chance of finding bugs. A simple candidate for such relation is the following: $p = \pi(c) = 1 - e^{-\theta c}$, where θ represents a measure of the efficiency of the investment: a larger θ corresponds to a higher efficiency of the investment. As the level of investment increases to infinity, the probability of discovery of each bug asymptotically approaches unity. The two firms may be different in how "efficient" they are in their investment. A firm with more prepared talents can expect higher chances of discovery with less investment. To capture the potential heterogeneity in the investment efficiencies, we consider two potentially different θ_i and θ_j. Our investment-discovery probability relation has the extra property that the relative efficiency of the investment stays constant for all investment values, specifically: $(\partial \pi_i / \partial c)/(\partial \pi_j / \partial c) = \theta_i/\theta_j$. This relation can also be equivalently represented in its inverse form: $c_i(p_i) = -\ln(1 - p_i)/\theta_i$ for $p_i \in [0,1)$, and likewise for j. Note that the condition of Proposition 1 $p_i, p_j < 1$ is automatically satisfied when $\lim_{p\to 1} c(p) \to \infty$, as is the case in our example.

To analyse this stage, we note that Proposition (1) fully determines (s_i^*, s_j^*) for each profile of (p_i, p_j). This allows us to treat the first stage as a "one-shot" game of investment with action profiles of the form (p_i, p_j).

3.3   The Case of $\delta < \tau$

For the case of $\delta < \tau$, from Proposition 1, the dominant strategy of both players is to share *all* of their findings, i.e., $s_i(p_j, n_i) = n_i$ and $s_j(p_i, n_j) = n_j$ for all $n_i, n_i \in \mathbb{N}^+$. Then, the second and third terms in (3) become zero, and we get:

$$\mathbb{E}[u_i(\omega, (p_i, s_i^*), (p_j, s_j^*))] = -c_i(p_i) + (-\tau - l)\mathbb{E}[B - N_i - N_j + N_{ij}]$$
$$= -c_i(p_i) + (-\tau - l)\lambda(1 - p_j)(1 - p_i)$$

The best response p_i^{BR} as a relation over p_j is hence:

$$p_i^{BR}(p_j) = [c_i'^{-1}(\kappa(1 - p_j))]^+, {}^9 \qquad\qquad \text{where } \kappa := \lambda(\tau + l). \qquad (6)$$

Note that when $p^{BR} > 0$, $\partial p_i^{BR}/\partial p_j = -\kappa/c_i''(p_i^{BR}) < 0$, i.e., more investment by the other player leaves *less* incentive for a player to invest. Similarly, we have: $\mathbb{E}[u_i(\omega, (p_i, s_i^*), (p_j, s_j^*))] = -c_j(p_j) + (-\tau - l)(1 - p_i)\lambda(1 - p_j)$, and hence: $p_j^{BR}(p_i) = [c_j'^{-1}(\kappa(1 - p_i))]^+$. The fixed points of the best response correspondence $(p_i, p_j) \Rightarrow ([c_i'^{-1}(\kappa(1 - p_j))]^+, [c_j'^{-1}(\kappa(1 - p_i))]^+)$ constitute the outcome of the first stage. For our example cost function $c = -\ln(1 - p)/\theta$, the simultaneous best response must hence satisfy the following (Fig. 2a):

$$p_i^{BR}(p_j) = [1 - \frac{1}{\theta_i \kappa (1 - p_j)}]^+, \qquad p_j^{BR}(p_i) = [1 - \frac{1}{\theta_j \kappa (1 - p_i)}]^+.$$

This, together with Proposition 1, lead to the following result:[10]

---

[10] The exact values of the investments depend on the cost function adopted, however, the qualitative observations hold for a wide class of such functions.

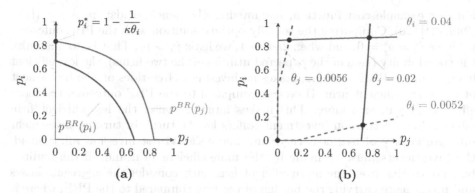

Fig. 2. (a) Example best response curves for the case of $\delta < \tau$, investigated in §3.3. In the figure $\theta_i > \theta_j$. The intersection gives the simultaneous best response pair in the first stage of the game as: $(p_i^*, p_j^*) = ([1 - (\kappa\theta_i)^{-1}]^+, 0)$. The parameters used are: $\lambda = 100$, $\tau = 0.5$, $l = 1$, $\theta_i = 0.04$, $\theta_j = 0.02$. (b) Example best response curves for the case of $\delta < \tau$ and different θ_is and θ_js.

Proposition 2. *If $\delta < \tau$ and $\theta_i > \theta_j$, the Perfect Bayesian Equilibrium (PBE) of the two-stage game is $((p_i^*, s_i^*(p_j, n_i)), (p_j^*, s_j^*(p_i, n_j))) = (([1 - \frac{1}{\kappa\theta_i}]^+, n_i), (0, n_j))$ for all $n_i, n_j \in \mathbb{N}^+$ and all $p_i, p_j \in [0,1)$, where $\kappa := \lambda(\tau + l)$. That is, only the more efficient firm invests in discovery of the bugs – to achieve discovery probability of $[1 - (\kappa\theta_i)^{-1}]^+$ – and all the findings are then shared.[11]*

Discussion The less efficient firm free-rides on the bug discovery investment of the more efficient company, knowing that all the findings will be shared. This might leap the reader to the conclusion that the PBE outcome is socially inefficient simply because of the existence of "free-riding". However, a social planner may also prefer that the investment is done by the more efficient firm as opposed to distributing the investment among both, hence garnering a higher social return on the aggregate investments. In what follows, we will evaluate the social utility and the socially efficient outcome and compare the two.

Investigating Social Welfare: Let W represent the expected (utilitarian) *social utility*, defined simply as the sum of the expected utilities of the two firms, i.e., $W := u_i + u_j$.[12] First off, it is straightforward to argue that in the socially optimal outcome, all the findings are shared (the social utility can only be improved by sharing the findings, as the investment decisions are now disentangled from the sharing decisions). The expected social utility is hence as follows:

$$\mathbb{E}W = -c_i - c_j - 2(\tau + l)\mathbb{E}B_{\neg i, \neg j} = -c_i(p_i) - c_j(p_j) - 2\kappa(1 - p_i)(1 - p_j) \quad (7)$$

---

[11] When $\theta_i = \theta_j = \theta$, i.e., the two firms are homogeneous in terms of their efficiencies of bug discovery investments, the equilibrium point is not unique and becomes the set: $\{(p_i^*, p_j^*) \in [0,1]^2, p_i^* = [1 - (\theta\kappa(1 - p_j))^{-1}]^+\}$.

[12] Other notions of social welfare exist, e.g., the egalitarian objective $W := \min(u_i, u_j)$.

For our example cost function, maximizing $\mathbb{E}W$ hence yields: $(\hat{p}_i, \hat{p}_j) = ([1 - (2\kappa\theta_i)^{-1}]^+, 0)$. Comparing the socially optimal solution with the PBE outcome, we have $\hat{p}_j = p_j^* = 0$, and when $2\kappa\theta_i > 1$, we have: $\hat{p}_i > p_i^*$. That is, to maximize the social utility (sum of the expected utilities of the two firms), the less efficient firm, as in the PBE outcome, makes no investment free-rides on the investment of the more efficient firm. However, compared to the PBE outcome, the more efficient firm invests more. This makes intuitive sense: the less efficient firm offers a lower return on investment (offers less "return" in turning investment into probability of bug discovery) and hence should not invest at all. Instead, the investments must be made by the more efficient firm and all the findings be shared. Moreover, the more efficient firm must consider the aggregate losses and invest more carrying the burden of the two, compared to the PBE, where it only considers the effect of its investment on its own losses. Note that even when the players are homogeneous in terms of their efficiencies, i.e., when $\theta_i = \theta_j$, the socially optimal investment turns out to choose only one of the firms to invest. This is because it will prevent from discovery of the same bugs by both players. The value of the optimum social welfare is:

$$W(\hat{p}_i, \hat{p}_j) = -\ln(2\kappa\theta_i)/\theta_i - 1/\theta_i \text{ for } \kappa\theta_i > 1/2, \text{ and: } -2\kappa \text{ for } \kappa\theta_i \leq 1/2. \quad (8)$$

The social welfare that is achieved at the equilibrium outcome of the game is:

$$W(p_i^*, p_j^*) := -\ln(\kappa\theta_i)/\theta_i - 2/\theta_i \text{ for } \kappa\theta_i > 1, \text{ and: } -2\kappa \text{ for } \kappa\theta_i \leq 1. \quad (9)$$

An example comparison between the two is depicted in Fig. 3a.

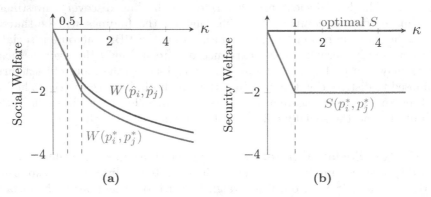

(a)   (b)

Fig. 3. (3a): Example depiction of the optimal and achieved social welfare (3a) and security utility (3b) for the case of $\delta < \tau$ as functions of $\kappa = \lambda(\tau + l)$

Here, we define another metric of social welfare in the context of economics of network security. Let the *security utility* u^S of a player be the negative of the costs of security attacks. Security utility, such defined, is related to the utility of a player as $u^S = u + c$: it includes all the secueity damages but excludes the investment cost. Now, let the *security welfare* S, as a metric of the aggregate security of the two firms, be the sum of their security utilities: $S := u_i^S + u_j^S$. The security utility is related to the utilitarian social welfare in the following way: $S = W + c_i(p_i) + c_j(p_j)$. The optimal S is achieved by picking $p_i = 1$ and sharing all the findings, which yields $S = 0$. Fig. 3b illustrates a comparison between the achieved security utility at the equilibrium and the optimal S.

Comparative Statics. [13] Recall from Proposition (2), that for $\delta < \tau$, in part we have: $(p_i^*, p_j^*) = ([1 - 1/(\kappa\theta_i)]^+, 0)$. Hence, as long as $\delta < \tau$, $\theta_i > \theta_j$ and $p_i^* > 0$ (i.e., for $1 < \kappa\theta_i$), we have the following straightforward observations:

$$\frac{\partial p_i^*}{\partial \tau}, \frac{\partial p_i^*}{\partial l}, \frac{\partial p_i^*}{\partial \lambda}, \frac{\partial p_i^*}{\partial \theta_i} > 0, \qquad \frac{\partial p_j^*}{\partial \tau}, \frac{\partial p_j^*}{\partial l}, \frac{\partial p_j^*}{\partial \lambda}, \frac{\partial p_j^*}{\partial \theta_j} = 0.$$

We also have $\partial p_i^*/\partial \theta_j = 0$, and perhaps most interesting of all $\partial p_i^*/\partial \delta = 0$; intuitively, player i shares all of its findings and thus removes any dependence of its utility (and hence its best strategy) on δ. Also, note that even though $\partial p_i^*/\partial \theta_i > 0$, i.e., more efficiency in investment means higher choice of probability of discovery, this does not necessarily translate to higher choice of investment. In fact, we have: $\partial c_i(p_i^*)/\partial \theta_i < 0$ for $1 < \kappa\theta_i < e$, and $\partial c_i(p_i^*)/\partial \theta_i > 0$ for $\kappa\theta_i > e$. Moreover, from (9), for $p_i^* > 0$ we have: $W^* := W(p_i^*, p_j^*) = -\ln(\kappa\theta_i)/\theta_i - 2/\theta_i$ and $S^* := S(p_i^*, p_j^*) = -2/\theta_i$. Hence, when $\delta < \tau$, $\theta_i > \theta_j$ and $1 < \kappa\theta_i$, we have:

$$\frac{\partial W^*}{\partial \tau}, \frac{\partial W^*}{\partial l}, \frac{\partial W^*}{\partial \lambda} < 0, \quad \frac{\partial W^*}{\partial \theta_i} > 0, \qquad \frac{\partial S^*}{\partial \tau}, \frac{\partial S^*}{\partial l}, \frac{\partial S^*}{\partial \lambda} = 0, \quad \frac{\partial S^*}{\partial \theta_i} > 0.$$

3.4   The Case of $\delta > \tau$

Following Proposition 1, the dominant strategy of the players in the second stage is to share *none* of their findings, i.e., $s_i(p_j, n_i) = 0$ and $s_j(p_i, n_j) = 0$ for all $n_i, n_i \in \mathbb{N}^+$ and all $p_i, p_j \in [0, 1)$. Then from (3), we obtain:

$$\mathbb{E}[u_i(\omega, (p_i, s_i^*), (p_j, s_j^*))] = -c_i(p_i) + (\delta - \tau)\lambda p_i(1 - p_j)$$
$$+ (-\delta - \tau - l)p_j\lambda(1 - p_i) + (-\tau - l)(1 - p_j)\lambda(1 - p_i) \quad (10)$$

The best response relation for player i is therefore:

$$p_i^{BR}(p_j) = [c_i'^{-1}(\lambda(\delta + l + p_j\tau))]^+.$$

A point to observe is that for $p_i^{BR} > 0$, we have: $\partial p_i^{BR}/\partial p_j = \lambda\tau/c_i''(p_i^{BR}) > 0$, i.e., more investment by the other player leads to *more* investment by a player. This is in sharp contrast to the the previous case of $\delta < \tau$. Similarly: $p_j^{BR}(p_i) = [c_j'^{-1}(\lambda(\delta + l + p_i\tau))]^+$. For our example cost function, the simultaneous best response is therefore the solution the following system (Fig. 2b):

$$p_i^{BR}(p_j) = [1 - \frac{1}{\theta_i\lambda(\delta + l + p_j\tau)}]^+, \qquad p_j^{BR}(p_i) = [1 - \frac{1}{\theta_j\lambda(\delta + l + p_i\tau)}]^+. \quad (11)$$

---

[13] In economics, *comparative statics* is the study of the change in the "equilibrium" outcome when a change in a parameter is/would be introduced.

Straightforward algebraic investigation reveals that the solution is unique and given as follows:

$$
\text{If } \Delta \geq 0: \quad
\begin{cases}
p_i^* = \dfrac{\left[-\lambda \theta_i \theta_j ((\delta + l)^2 - \tau^2) + \tau(\theta_i - \theta_j) + \sqrt{\Delta} \right]^+}{2\tau \theta_i \theta_j (\delta + l + \tau)} \\[4mm]
p_j^* = \dfrac{\left[-\lambda \theta_i \theta_j ((\delta + l)^2 - \tau^2) - \tau(\theta_i - \theta_j) + \sqrt{\Delta} \right]^+}{2\tau \theta_i \theta_j (\delta + l + \tau)}
\end{cases}
, \qquad (12)
$$

and if $\Delta < 0$: $(p_i^*, p_j^*) = (0,0)$, where $\Delta := \left(\tau(\theta_i + \theta_j) - \lambda \theta_i \theta_j (\delta + l + \tau)^2 \right)^2 - 4\tau^2 \theta_i \theta_j$. This, along with Proposition 1, fully determines the PBE:

Proposition 3. *When $\delta > \tau$, the Perfect Bayesian Equilibria (PBE) of the security information sharing game is unique, in which (p_i^*, p_j^*) are provided in (12), and $(s_i^*(p_j, n_i), s_j^*(p_i, n_j)) = (0,0)$ for all $n_i, n_j \in \mathbb{N}^+$ and all $p_i, p_j \in [0,1)$. That is, both of the firms may invest – to achieve discovery probabilities as given in (12) – and none of the consequent findings are shared.*

Discussion When $\delta > \tau$, the competitive gain outweighs the market shrinkage of not sharing the found bugs. Knowing that the found bugs will not be shared, both players, notably even the less efficient player, invest in discovery of the bugs on their own. This is because of two facts: 1- Since the findings are not shared, the firm would be exposed in its bugs if it does not discover and rectify them if it does not invest. 2- Since the other firm invests and expectedly discovers some bugs, the firm will further suffer through the competitive effect of being the sole victim of such bugs if it does not invest.

Comparison to Socially Optimal Outcome: The social optimal outcome certainly shares the found bugs. Compared to the case of $\delta < \tau$, both players invest strictly more in discovery of the bugs. The social inefficiency of the outcome for the case of $\delta < \tau$ was due to underinvestment. Here, it is primarily due to lack of sharing of the found bugs: if a player would receive information of a bug that has not discovered itself, the social utility would have improved by preventing the potential direct losses in that player as well as the market shrinkage losses in both players. Another source of social inefficiency is the fact that "both" players make discovery investment: there is a positive probability that the same bug can be discovered independently by both firms. The investment could have been more efficient by preventing such cases of "duplicate effort", if directed to only one player and the subsequent findings are shared. Anther source of social inefficiency, which is again rooted in lack of information sharing of the players, is the possibility of "over-investment" in bug discovery. The optimal expected social utility is the same as was computed in (8). Note in particular that it does not depend on the value of δ. Sharing the information in the social optimal removes the competitive effect of δ. However, in the case of $\delta > \tau$, the investment value of both players increases with δ. This means that the threat of competitive losses due to being the sole victim of a security attack can drive both firms to invest inefficiently large values in bug discovery, when they know the discoveries, as competitive advantages, will not be shared. A combination of all of these three effects is responsible for a high social inefficiency in this case.

Comparative Statics. Given $\delta > \tau$ and our example cost functions, we note that players' best response functions as in (11) are increasing and concave. Investigating the best-response expressions in (11) further reveals:

$$\frac{\partial p_i^{BR}}{\partial \tau}, \frac{\partial p_i^{BR}}{\partial l}, \frac{\partial p_i^{BR}}{\partial \lambda}, \frac{\partial p_i^{BR}}{\partial \theta_i}, \frac{\partial p_i^{BR}}{\partial \delta} > 0, \qquad \frac{\partial p_j^{BR}}{\partial \tau}, \frac{\partial p_j^{BR}}{\partial l}, \frac{\partial p_j^{BR}}{\partial \lambda}, \frac{\partial p_j^{BR}}{\partial \theta_j}, \frac{\partial p_j^{BR}}{\partial \delta} > 0.$$

This means that player i is willing to invest more as any of the following parameters increases: τ, l, λ, θ_i, and similarly for player j (with θ_i replaced by θ_j). Investigating the effect on the equilibrium point is a bit trickier. For simplicity of exposition, we illustrate the "shift" in the equilibrium pair pictorially. In Fig. 4, the effect of increasing δ is depicted. Note that, on the "p_i–p_j" plane, $p_i^{BR}(p_j)$ shifts "up" and $p_j^{BR}(p_i)$ shifts "right" as the value of δ increases. Hence, the intersection, which indicates the equilibrium, moves towards up and right. The algebraic details of the analysis is removed for brevity. Analysing the effect of each parameter in turn reveals:

$$\frac{\partial p_i^*}{\partial \tau}, \frac{\partial p_i^*}{\partial l}, \frac{\partial p_i^*}{\partial \lambda}, \frac{\partial p_i^*}{\partial \delta}, \frac{\partial p_i^*}{\partial \theta_i}, \frac{\partial p_i^*}{\partial \theta_j} \geq 0, \qquad \frac{\partial p_j^*}{\partial \tau}, \frac{\partial p_j^*}{\partial l}, \frac{\partial p_j^*}{\partial \lambda}, \frac{\partial p_j^*}{\partial \delta}, \frac{\partial p_j^*}{\partial \theta_j}, \frac{\partial p_j^*}{\partial \theta_i} \geq 0.$$

In words, the above inequalities indicate that if any of the following parameters increases, then firms would invest more: τ, l, λ, and δ. Indeed, the higher these parameters, the more severe impacts of security incidents would be, and thus both firms have to secure themselves, especially when they receive no aid from the other. An interesting result is the effect of improvement in the investment efficiency of the competitor: If θ_j is improved, then firm i invests more in vulnerability research. Intuitively, this is due to the fact that an improvement in the discovery probability of the competitor firm j means more competitive pressure on firm i. This is because each bug that is discovered exclusively by firm j brings it a net advantage of $\delta - \tau$ at the cost of firm i. Thus the increase in efficiency of firm j forces firm i to also improve its probability of discovery, which happens by increasing its investment. This means that the utility of player i decreases as the result of an improvement in player j's efficiency. Specifically, $\partial u_i(p_i^*, p_j^*)/\partial \theta_j < 0$. This is while, $\partial u_j(p_i^*, p_j^*)/\partial \theta_j > 0$. Due to these opposing effects of efficiencies on individual utilities, in general, the equilibrium social welfare, $W(p_i^*, p_j^*)$, which is the sum of the two utilities at the equilibrium, may increase or decrease as θ_i or θ_j is improved. Note, however, that the equilibrium security welfare, $S(p_i^*, p_j^*)$, always improves when θ_i or θ_j increases.

4   Mediation: Encouraging Information Sharing

Our analysis in the previous section characterized the players' behaviour in equilibria. For the case of $\delta < \tau$, which pertain to a the case where security acts

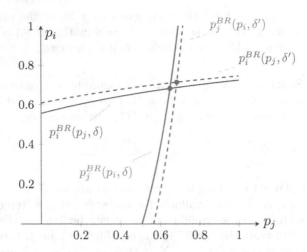

Fig. 4. Example illustration of the comparative statics for the case of $\delta > \tau$. The parameters used are $\lambda = 1.5$, $l = 0.5$, $\theta_i = 1$, $\theta_j = 0.9$, $\tau = 0.9$, and the value of δ is increased from $\delta = 1$ to $\delta' = 1.2$. Notice the shift in the equilibrium value towards "up" and "right" as a result.

effectively as a "common good", sharing of security findings becomes inevitable, and exactly because of that, free-riding emerges, which in turn leads to underinvestment. In contrast, when $\delta > \tau$, which represents cases where security effectively becomes a "competitive advantage", firms would individually strive for their security and refrain from sharing their findings. We observed that none of these outcomes are in line with desirable social planning.

In this section, we make a preliminary attempt to remedy one of the sources of social inefficiency, specifically, failure in information sharing in the "competitive advantage" case. We develop a mediation mechanism that partially removes the negative incentives of sharing the information while allowing the players to gain from its positive effects. Informally put, our mediation plan states that if a firm wants to be informed about n bugs that it failed but the other firm succeeded to discover, it must reveal in exchange n bugs that the other firm is not aware of. Note that this was not possible in the previous sections, as there was no means of making the sharing actions of a firm "contingent" on the action of the other. The mediator effectively ensures that no net "competitive advantage" is lost by sharing the vulnerability findings, as any leakage of an "exclusive" discovery is *matched* by an "exclusive" discovery of the competitor. We will hence refer to our mediation plan as "matched sharing".

Matched sharing operates in two steps: (i) each player/firm submit its set of found bugs to the mediator, along with a specification of a "threshold" as the maximum number of bugs it is willing to exchange with the other firm. (ii) Subsequently, based on the reported sets and the players' thresholds, the mediator moderates the exchange of as many bugs as possible in the following manner: the mediator marks the bugs that are exclusive to each player, i.e., that the other player has not discovered them. Then the information of a bug is transferred from player i to player j iff a) there is an exclusive bug to *match*, i.e., to transfer from player j to i, and b) if the total number of bugs transferred so far does not exceed either one of the players' requested maximum threshold. Note that the mediator is not a strategic player, and its behaviour is known to and trusted by both players.

From the above description, a sharing action of a player entails the selection of the threshold on exchange number. Note specifically, that we can without loss of generality assume that both players submit all of their findings to the mediator.[14] This is because the players can restrict the sharing of their findings by specifying the threshold. For instance, no sharing corresponds to requesting a threshold of "zero". Note that due to the nature of the Bayesian game, each player must pick this bound for every realisation of bugs it discovers (given the investment decisions). Formally, we can reuse the notations $s_i(p_j, n_i)$ and $s_j(p_i, n_j)$ to represent the sharing strategies, with the different interpretation that s_i and s_j denote the threshold, i.e., the maximum number of their bugs to be shared by the mediator to the other player. Hence, the expressions in (2) in the presence of the mediator and the new interpretation of the strategies become:

$$\mathbb{E}[B_{i,\neg j}|\omega, s_i, s_j] = n_i - n_{ij} - \min\{s_i(p_j, n_i), s_j(p_i, n_j), n_i - n_{ij}, n_j - n_{ij}\}$$
$$\mathbb{E}[B_{\neg i,j}|\omega, s_i, s_j] = n_j - n_{ij} - \min\{s_i(p_j, n_i), s_j(p_i, n_j), n_i - n_{ij}, n_j - n_{ij}\}$$

and, as before, $\mathbb{E}[B_{\neg i,\neg j}|\omega, s_i, s_j] = b - n_i - n_j + n_{ij}$. In words, the term represented by the min function determines the number of bugs that are exchanged between the players, which should be no more than the bounds set by both firms, as well as what each firm individually has to offer. This in turn gives:

$$u_i(\omega, \sigma_i, \sigma_j) = - c_i(p_i) + \delta(n_i - n_j) - \tau(b - n_{ij}) - l(b - n_i)$$
$$+ (2\tau + l)\min\{s_i(p_j, n_i), s_j(p_i, n_j), n_i - n_{ij}, n_j - n_{ij}\} \qquad (13)$$

As we can see, the only term that involves $s_i(p_j, n_i)$ is the last term. Maximization of the expected utility of player i given the strategy of player j therefore translates to maximizing $\min\{s_i(p_j, n_i), s_j(p_i, n_j), n_i - n_{ij}, n_j - n_{ij}\}$. Hence, we have the following result:

Proposition 4. *Suppose $p_i, p_j < 1$. The weakly dominant pure Bayesian Nash Equilibrium of the second stage of the game is $(s_i^*(p_j, n_i), s_j^*(p_i, n_j)) = (n_i, n_j)$ for all $n_i, n_j \in \mathbb{N}^+$ and $p_i, p_j \in [0,1)$, i.e., asking the mediator to share the maximum number of exclusive bugs. This proposition holds irrespective of the distribution of the total number of bugs, or correlation in the discovery of bugs.*

Proof. First, note that irrespective of the choice of s_j, $s_i(p_i, n_i) = n_i$ maximizes the expression $\min\{s_i(p_j, n_i), s_j(p_i, n_i), n_i - n_{ij}, n_j - n_{ij}\}$, and likewise for $s_j(p_i, n_j) = n_j$. Hence $(s_i(p_j, n_i), s_j(p_i, n_j)) = (n_i, n_j)$ for all $n_i, n_j \in \mathbb{N}^+$ and $p_i, p_j \in [0, 1)$ belongs to the set of pure Bayesian Nash equilibria of the second stage of the game. To see the weak dominance, consider the cases where

---

[14] Assuming that both parties have established trust with the mediator.

$n_j > n_i > 0$ and $n_{ij} = 0$. Note that $\Pr[N_j > n_i \wedge N_{ij} = 0 \mid N_i = n_i] > 0$. Consider the strategy of player j as $s_j(p_i, n_j) = n_j$ for all $n_j \in \mathbb{N}^+$. Then $u_i(\omega, (p_i, n_i), (p_j, s_j)) > u_i(\omega, (p_i, s_i'), (p_j, s_j))$ for any $s_i'(p_j, n_i) < n_i$, because: $\min\{n_i, s_j(p_i, n_j), n_i - n_{ij}, n_j - n_{ij}\} > \min\{s_i'(p_j, n_i), s_j(p_i, n_j), n_i - n_{ij}, n_j - n_{ij}\}$ for any $s_i'(p_j, n_i) < n_i$ when $n_j > n_i$, $n_{ij} = 0$ and $s_j(p_i, n_j) = n_j$.

4.1   Game's First Stage: Investment in the presence of the Mediator

Given the weakly dominant equilibrium in Proposition 4, $\min\{s_i^*(p_j, N_i), s_j^*(p_i, N_j), N_i - N_{ij}, N_j - N_{ij}\} = \min\{N_i, N_j\} - N_{ij}$. Hence, utility of player i in (13) becomes:

$$\mathbb{E}u_i(\omega, p_i, p_j, s_i^*, s_j^*) = -c_i(p_i) + \delta\mathbb{E}[N_i - N_j] - \tau\mathbb{E}[B - N_{ij}] - ql\mathbb{E}[B - N_i]$$
$$+ (2\tau + l)(\mathbb{E}[\min\{N_i, N_j\}] - \mathbb{E}[N_{ij}])$$
$$= -c_i(p_i) + \lambda\delta(p_i - p_j) - \lambda\tau(1 - p_i p_j) - \lambda l(1 - p_i) + (2\tau + l)(\mathbb{E}[\min\{N_i, N_j\}] - \lambda p_i p_j)$$

The term $\mathbb{E}[\min\{N_i, N_j\}]$ depends on the specific distribution of the total number of bugs. A good candidate is the Poisson distribution. The presence of this term in the utility function prevents a closed-form solutions for the best responses and the equilibrium points. Instead, we pictorially illustrate in Fig. 5 the potential usefulness of the mediator when $\delta > \tau$, i.e., when players are motivated more by competition than aggregate security. Fig. 5a depicts the equilibrium points of players' investments in two cases: sharing in the absence of the mediator (which leads to no sharing) and our "matched sharing". These are set in the context of low security damage (l) compared to competitive advantage (δ) and inefficient investment ($\theta_i = \theta_j = 0.1$). The end result is that with matched sharing, both players invest more in finding vulnerabilities, which guarantee a a better security for both. However, the social welfare, as well as the individual utilities of both players, worsens with the introduction of the matched sharing, as it exacerbates the already inefficiently high investments of the players in this example.

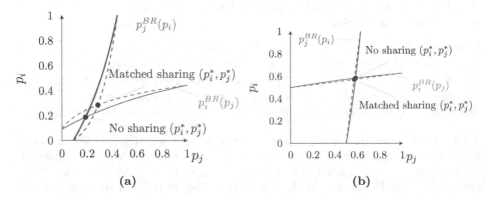

Fig. 5. Illustration of opportunistic sharing vs. matched sharing when $\delta > \tau$, with $\delta = 10$, $\tau = 1$, $\theta_i = \theta_j = 0.1$, with (a) $l = 1$ and (b) $l = 10$

In contrast, Fig. 5b shows the effect of our mediator plan in situations with either a significant security damage value (large l) or efficient investments (high θ_i, θ_j), or both. In such scenarios, equilibrium points of the two cases are relatively close to each other, i.e., they make similar levels of investments. With the help of the mediator, players would share their intelligences and thus gain extra value in security, making mediation a superior solution to opportunistic sharing. This suggests the potential of our matched sharing mediation scheme, and that it should be in the interest of the social planner to monitor environment parameters and establish trusted mediation among firms whenever appropriate for players/societal benefits.

5   Conclusion

In this work, we focused on the problem of sharing cybersecurity information, as an envisioned pillar of cybersecurity planning for a more secure infrastructure. We analysed the strategic decisions of two competing firms with regards to investment for discovery of security vulnerabilities (generating valuable cyber-intelligence) and subsequently, to share their findings. We showed that sharing becomes a dominant strategy when security tends to behaves as a common good, i.e., when the common losses as a result of security attacks outweigh the competitive gains of being protected. We analysed how in turn this leads to free-riding of less efficient firm and the under-investment of the more efficient firm. We also established that when security effectively becomes a competitive advantage, i.e., when there is a net positive gain when a competitor is a sole victim of an attack, then sharing no information becomes the dominant strategy, with negative implication on the social efficiency. Finally, we provided a monetary-free light-weight mediation mechanism that (partially) enables sharing of the found vulnerabilities in cases where they fail to achieve any sharing on their own.

Future Research This work has the potential to be extended in many directions. We have already made some grounds in extending our results to the multi-player situation. An interesting addition is considering "features" for the found bugs, such as severity (seriousness of the potential damage), sophistication (exploitability), etc., and hence letting the sharing strategies depend on the type of the found bug as well. Investigating the behaviour of risk-averse players – as opposed to risk-neutral in this work – is another problem. Identifying other types of "security information" to share is another interesting direction, for instance, revealing past incidents of successful attacks and resultant losses carries some market implications that sharing merely discovered security vulnerabilities does not. Also, we assumed that both firms use a common implementation (the "platform"). If instead, for instance, the firms are using a common protocol but with their private implementations of it, then "some" of the discovered bugs may be just exclusive to that party's implementation. Sharing found bugs now requires a modified analysis. Investigating other means of encouraging sharing is another important direction. An example is "bargaining": A player starts by sharing one bug, then the other player matches with a bug of its own findings, and so on, until one stops. Another example is a generalisation of the "matched sharing" mechanism in this work by allowing unequal number of matching that may involve some randomisation as well. An exchange market of vulnerabilities is another idea, although it may suffer from adverse selection and moral hazard.

References

1. Press-Release: Government launches information sharing partnership on cyber security (March 27, 2013), http://www.gov.uk
2. of Homeland Security Department.: National cybersecurity and communications integration center, http://www.us-cert.gov/nccic (accessed, June 2014)
3. Cavusoglu, H., Mishra, B., Raghunathan, S.: The effect of internet security breach announcements on market value: Capital market reactions for breached firms and internet security developers. Intl. J. of Electronic Commerce 9(1), 70–104 (2004)
4. Campbell, K., Gordon, L.A., Loeb, M.P., Zhou, L.: The economic cost of publicly announced information security breaches: Empirical evidence from the stock market. Journal of Computer Security 11(3), 431–448 (2003)
5. Goel, S., Shawky, H.A.: Estimating the market impact of security breach announcements on firm values. Information & Management 46(7), 404–410 (2009)
6. Lovells, H.: DOJ and FTC clarify antitrust implications of cybersecurity information sharing (April 22, 2014), http://www.hoganlovells.com/
7. Netcraft: Half a million widely trusted websites vulnerable to heartbleed bug (April 08, 2014), http://www.news.netcraft.com
8. Gordon, L.A., Loeb, M.P., Lucyshyn, W.: Sharing information on computer systems security: An economic analysis. Journal of Accounting and Public Policy 22(6), 461–485 (2003)
9. Gal-Or, E., Ghose, A.: The economic incentives for sharing security information. Information Systems Research 16(2), 186–208 (2005)
10. Hausken, K.: Income, interdependence, and substitution effects affecting incentives for security investment. J. of Accounting and Public Policy 25(6), 629–665 (2006)
11. Hausken, K.: Information sharing among firms and cyber attacks. Journal of Accounting and Public Policy 26(6), 639–688 (2007)
12. Liu, D., Ji, Y., Mookerjee, V.: Knowledge sharing and investment decisions in information security. Decision Support Systems 52(1), 95–107 (2011)
13. Liu, C.Z., Zafar, H., Au, Y.A.: Rethinking fs-isac: An it security information sharing network model for the financial services sector. Communications of the Association for Information Systems 34(1), 2 (2014)
14. Xiong, Q., Chen, X.: Incentive mechanism design based on repeated game theory in security information sharing. In: 2nd International Conference on Science and Social Research (ICSSR 2013). Atlantis Press (2013)
15. Gao, X., Zhong, W., Mei, S.: Security investment and information sharing under an alternative security breach probability function. Inf. Systems Frontiers, 1–16
16. Gal-Or, E.: Information sharing in oligopoly. Econometrica: Journal of the Econometric Society, 329–343 (1985)
17. Shapiro, C.: Exchange of cost information in oligopoly. The Review of Economic Studies 53(3), 433–446 (1986)
18. Vives, X.: Trade association disclosure rules, incentives to share information, and welfare. RAND Journal of Economics 21(3), 409–430 (1990)
19. Katz, M.L., Shapiro, C.: R and d rivalry with licensing or imitation. The American Economic Review, 402–420 (1987)

Optimal Contracts for Outsourced Computation*

Viet Pham, M.H.R. Khouzani, and Carlos Cid

Information Security Group
Royal Holloway, University of London
Egham, Surrey, TW20 0EX, UK
{viet.pham.2010,arman.khouzani,carlos.cid}@rhul.ac.uk

Abstract. While expensive cryptographically verifiable computation aims at defeating malicious agents, many civil purposes of outsourced computation tolerate a weaker notion of security, i.e., "lazy-but-honest" contractors. Targeting this type of agents, we develop optimal contracts for outsourcing of computational tasks via appropriate use of rewards, punishments, auditing rate, and "redundancy". Our contracts provably minimize the expense of the outsourcer (principal) while guaranteeing correct computation. Furthermore, we incorporate practical restrictions of the maximum enforceable fine, limited and/or costly auditing, and bounded budget of the outsourcer. By examining the optimal contracts, we provide insights on how resources should be utilized when auditing capacity and enforceability are limited. Finally, we present a light-weight cryptographic implementation of the contracts to mitigate the double moral hazard problem between the principal and the agents.

1 Introduction

The idea of outsourcing complex computation tasks has been proposed and implemented in a variety of applications. Research projects involving complex analysis on a huge multitude of data have utilized parallel processing of their computations on the processors of millions of volunteering Internet users. These include search for extra-terrestrial life (*SETI@Home*), investigation of protein folding and computational drug design (*Folding@Home* and *Rosetta@home*). Businesses from different sections including finance, energy infrastructure, mining and commodities transport, technology and innovation [7] have also realized the benefits of outsourcing their data and computation, and "moving to the cloud". The cloud, as a dedicated infrastructure with specialized man-force and powerful computing capabilities, along with the ability to pool demands from different clients and dynamic assignment of the resources can reduce the cost of computation. Meanwhile, the outsourcer is also relieved from maintaining a dedicated computing infrastructure and in addition, has the total flexibility of pay-per-use paradigm, to flex-on or to flex-off services effortlessly [7]. This growing trend has made possible small virtualised computers and smart devices with powerful computational power, applicable to critical mission scenarios and everyday use.

---

* The full version of this paper with formal proofs for all the propositions is accessible via https://eprint.iacr.org/2014/374.pdf.

R. Poovendran and W. Saad (Eds.): GameSec 2014, LNCS 8840, pp. 79–98, 2014.
© Springer International Publishing Switzerland 2014

In all of these scenarios, there is a concern for the outsourcer (client) about the correctness of the returned results. The provider of computation services (the servers) have an economic incentive to return guessed results as opposed to performing the computation completely and honestly, and thereby save on the computation work. Hence, to make this paradigm viable and guarantee soundness of the results, there must be an auditing mechanism in place. The auditing, however, is not free: it either creates computational overhead for the client, the server, or both. The auditing can be done by the outsourcer itself or through a trusted third party for a fee, say, through re-computation. Alternatively, a *redundancy* scheme can be employed in which the same job is outsourced to multiple servers and the results are checked against each other.

Irrespective of the auditing mechanism, the outsourcer can set an extremely large fine for detected wrong results, and make cheating theoretically impossible even for the lowest probability of cheat detection. However, in practice, an extremely large fine is a non-credible threat. A more reasonable assumption is a cap on the maximum enforceable fine, with the special interesting case where the cap is zero. In this paper we provide a concrete and general approach based on Principal-Agent modelling from game theory to optimal contract designs for outsourcing from the client (principal) to the servers (agents). Specifically, we assume a general maximum enforceable fine, maximum budget, and costly and/or limited auditing rate. We formulate the utilities of both the principal and the agents, as well as essential constraints that guarantee honest computation (incentive compatibility) along with their acceptance of the offer (participation). This allows us to systematically compute the optimal contract such that the principal's expense is minimized. Our work hence potentially provides a benchmark enabling comparison among different deployments of outsourcing.

The paper is structured as follows: In Section 2, we review previous results and describe our contributions. This is followed by a detailed motivation of our contract model in Section 3, along with descriptions of important constraints that make the problem non-trivial. In Section 4, we compute optimal contracts involving only one agent, and explore related improvements. In Section 5, we allow the principal to also potentially outsource the same task to multiple non-colluding agents as an alternative means of auditing and develop optimal hybrid contracts. We further establish the global optimality of our hybrid two-agent contracts among all possible contracts involving any number of non-colluding agents with respect to the notion of Nash Equilibria. In Section 6, we comment on cryptographic implementation of our contracts, i.e., how to enforce the terms and policies in an automated way. Finally, in Section 7, we conclude the paper with a summary of the results and remark on some potential future directions.

2   Related Work

A line of research is focused on designing reliable verification techniques for outsourcing of special-purpose computations. For instance, [17] investigates outsourcing of linear optimizations. Other notable examples are queries on outsourced databases, including typical queries [1,5] and aggregation [18]. Their main paradigm is for the querier to rely on trusted information directly given by the data owner (outsourcer) to verify the results returned by the servers.

Verification methods for general-purpose computing also appear in several remarkable works. In [12] verification is performed by re-executing parts of the computation. A variation is presented in [3] in which the authors utilize redundancy over multiple agents, assuming that at least one of them is honest. Outsourced computation has also caught attraction in cryptographic research: in a seminal work, the authors of [8] formally define verifiable computation and give a non-interactive solution. Their solution uses Yao's garbled circuits to represent the computation and homomorphic encryption to hide such circuits from the agents. More efficient but interactive solutions that use *probabilistically-checkable proofs* (PCPs) have since been developed such as PEPPER [15] and GINGER [16].

Incentive-based solutions such as [2,13] have studied contracts that the outsourcer may offer to the agents and through a combination of auditing, fines and rewards, honest computation is enforced. All of these verification techniques are, however, costly in terms of computation, memory, incentive rewards, etc., either to the prover or the verifier, or both. For example, the scheme in [12] requires partial re-execution of the tasks, and the verification in [3] incurs cost in the redundancy of the number of computing agents. Also, efficient protocols like PEPPER still incurs a cost in the order of m^3 [15] on the principal, where m is the size of the problem. The cost of employing verifiable computing across these different schemes hence raises the important question of how to use them economically, especially when there is a flexibility in parameters that govern the overall cost to the outsourcer. Motivated by this, we abstract verification techniques as an auditing tool with a exogenous cost and provide incentive-based contracts that minimise the expected cost of the principal. Our contributions generalize the results in [2,13] by (1) extending the feasibility of honesty enforcing schemes for *any* bound on the enforceable fines and *any* auditing capacity; (2) explicitly accounting for the cost of auditing and treating the auditing rate as one of the choice variables; and (3) providing optimal contract that minimize the aggregate cost of the principal as a combination of incentive payments and auditing costs. In short, our work explicitly extends both applicability and efficiency of incentive-based solutions based on a general abstraction of the verification method employed. For readers' interests, we also study in [10] the coalition among agents that may give them advantages in cheating the principal.

3   Problem Definition: General Setup

In this section, we describe the general setting of the problem and basic assumptions behind our model. A list of notations is provided in Table 1 for reference.

The outsourcer, which we refer to as the *principal*[1] has a deterministic *computation task* to be executed to obtain the output (result). Instead of executing the task itself, the principal hires a set of *agents*[2] to do this. The principal

---

[1] Also called the *boss* [2], *master* [6], *outsourcer* [4], *client* [8], *data owner* [13], etc.

[2] Also referred to as the *workers, servers, clouds,* or *contractors.*

aims to enforce *fully honest* computation of the task through setting a contract, involving rewards, auditing, and punishments (fines).

The principal and the agents are each selfish non-cooperative expected utility maximizers. Initially, we assume that everybody is risk-neutral, i.e., they have no strict preference between their expected utility and their utility of expected reward, and hence [9, ch.2.4], their utilities are linear function of the costs (with negative sign) and the rewards (with positive sign). Moreover, we assume that agents are "lazy but not malicious", that is, they do not have any interest in potentially reporting dishonest computations other than saving in their computation cost. Suppose the range and the probability distribution of the computation result is known. Generating a guessed output according to this distribution has zero computation cost and accuracy probability of q_0 (which can be negligibly small if the range of the output is large). For the sake of generality, as in [2], suppose each agent also has access to a private and independent *tricky algorithm* Alg that generates the correct output with probability q_1, where $q_0 < q_1 < 1$, at the cost of $c(q_1) \geq c(q_0) = 0$. The cost of honest computation is $c(1)$, which is strictly greater than $c(q_1)$. To enforce honesty of the agents, the principal *audits* the returned result with probability λ. We assume that auditing is perfect, i.e., if the output is indeed correct, the audit definitely confirms it (no "false positives"), and if the output is incorrect, the audit surely detects it (no "false negatives"). In the most basic contract, the principal decides on an auditing rate λ, sets a penalty (fine) f for detected erroneous answers and reward r otherwise. What make the problem non-trivial are the following observations:

1. *Costly detectability of cheating*: that auditing *all* of the results is either *infeasible* or *undesirable*. Regarding the infeasibility, suppose that in the long run the principal has a continuous demand (e.g. the Folding@Home project) of tasks awaiting computation, appearing at a rate ρ tasks per unit time. Also, suppose that each audit takes the principal ν machine cycles, and the computation capacity of the principal's machine is κ cycles per unit time. Then the maximum feasible rate of verification is $\frac{\kappa}{\nu\rho}$.[3] Moreover, auditing (e.g. through re-computation) may be costly as it will consume the computation power of the principal's machine and slow it down, or it will require obtaining additional hardware. The principal chooses the probability of auditing of a task $\lambda \in [0, \Lambda]$, where $0 < \Lambda \leq 1$ is associated with the computational capacity of the principal. The principal incurs the cost $\Gamma(\lambda)$ which is nondecreasing in λ. For simplicity of exposition, we assume a linear relation: $\Gamma(\lambda) = \gamma\lambda$ for a given $\gamma \geq 0$. An alternative to the occasional redoing of the whole computation by the principal can be using a third-party cloud that

---

[3] Note that even when the principal is verifying at full capacity, it should not pick the next immediate task to verify after finishing the previous one, since it may create a "learnable" pattern of audited tasks, which the agent can use to only be honest when computing them. This however can be avoided if the principal picks uniformly randomly tasks at the rate of $\frac{\kappa}{\nu\rho}$ and store them in a queue. However, the practical buffer has a storage limit. Consequently, the maximum feasible auditing rate with no essential pattern is strictly less than the full capacity rate $\frac{\kappa}{\nu\rho}$.

is highly reliable but costly (with per access cost of γ). For this scenario, the maximum auditing rate Λ is one, i.e., all of the tasks could be audited, albeit at an excessive cost.

2. *Limited enforceability of the fines*: The problem of verifiable computing could become trivial if there is no bound on the fine that can be practically levied on a wrongdoer: as long as there is even a tiniest probability of detection, then the principal can make the expected utility of the smallest likelihood of cheating become negative by setting the fine for erroneous results large enough. The issue with this argument is that such a fine may be extremely large and hence, become an *incredible threat*, in that, if the cheating of an agent is indeed caught, the fine is practically or legally non-collectable. Thus, existence (feasibility) results of honesty enforcement that rely on choosing a "large enough" fine are rather straightforward and uninteresting. In particular, such approaches leave unanswered the question of whether honest computation is still attainable for a bounded enforceable fine below their prescriptive threshold. Moreover, such results do not provide a good metric of comparison between alternative incentive schemes, or across different choices of parameters for a particular scheme. We will explicitly introduce $F \geq 0$ in our model to represent the maximum enforceable fine and obtain the optimal contracts subject to $f \leq F$. This can be the "security deposit", prepaid by the agent to the principal, that is collectible upon a provable detection of an erroneous result. A special case of interest is $F = 0$, i.e., when the only means of punishment is refusal to pay the reward.

3. *Limited budget*: As with the maximum enforceable fine to make it a credible threat, the maximum instantaneous "budget" of the principal leads to a bound on the reward to make it a credible promise. Let the maximum instantaneous payable reward by the principal be R. Thus, we require: $r \leq R$.

4   Contracts for Single Agent

In this section, we consider the case where the contract is designed for and proposed to only one computing agent. We provide the optimal contract for the basic model in subsection 4.1. In subsection 4.2, we investigate what happens if the risk-neutrality assumption of the agents is relaxed. Next in subsection 4.3, we comment on moderating against using tricky algorithms and clever guesses. Subsequently, in subsection 4.4, we discuss the optimal choice of the principal in the light of the optimal contracts theretofore developed. We close the case of single-agent in subsection 4.5 by generalising our results to contracts in which the principal is allowed to reward unaudited and verified tasks potentially differently. In Section 5, we will investigate the multi-agent case.

The action of the agent, given the parameters of the contract set by the principal, is first whether to accept it, and if so, which (probabilistic) algorithm to choose for computation of the assigned task. Since a naive random guess is correct with probability q_0, we assume that the agent's algorithm is correct with probability $q \in [q_0, 1]$. Let u_A denote the expected utility of the agent after accepting the contract. With correctness probability of q, the agent is

Table 1. List of main notations

parameter	definition
λ	probability of auditing an outsourced computation by the principal
Λ	the physical upper-bound on λ
γ	cost of auditing (incurred by the principal)
q	probability of a correct computation by the agent
q_0	the correctness probability of a random guess from the output space
$c(q)$	the expected cost of computation to an agent for the correctness level of q
$c(1), c$	cost of an honest computation to an agent
f	fine collected from agent upon detection of an erroneous computation
F	the maximum enforceable fine
r	reward to the agent for an unaudited or audited and correct computation
R	the maximum feasible reward
z	the reserve utility (a.k.a., fallback utility or aspiration) of the agent
H	auxiliary coefficient defined as $c(1) + z$ (§4)
K	auxiliary coefficient defined as $(c(1) - c(q_1))/(1 - q_1)$ (§4)
$\mathcal{C}$	the expected cost of the contract to the principal
α	probability of using two agents for the same computation (§5.1)
F_0	auxiliary coefficient defined as $c/\Lambda - c$ (Proposition 5, §5.1)
F_1	auxiliary coefficient defined as $c[c - \gamma]^+/[2\gamma - c]^+$ (Proposition 5, §5.1)

caught (and fined) with probability $(1-q)\lambda$. Hence, u_A is composed of expected reward $[1 - (1 - q)\lambda]r$, minus the expected cost composed of the cost $c(q)$ of the agent's algorithm and the expected fines $(1 - q)\lambda f$. Hence: $u_A(q) = [1 - (1 - q)\lambda]r - c(q) - (1 - q)\lambda f$. The agent may be able to achieve the same level of correctness, i.e., q, with different randomizations between the tricky algorithm Alg, the honest algorithm and random (naive) guessing. It is straightforward to make the following observation: For any q, the best $c(q)$ is achieved as follows:

a) If $[c(1) - c(q_1)]/(1 - q_1) > c(1)/(1 - q_0)$, then: $c(q) = \begin{cases} L_{q_0,q_1}(q) & q_0 \leq q \leq q_1 \\ L_{q_1,1}(q) & q_1 \leq q \leq 1 \end{cases}$;

b) If $[c(1) - c(q_1)]/(1 - q_1) < c(1)/(1 - q_0)$, then: $c(q) = L_{q_0,1}(q)$, where in both cases,

$L_{x,y}(z) := c(x) + \dfrac{c(y) - c(x)}{y - x}(z - x)$, i.e., the linear combination of the costs of the corresponding two end points.

Note that in case-(b), the risk-neutral agent would never use Alg, since the cost of using it can be undercut (in expected value) by randomizing between honest computation and random guessing. Hence, we only consider case-(a) for now and revisit case-(b) in §4.3.

4.1   Optimum Contract for a Single Agent

The principal chooses the contract by setting the rate of auditing and reward and punishment values, in order to maximize its own utility and ensure *fully honest* computation. Hence, the reward and punishments, r and f, should be chosen such that honest computation is the optimal course of action for the agent, if the contract is accepted. This means ensuring: $1 = \arg\max u_A(q)$. Following the

Principal-Agent modelling in game theory (e.g. [9, ch.7] or [14, ch.6]), we will refer to this as the *incentive compatibility* constraint. For case (a), this becomes:

$$u_A(1) = r - c(1) \geq u_A(q_1) = [1 - (1 - q_1)\lambda]r - c(q_1) - (1 - q_1)\lambda f \qquad (1)$$

The agent accepts the contract if its expected utility is larger than its *reserve utility*, $z \geq 0$.[4] Given incentive compatibility, this *participation constraint* is hence:[5]

$$u_A(1) = r - c(1) \geq z. \qquad (2)$$

The principal wants to get away with the least reward and auditing rate. Therefore, the *optimal* contract for the single agent reduces to solution of the following optimization:

$$\min_{r,f,\lambda} \mathcal{C} := r + \gamma\lambda \qquad (3a)$$

$$s.t. \quad r \leq R, \ 0 \leq f \leq F, \ 0 \leq \lambda \leq \Lambda, \qquad (3b)$$

$$r \geq H, \ r\lambda + f\lambda \geq K \qquad (3c)$$

where (3c) is derived from (1) and (2) in which we have used the auxiliary coefficients $H := c(1) + z$ and $K := [c(1) - c(q_1)]/(1 - q_1)$ for brevity. Then:

Proposition 1. *With the parameters given in Table 1, the contract that enforces honest computation and is accepted by the agent, and minimizes the cost of the principal is by setting $f^* = F$ and choosing λ^*, r^* as given by the following.*[6]

$$\gamma \leq \frac{K}{\Lambda^2} : \begin{cases} [\frac{K}{\Lambda} - H]^+ \leq F : & \lambda^* = \frac{K}{H + F}, \ r^* = H, \ C^* = H + \frac{\gamma K}{H + F} \\ [\frac{K}{\Lambda} - R]^+ \leq F < [\frac{K}{\Lambda} - H]^+ : & \lambda^* = \Lambda, \ r^* = \frac{K}{\Lambda} - F, \ C^* = \frac{K}{\Lambda} + \gamma\Lambda - F \end{cases}$$

$$\gamma > \frac{K}{\Lambda^2} : \begin{cases} [\sqrt{K\gamma} - H]^+ \leq F : & \lambda^* = \frac{K}{H + F}, \ r^* = H, \ C^* = H + \frac{\gamma K}{H + F} \\ [\sqrt{K\gamma} - R]^+ \leq F < [\sqrt{K\gamma} - H]^+ : & \lambda^* = \sqrt{\frac{K}{\gamma}}, r^* = \sqrt{K\gamma} - F, C^* = 2\sqrt{K\gamma} - F \\ [\frac{K}{\Lambda} - R]^+ \leq F < [\sqrt{K\gamma} - R]^+ : & \lambda^* = \frac{K}{R + F}, \ r^* = R, \ C^* = R + \frac{\gamma K}{R + F} \end{cases}$$

For $F < [\frac{K}{\Lambda} - R]^+$, the optimization is infeasible, i.e., there is no honesty-enforcing contract that is also accepted by the agent.

---

[4] The reserve utility (also referred to as the *fall-back utility* or *aspiration wage*) is the minimum utility that the agent aspires to attain or can obtain from other offers. Naturally, $z \geq 0$. Note that an implicit assumption here is that the agent is replaceable by any other agent with the same fall-back utility, i.e., there are many agents available with the same reserve utility. Without this assumption, the agent has negotiation power by refusing the contract knowing that it cannot be replaced. Alternatively, z can be thought as to (exogenously) capture the negotiation power of the agents. This is an assumption we make throughout the paper.

[5] Participation constraint is sometimes also called Individual Rationality constraint.

[6] The notation $x^+ := \max\{0, x\}$.

Discussion. The first observation is that the optimal contract should fully utilize the maximum enforceable fine and punish at no less than F. For large values of enforceable fines, we note that r^* is at H, the minimum value to ensure participation, and $\lim_{F\to\infty} \lambda^* = 0$, which yields $\lim_{F\to\infty} C^* = H$. These are compatible with intuition as a huge fine implies that honesty can be enforced with minimum compensation and minuscule rate of inspection. When auditing is cheap ($\gamma \leq K/\Lambda^2$), increasing the auditing rate is the better option to compensate for lower values of F to maintain incentive compatibility (honest computation). This is unless the auditing rate is at its maximum Λ, in which case, reward must increase above H to maintain incentive compatibility and compensate for the low value of F. Note that in this case, the participation constraint is not active and is satisfied with a slack, while the incentive compatibility constraint is satisfied tightly. For yet lower values of enforceable fine F, even maximum reward $r = R$ and auditing rate $\lambda = \Lambda$ might not impose a strong enough threat against cheating, hence the infeasibility region. When auditing is expensive ($\gamma > K/\Lambda^2$), in order to retain incentive compatibility in the situation of very low fine F, the principal should increase reward, and only consider more frequent auditing if the reward budget R has been reached. Fig. 1 depicts the optimal parameters of the contract versus the maximum enforceable fine for the latter case ($\gamma > K/\Lambda^2$).

Note that the infeasible region does not necessarily exist. Specifically, when the principal's instantaneous budget R is larger than K/Λ, then there is always a feasible contract. Then even for $F = 0$, i.e., no enforceable fine, a contract that enforces honest computing is feasible, albeit by using high values of reward and/or auditing rate. In such cases, the principal "punishes" audited erroneous computations only through not rewarding the agent. However, it is clear that honesty cannot be enforced with zero auditing rate, and hence the case of $\Lambda = 0$ trivially leads to infeasibility. Moreover, to satisfy the participation constraint at all, R has to be at least as large as H. Hence, for $R < H$, likewise, there exists no feasible contract for any F. We also show that except for the special case of $\gamma = 0$, the optimal contract has the feature that it is *unique*. Figures 2a and 2b depict

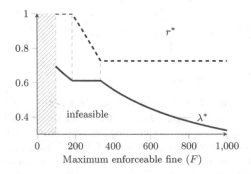

Fig. 1. Change of contract parameters r^*, λ^* w.r.t. the maximum enforceable fine F (Prop. 1, case of $\gamma > \frac{K}{\Lambda^2}$), where $K = 450$, $\gamma = 1200$, $\Lambda = 0.7$, and $c = 400$

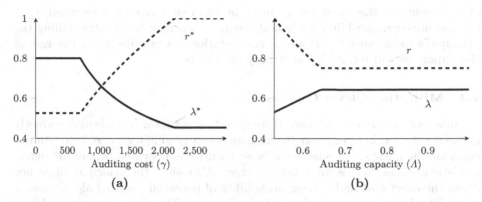

Fig. 2. Optimal contract parameters w.r.t (a) the auditing cost γ, with $K = 450$, $\Lambda = 0.8$, $c = 400$, and (b) auditing capacity Λ, with $K = 450$, $\gamma = 450$, $c = 450$

the change in the structure of the optimal contract versus varying auditing cost γ and the maximum auditing capacity, respectively. From Fig. 2a, we can see that for larger values of γ, the optimal contract utilizes lower values of inspection rate λ^* while using higher values of reward r to enforce honest computation. This transition progress culminates when the payment reaches its threshold R, after which the contract remains unchanged. In contrast, Fig. 2b shows how increasing the maximum auditing capacity affects the optimal contract in the opposite trend: as the principal is more capable of auditing, it should consider more frequent auditing and lessen the reward for honest computation. The payment, however, can never be lowered below H to maintain participation.

4.2 A Risk-Averse Agent

So far, we modelled the agent as risk-neutral, i.e., one that is indifferent between its expected utility and utility of expectation, leading to a linear utility function. However, empirically, individuals tend to show risk-aversion regarding decisions that affect their income. By definition, (strict) risk aversion is (strict) preference of expected utility over utility of expectation. Following *Jensen*'s inequality, this is equivalent to assuming a (strictly) concave utility function (ref. e.g. [9, ch.2.4]). We have the following simple but re-assuring result:

Proposition 2. *The optimal contract given in Proposition 1 developed for a risk-neutral agent stays feasible for any risk-averse agent as well.*

Note that even though the feasibility of our contract is guaranteed, its optimality might no longer hold. This is because a lower value of fine and/or rewards could potentially maintain incentive compatibility, as intuitively, cheating with a chance of getting caught can be seen as a lottery. However, because the level of risk-averseness of an agent is unknown, we argue that it is best practice to design the optimal contract for the worst case with respect to risk, i.e., risk neutrality. Specially, if a contract is designed assuming a particular degree of risk-aversion of the agent but the agent turns out to be less risk-averse than assumed, then the incentive-compatibility for honest computation may be violated, failing the principal's intolerance of erroneous computations. Accordingly, for the rest of the paper, we will retain risk-neutrality for agents.

4.3   Mitigating Clever Guesses

An inherent problem of outsourced computation is that a (not always) correctly guessed output is indistinguishable from an honestly computed one. For instance, consider the question of whether a large natural number is a prime: the deterministic guess of "no" is most likely correct. Also, since the principal might not know the exact cost and success probability of potential guessing algorithms, it is hard to design a contract that enforces honesty. Therefore, the principal may prefer to avoid identifying the parameters of guessing algorithms altogether.

One way to mitigate the possibility of "clever" guesses is to enlarge the output range by requiring the agent to return not just the final computation output, but also snapshots of intermediate steps of the computing process [2]. This will reduce the correctness probability of a naive guess down to $q_0 = $ negl. Moreover, requiring snapshots of the intermediate steps makes guessing of the correct output more costly. Let $c(q_1)$ be the cost of a tricky algorithm that tries to produce the expanded output with the intermediate steps of the honest computation, where it succeeds with probability q_1. We make the assumption that now $c(q_1) > q_1 c(1)$, so that any guessing algorithm with cost $c(q_1)$ can be replaced with a randomization between naive guess (with weight $1 - q_1$) and honest computation (with weight q_1). Thus, for incentive compatibility, we only need to make sure that the agent's utility from honest computation is better than a naive guess that succeeds with negligible probability $q_0 = $ negl. To avoid distraction in our analysis, we assume $q_0 = 0$, as the results can easily be realized for $q_0 = $ negl. Our simplified constraints for the contract become:

$$\text{participation}: \ r \geq c(1) + z, \ \text{incentive compatibility}: \ r \geq \frac{1}{\lambda}c(1) - f. \quad (4)$$

Comparing to the constraints in (3c), this translates to changing K to $c(1)$. This in turn implies that the new incentive compatibility constraint requires a strictly lower fine value. Intuitively, as guessing becomes more difficult, cheating becomes less attractive and hence can be deterred with a smaller fine. Hereafter, we assume that the principal is employing this technique and use the above incentive compatibility constraint. Moreover, for simplicity of exposition, we assume that the reserve utility z is zero, and hence H becomes $c(1)$, which we will abbreviate as c.

4.4   Optimal Choice for the Principal

So far we have considered auditing as a blackbox and only included its cost and capacity into the model. However, when auditing is via redoing the computation (at the cost of γ) it might be optimal for the principal to not offer any contract at all. Indeed, when $\Lambda = 1$, the principal can potentially audit all computations by redoing them. Specifically, if the optimal contract costs $C^* \geq \gamma$, then it is optimal for the principal to do the computation itself, as that only costs $\gamma\Lambda = \gamma$. In case $\Lambda < 1$, the principal cannot *do* all the computations, and must outsource a portion of it. Interestingly, the following proposition establishes that the principal's optimal choice is either to not outsource at all, or fully outsource its computation.

Proposition 3. *Consider the case where auditing is through redoing the computation. Let x be the probability that the principal computes the tasks itself. Then, either $x^* = 0$ and the optimal contract is as per Proposition 1, or $x^* = \Lambda = 1$ and there should be no outsourcing.*

The proposition has this important corollary:

Corollary 1. *When $\Lambda < 1$, the optimal choice for the principal is to use the optimal contact given by Proposition 1. When $\Lambda = 1$, the optimal choice of the principal is to compare the expected cost achieved by the optimal contract in Proposition 1 (for the value of maximum enforceable fine at hand) against γ, and accordingly decide to outsource or independently compute all of the tasks.*

4.5   Optimal Contract for a Single Agent: Two-Level Reward

In our contracts so far, verified correct results and unaudited results are rewarded identically at r. Suppose, alternatively, that the principal rewards r_0 for accepted but not audited results and r_1 for corroborated correct answers, and as before, penalizes f for detected wrong computations. This way, the principal may hope to save significantly by, for example, not paying for unaudited computations. The new incentive compatibility and participation constraints are: $(1-\lambda)r_0 + \lambda r_1 - c \geq (1-\lambda)r_0 - \lambda f$ and $(1-\lambda)r_0 + \lambda r_1 - c \geq 0$, respectively. The optimization of (3) for a contract with two-level reward changes to:

$$\min_{r_0, r_1, f, \gamma} C := r_1\lambda + r_0(1-\lambda) + \gamma\lambda$$

$$s.t. \quad r_0, r_1 \leq R, \ f \leq F, \ 0 \leq \lambda \leq \Lambda, \ r_1\lambda + r_0(1-\lambda) \geq c, \ r_1 \geq \frac{c}{\lambda} - f.$$

Proposition 4. *For $F \geq [c/\Lambda - R]^+$, the optimal single-agent contract for two-level rewarding is given as: $f^* = F$, $\lambda^* = c/(F+R)$, $r_1^* = R$, $r_0^* = Fc/(R-c+F)$, $C^* = c(1 + (\gamma + c - R)/(F+R))$. For $F < [c/\Lambda - R]^+$, the contract is infeasible.*

Discussion of the two level reward contract. First, note that there is no improvement in terms of the infeasibility region compared with the single-level reward contract. However, the achieved cost is always better. This was to be expected as the single-level rewarding can be thought of as a special case of two-level. However, the behaviour of the optimal contract now does not depend on the value of the auditing cost γ. This is where the strength of the two-level rewarding lies: for high values of γ, the two-level contract increasingly outperforms the single reward-level contract.

Note that the optimal reward for audited and correct results r_1 is at the principal's maximum budget R irrespective of the value of F. The value of reward for unaudited results r_0 is always strictly less than c, i.e., the cost of honest computation (and hence strictly less than r_1 as well). The value of r_0, unlike r_1, depends on F: For higher values of maximum enforceable fine, in fact somewhat unexpectedly, the optimal contract chooses increasing values of reward r_0^*. Still intuitively, a larger threat allows less necessity for auditing, and thus the contract starts to behave as a "lottery", in which the low-chance "winner" receives $r_1^* = R$ and the "loser" $r_0 < c < R$.

5   Optimal Contracts for Multiple Agents

When there are more than one agent available, the set of possible contracts gets extended. Specifically, as e.g. [2] and [13] discuss, the principal has the option of submitting the same task to multiple agents and comparing the outcomes. We will refer to this option as the *redundancy* scheme. If the returned results do not match, it is clear that at least one agent is cheating. Furthermore, as [13] assumes, if the agents are non-colluding, and returning the intermediate steps along with the computation result is required, then the probability that the results produced by cheating will be the same will be negligible, which we again assume to be zero (for simplicity). Hence, the returned results are correct *if and only if* they are the same.

In the next subsection, we develop optimal contracts considering two agents. Subsequently, we establish the global optimality of two-agent contracts among any number of agents with respect to the notion of Nash Equilibrium.

5.1   Optimal Contracts for Two Agents

Consider the case that there are two agents available: agent 1 and 2. As in the single-agent case, consider a principal that has a computation task and a maximum auditing rate of Λ. Then, in general, a principal can use a hybrid scheme: it may choose to send the same job to both of the agents sometimes, and to one randomly selected agents the rest of the time. Sending the same task to two agents provides a definite verification, however, at the cost of paying twice the reward, since both agents must be rewarded for honest computation. Hence, an optimal choice of redundancy scheme is not immediately clear, even less so if this schemes is randomized with just choosing one agent and doing independent audits. In this section, we investigate optimal contracts among all hybrid schemes.

Besides lack of collusion, we assume the agents do not communicate either. Therefore, on the event that any of the agents receives a task, it has no information about the busy/idle state of the other agent. The action of each agent is selection between honest computation, which we represent by $\mathcal{H}$, and cheating, which we denote by $\mathscr{C}$. Since the agents have no information about the state of the other agent, the set of their (pure) strategies and actions are the same.

The expected utility of each agent depends in part on the action of itself and of the other agent. Let $u_A(a_1, a_2)$ represent the utility of agent 1 when it chooses action a_1 and agent 2 chooses a_2, where $a_1, a_2 \in \{\mathscr{H}, \mathscr{C}\}$. The principal wants to enforce honest computation with probability one. If $u_A(\mathscr{H}, \mathscr{H}) \geq u_A(\mathscr{C}, \mathscr{H})$, then given that agent 2 is going to be computing honestly, agent 1 will prefer to do the same too, and due to symmetry, likewise for agent 2. In the game theoretic lingo, this means that $(\mathscr{H}, \mathscr{H})$ is a (Nash) equilibrium. If, further, $u_A(\mathscr{H}, \mathscr{C}) \geq u_A(\mathscr{C}, \mathscr{C})$, then $(\mathscr{H}, \mathscr{H})$ will be the dominant (Nash) equilibrium, i.e., honest computation is the preferred action irrespective of the action of the other agent.

The principal utilizes the redundancy scheme with probability α or employs only one of the agents (selected equally likely)[7] with probability $1 - \alpha$. If the principal chooses only one agent, then it audits it with probability ρ. Since auditing only occurs when a single agent receives the task, the likelihood λ that the task will ever be audited is $\rho(1 - \alpha)$. As in the single-agent single-reward scenario, if only one agent is selected, the agent is rewarded r if there is no indication of wrongdoing, and is punished f if audited and caught wrong. When the redundancy scheme is selected and the returned results are equal, both agents are rewarded r. Otherwise, both are fined at f. With the model so described, the expected utilities of an agent are computed as follows:[8]

$$u_A(\mathscr{H}, \mathscr{H}) = r - c, \qquad u_A(\mathscr{C}, \mathscr{H}) = (1 - \alpha - \lambda)r/2 - (\alpha + \lambda/2)f.$$

Hence, the condition $u_A(\mathscr{H}, \mathscr{H}) \geq u_A(\mathscr{C}, \mathscr{H})$ becomes: $r \geq (1 + \alpha)c/(\lambda + 2\alpha) - f$. Subject to making $(\mathscr{H}, \mathscr{H})$ an equilibrium, the contract is accepted if the expected utility of it to the agents is above their reserve utility, which we assume here too to be zero for simplicity: $r - c \geq 0$. Then the expected cost of the contract to the principal is:

$$C = 2r\alpha + \gamma\lambda + r(1 - \alpha) = (1 + \alpha)r + \gamma\lambda.$$

The principal chooses λ, α, f, r such that honest computation is enforced, the contract is accepted, and the expected cost of the principal is minimized. λ and α must satisfy the structural condition $0 \leq \alpha \leq 1$, $0 \leq \lambda \leq \Lambda$ and $\alpha + \lambda \leq 1$. The instantaneous budget of the principal imposes $r \leq R$ if $\alpha = 0$, and $2r \leq R$ if $\alpha > 0$. We assume $R \geq 2c$, since otherwise, the principal can never employ both of the agents without violating its instantaneous budget constraint, and hence, the problem reduces to the single agent problem. Then, the budget constraint

---

[7] We will formally show through the proof of proposition 6 that equal randomization is the best option. Intuitively, this removes any information that the agents may infer upon receiving a task.

[8] Since the only information state to an agent is whether it receives the job, the ex-ante and ex-post analysis, i.e., before and after reception of the task, become equivalent. We present the ex-ante view for simplicity.

simplifies to $r \leq R/2$. Therefore, the optimal contracts for two agents that make $(\mathcal{H}, \mathcal{H})$ an equilibrium are solutions of the optimization problem of:

$$\min_{r,f,\alpha,\lambda} r(1+\alpha) + \gamma\lambda \text{ subject to:}$$

$$r \leq R/2, \ f \leq F, \ 0 \leq \lambda \leq \Lambda, \ \lambda \leq 1 - \alpha, \ \alpha \geq 0, \ r \geq c, \ r \geq \frac{c(1+\alpha)}{\lambda + 2\alpha} - f.$$

Note that the above optimisation only guarantees that $(\mathcal{H}, \mathcal{H})$ is a Nash equilibrium. Other strategy profiles might become equilibria, for example $(\mathcal{C}, \mathcal{C})$. However we notice that because agents are only rewarded when they are both honest, $(\mathcal{H}, \mathcal{H})$ is thus the most attractive equilibrium to agents both individually and socially. We therefore only care to ensure that $(\mathcal{H}, \mathcal{H})$ is an equilibrium. The optimal contracts are as follows:

Proposition 5. *Let $F_0 = c/\Lambda - c$ and $F_1 = c[c - \gamma]^+/[2\gamma - c]^+$,[9] the optimal one-level reward two-agent contract that makes $(\mathcal{H}, \mathcal{H})$ a Nash equilibrium is:*

$$\begin{cases} F_1 \leq F: & f^* = F, \ \alpha^* = \dfrac{c}{2F+c}, \ \lambda^* = 0, r^* = c, \ C^* = c(1 + \dfrac{c}{2F+c}) \\[2mm] F_0 \leq F < F_1: & f^* = F, \ \alpha^* = 0, \ \lambda^* = \dfrac{c}{c+F}, \ r^* = c, \ C^* = c(1 + \dfrac{\gamma}{F+c}) \\[2mm] F < \min(F_0, F_1): & f^* = F, \ \alpha^* = \dfrac{c - \Lambda(c+F)}{c+2F}, \ \lambda^* = \Lambda, \ r^* = c, \ C^* = \dfrac{c(c+F)(2-\Lambda)}{c+2F} + \gamma\Lambda \end{cases}$$

For $\Lambda = 1$, $(\mathcal{H}, \mathcal{H})$ is moreover the dominant Nash equilibrium.

Corollary 2. *If auditing is more expensive than the cost of honest computation $(\gamma \geq c)$, the optimal contract only uses the redundancy scheme. When $\gamma \leq c/2$, either there is no redundancy scheme $(\alpha = 0)$ or the whole auditing capacity is used $(\lambda^* = \Lambda)$.*

The first part of the corollary is quite intuitive: when $\gamma > c$, any instance of outsourcing to a single agent and performing independent auditing can be replaced by the redundancy scheme (job duplication) and strictly lower the cost by $\gamma - c$.

Further Discussion. First, note that in our optimal two-agent contract, as long as $R \geq 2c$, there is no infeasible region: there is always a contract that makes $(\mathcal{H}, \mathcal{H})$ an equilibrium. Moreover, the payment to any of the agents is never more than the cost of honest computation. Fig. 3a provides a pictorial representation of the proposition where $c/2 < \gamma < c$ and $\Lambda = 0.5$. When the enforceable fine is large, the redundancy scheme is preferable. This is despite the fact that the redundancy scheme is more expensive than auditing: it costs an extra c as opposed to $\gamma < c$. In other words, for high values of fine, the redundancy scheme is a more effective threat against cheating than independent auditing. When F is less than F_1, the independent auditing becomes the preferred method. For lower values of F, when the auditing capacity is all used up, the redundancy scheme is added to compensate for the low value of fine to maintain incentive compatibility. Fig. 3b depicts the effect of auditing capacity, Λ, on the optimal contract

---

[9] We adopt the convention that $x/0 = +\infty$ for $x > 0$.

where $c/2 < \gamma < c$. When $\Lambda = 0$, redundancy scheme is the only means to enforce honest computation. If furthermore no fine can be enforced ($F = 0$), then $\alpha = 1$: the job should be always duplicated. As Λ increases, there is a gradual transition from using redundancy scheme to independent auditing ($F < F_1$).

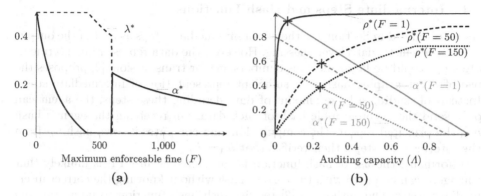

Fig. 3. Optimal contract (where $c = 400, \gamma = 250$) w.r.t. (a) max. enforceable fine F ($\Lambda = 0.5$); and (b) auditing capacity Λ ($F_1 = 600$). Recall $\rho = \frac{\lambda}{1-\alpha}$ is the conditional probability of auditing given the job is assigned to a single agent.

5.2   Global Optimality of Two-Agent Contracts

In developing the optimal contracts for two-agent case, we made a few critical assumptions: (a) the independent auditing is perfect; (b) the agents are non-colluding and non-communicating; (c) the range of intermediate steps is large enough that the probability of any two guessed results to be same, or the guessed result to be the correct result, is negligible; and (d) the agents are lazy but not malicious. It turns out that these assumptions are sufficient to warrant global optimality of two-agent contracts among contracts that engage any number of agents in the following notion:

Proposition 6. *The contract that hires at most two agents and chooses its terms according to proposition 5, is globally optimal, that is, it achieves the least cost to the principal among all contracts that employ any number of agents and aim to make honest computation a Nash Equilibrium.*

The above proposition shows that our contract for two agents is not just a special case solution of multiple agents, but it is indeed the solution involving any number of agents. In other words, given the stipulated assumptions, there is no advantage ever in hiring more than two agents. Incidentally, we also show that the best contracts makes the probability of any of the agents to be hired equal. This makes intuitive sense, as unequal probability of task assignment creates some "information" which the agents can potentially exploit to their benefit, and to the detriment of the principal.

6   Contract Implementation

For completeness of the solutions, in this section we discuss notable technical concerns on the implementation of our contracts.

6.1   Intermediate Steps and Hash Functions

As we discussed in Section 4.3, the use of intermediate steps as part of the output would prevent trivial/clever guessing. However, the data representing intermediate steps could be large and thus cumbersome for transmission. [2] proposes the use of cryptographic hash as a sufficient representation of intermediate steps: Instead of sending a large amount of data detailing these steps, the agent can only send the cryptographic hash of such data. On receiving the agent's hash h_A, the principal repeats the computation, and computes its own hash h_P from the intermediate steps, then verifies that $h_A = h_P$.

Informally, the use of hash function is considered secure if it is unlikely that the agent can come up with the correct hash without knowing the correct intermediate steps. The authors in [2] require such hash function to be a "random oracle", i.e., a function mapping in which each output is chosen uniformly randomly regardless of the input. While this is a sufficient condition, the notion of random oracle is rather impractical, and also an overkill. Indeed, we argue that for this purpose of hash checking, it is necessary and sufficient that the hash function is "collision resistant", that is, it should be difficult to find two different messages with the same hash.

Lastly, note that the process of hashing the intermediate steps may itself carry a considerable cost. For instance, if the computation task is to hash a large string, then the cost of hashing the intermediate steps (if the same hash function is used) would be at least as much as computation cost. Therefore, either the cost of hasing intermediate steps must be negligible compared to that of the original computation task, or it must enter the contract model.

6.2   Enforcing Contract Policies

With regards to legal enforcement of the contract, it is necessary that behaviours of contract participants are observable and verifiable. Actions such as "assigning a job" or "paying a reward" are of this type. However, probabilistic behaviours, e.g., "employing two agents with probability α", are usually unverifiable. Our contracts unfortunately rely on these probabilistic actions of the principal as explicitly stated in the terms and policies for auditing, task duplication and/or rewarding (the latter in two-level reward contracts of §4.5). It is critical to ensure (by both ends) that the principal in reality sticks to such actions, for two reasons. Firstly, the principal must establish to the agents its compliance to the contract so as to make the threats credible. Secondly, the agent needs an assurance that the principal cannot deviate from the contract and thus take away some of its benefits (in two-level rewarding). Without an appropriate security measure, this is usually not possible, e.g., the fact that the principal does not audit tells little about whether its auditing probability is indeed $\lambda = 0.3$ or $\lambda = 0.6$. This important implementation issue has not been discussed in previous works.

Usually this could be achieved cryptographically using multiparty computation (MPC) [11], in which a sampling function on the principal's behaviour is accurately and securely computed among the contract participants. However, MPC assumes pairwise secure communication among participants, which in this case implies a need for direct communication between the agents. This poses a potential threat to our model: if agents can freely communicate, they may as well collude and give identically incorrect result, thus fooling the principal. Therefore we seek a mechanism that requires no agent-to-agent communication. In Fig. 4, we propose a communication protocol between the principal and two agents that resolves this problem. Particularly, our security objective is to make sure that the principal gains negligible benefit by deviating from its prescribed behaviour as stated in the contract. To fulfil this objective, we rely on the fact that the contract can be legally enforced by an authority (e.g., a court), and thus punishment on the principal's cheating is guaranteed if there is enough evidence for the accusation. What remains is to ensure that each agent alone can prove the principal's deviation (from the contract) whenever the principal benefits from doing so.

Protocol ContractProtocol

Requirement A (non-interactive) commitment scheme (Setup, Commit, Open) and a trusted third party TTP. An optimal contract $\langle r, f, \alpha, \lambda \rangle$, a compu tation task J, and a security parameter $k > 0$.

Preparation The contract is signed by all parties with an additional term: if the principal is caught deviating from the below protocol, it must pay the worst possible cost [10]. TTP generates CK from Setup(k) and gives it to the principal (P), agent 1 (A_1) and agent 2 (A_2).

Protocol 1. A_1: generates $N_1 \leftarrow_\$ \{0,1\}^k$, computes $(c_1^*, d_1^*) = \text{Commit}_{CK}(N_1)$, then sends c_1^* to P.

2. A_2: generates $N_2 \leftarrow_\$ \{0,1\}^k$, computes $(c_2^*, d_2^*) = \text{Commit}_{CK}(N_2)$, then sends c_2^* to P.

3. P: sends (c_1^*, c_2^*) to both A_1 and A_2.

4. A_1: sends d_1^* to P.

5. A_2: sends d_2^* to P.

6. P: opens $N_1 \leftarrow \text{Open}_{CK}(c_1^*, d_1^*)$, $N_2 \leftarrow \text{Open}_{CK}(c_2^*, d_2^*)$, compute $\omega \leftarrow \text{Gen}_{\Delta(\Omega)}(N_1 \oplus N_2)$, and follows plan ω.

7. P: sends (d_1^*, d_2^*) to both A_1 and A_2.

8. A_1, A_2: open $N_1 \leftarrow \text{Open}_{CK}(c_1^*, d_1^*)$, $N_2 \leftarrow \text{Open}_{CK}(c_2^*, d_2^*)$, compute $\omega \leftarrow \text{Gen}_{\Delta(\Omega)}(N_1 \oplus N_2)$, and check if the principal follows ω.

Fig. 4. Communication protocol for the contract

In order to provably design such mechanism, we define the principal's action as a *plan*, which essentially captures the its deterministic choices for all possible decision-making situations which might arise while executing the contract. An example of such plan could be: give the task to both agents; if the result coming back is the same, then reward both. Another example is: give the task to agent 1, then audit on return. For convenience we denote the set of all possible plans as Ω, which also contains an element $\perp$ representing an invalid plan. The principal P is supposed to pick a plan $\omega \in \Omega$ according to a contract-specific probability distribution $\Delta(\Omega)$, but the agents do not know if P actually follows this distribution, or a different one to its eventual benefit. As a result, we decide to let such a plan be picked by the agents instead of the principal. The protocol for "picking plan" should satisfy the following properties:

- **Correctness**: Honest execution of the protocol must ensure that the plan is picked according to $\Delta(\Omega)$.
- **Hiding**: Before the contract is executed, the agents must know nothing about the plan they have picked for the principal.
- **Revealing**: After the contract is executed, there must be a secure way for the previously picked plan to be revealed to the agents.
- **No cheating**: Suppose that the agents execute the protocol honestly, then the principal receives no better benefit than being a honest principal.

We are now ready to construct our contract implementation protocol. For each probability distribution $\Delta(\Omega)$ assume that there exists a PPT *contract-generation* algorithm $\mathsf{Gen}_{\Delta(\Omega)}(\cdot)$ which efficiently samples $\Delta(\Omega)$, that is, there exists a negligible function ϵ_G and $k_1 > 0$ such that for all $k \geq k_1$:

$$\sup_{o \in \Omega} \left| \Pr\left[r \leftarrow_\$ \{0,1\}^k; \omega \leftarrow \mathsf{Gen}_{\Delta(\Omega)}(r) \right] - \Pr\left[\omega \leftarrow \Delta(\Omega) \right] \right| \leq \epsilon_G(k). \quad (7)$$

Whilst the protocol construction can be seen in Fig. 4, its security is described in Proposition 7. In words, the protocol involves the agents independently generate at uniformly random nonces N_1 and N_2, respectively. The agents then exchange these values using a commitment scheme via the principal P. The use of commitment ensures that even if the principal is able to modify the messages, it must not be able to convince each agent A_i of a nonce from the other which is dependent of N_i. This ensures that when A_i perform $N_1 \oplus N_2$ it would get a uniformly random value. Given the above property of $\mathsf{Gen}_{\Delta(\Omega)}$ the agent would receive a plan ω in the same distribution implied by the contract, thus avoid meaningful cheating by the principal.

Proposition 7 (informal). *Suppose all participants in* ContractProtocol *are PPT algorithms. Suppose that* (Setup, Commit, Open) *is a secure non-malleable commitment scheme, and that contract terms can be legally enforced and that both agents are honest, then* ContractProtocol *satisfies the following properties: correctness, hiding, revealing, and no cheating.*

---

[10] Here the worst possible cost (including what has been spent) is $\max(2r, r + \gamma)$, and it could either be distributed to the agents, or paid to the court as fine.

7   Conclusion

In this paper, we provide an incentive analysis of outsourced computation with non-malicious but selfish utility-maximising agents. We design contracts that minimise the expected cost of the outsourcer whilst ensuring participation and honesty of computing agents. We incorporate important real-world restrictions, in that the outsourcer can only levy a restricted fine on dishonest agents and that auditing can be costly and/or limited. We allow partial outsourcing, direct auditing and auditing through redundancy, i.e., employing multiple agents and comparing the results, and optimized the utility of the outsourcer among all hybrid possibilities.

We observe that outsourcing all or none of the tasks is optimal (and not partial outsourcing). We show that when the enforceable fine is restricted, achieving honest computation may still be feasible by appropriately increasing the reward above the sheer cost of honest computation. We demonstrate that when auditing is more expensive than the cost of honest computation, redundancy scheme is always the preferred method, and when the auditing cost is less than half of the cost of honest computation, independent auditing is preferable. When the cost of auditing is between half and the full cost of honest computation, the preferred method depends on the maximum enforceable fine: for large enforceable fines, redundancy scheme is preferred despite the fact that it is more expensive "per use" than independent auditing, since owing to its higher effectiveness, it can be used more sparingly. We establish the global optimality of contracts involving at most two agents among any arbitrary number of agents as far as implementing honesty as a Nash Equilibrium is aimed for. Finally, we present a light-weight cryptographic implementation of our contracts that provides mutual affirmation on proper execution of the agreed terms and conditions.

Acknowledgement. This research was partially sponsored by US Army Research laboratory and the UK Ministry of Defence under Agreement Number W911NF-06-3-0001. The views and conclusions contained in this document are those of the authors and should not be interpreted as representing the official policies, either expressed or implied, of the US Army Research Laboratory, the U.S. Government, the UK Ministry of Defense, or the UK Government. The US and UK Governments are authorized to reproduce and distribute reprints for Government purposes notwithstanding any copyright notation hereon.

References

1. Atallah, M.J., Cho, Y.S., Kundu, A.: Efficient data authentication in an environment of untrusted third-party distributors. In: IEEE ICDE (2008)
2. Belenkiy, M., Chase, M., Chris Erway, C., Jannotti, J., Küpçü, A., Lysyanskaya, A.: Incentivizing outsourced computation. In: NetEcon. ACM (2008)
3. Canetti, R., Riva, B., Rothblum, G.N.: Practical delegation of computation using multiple servers. In: ACM CCS (2011)

4. Carbunar, B., Tripunitara, M.V.: Payments for outsourced computations. IEEE Transactions on Parallel and Distributed Systems 23(2) (2012)
5. Chen, H., Ma, X., Hsu, W., Li, N., Wang, Q.: Access control friendly query verification for outsourced data publishing. In: Jajodia, S., Lopez, J. (eds.) ESORICS 2008. LNCS, vol. 5283, pp. 177–191. Springer, Heidelberg (2008)
6. Christoforou, E., Anta, A.F., Georgiou, C., Mosteiro, M.A., Sánchez, A.: Applying the dynamics of evolution to achieve reliability in master–worker computing. Concurrency and Computation: Practice and Experience (2013)
7. Fullbright, N.R.: Outsourcing in a brave new world: An international survey of current outsourcing practice and trends. Technical report (2011)
8. Gennaro, R., Gentry, C., Parno, B.: Non-interactive verifiable computing: Outsourcing computation to untrusted workers. In: Rabin, T. (ed.) CRYPTO 2010. LNCS, vol. 6223, pp. 465–482. Springer, Heidelberg (2010)
9. Gintis, H.: Game Theory Evolving: A Problem-Centered Introduction to Modeling Strategic Interaction. Princeton University Press (2009)
10. Khouzani, M.H.R., Pham, V., Cid, C.: Incentive engineering for outsourced computation in the face of collusion. In: Proceedings of WEIS 2014 – 13th Annual Workshop on the Economics of Information Security (2014)
11. Maurer, U.: Secure multi-party computation made simple. In: Cimato, S., Galdi, C., Persiano, G. (eds.) SCN 2002. LNCS, vol. 2576, pp. 14–28. Springer, Heidelberg (2003)
12. Monrose, F., Wyckoff, P., Rubin, A.D.: Distributed execution with remote audit. In: NDSS (1999)
13. Nix, R., Kantarcioglu, M.: Contractual agreement design for enforcing honesty in cloud outsourcing. In: Grossklags, J., Walrand, J. (eds.) GameSec 2012. LNCS, vol. 7638, pp. 296–308. Springer, Heidelberg (2012)
14. Rasmusen, E.: Games and information: An introduction to game theory (1994)
15. Setty, S., McPherson, R., Blumberg, A.J., Walfish, M.: Making argument systems for outsourced computation practical (sometimes). In: NDSS (2012)
16. Setty, S., Vu, V., Panpalia, N., Braun, B., Blumberg, A.J., Walfish, M.: Taking proof-based verified computation a few steps closer to practicality. In: USENIX Security (2012)
17. Wang, C., Ren, K., Wang, J.: Secure and practical outsourcing of linear programming in cloud computing. In: INFOCOM 2011 (2011)
18. Yi, K., Li, F., Cormode, G., Hadjieleftheriou, M., Kollios, G., Srivastava, D.: Small synopses for group-by query verification on outsourced data streams. In: ACM TODS (2009)

A Supervisory Control Approach to Dynamic Cyber-Security

Mohammad Rasouli, Erik Miehling, and Demosthenis Teneketzis

Department of Electrical Engineering and Computer Science
University of Michigan, Ann Arbor, MI, USA

Abstract. An analytical approach for a dynamic cyber-security problem that captures progressive attacks to a computer network is presented. We formulate the dynamic security problem from the defender's point of view as a supervisory control problem with imperfect information, modeling the computer network's operation by a discrete event system. We consider a min-max performance criterion and use dynamic programming to determine, within a restricted set of policies, an optimal policy for the defender. We study and interpret the behavior of this optimal policy as we vary certain parameters of the supervisory control problem.

Keywords: Cyber-Security, Computer Networks, Discrete Event Systems, Finite State Automata, Dynamic Programming.

1 Introduction

Cyber-security has attracted much attention recently due to its increasing importance in the safety of many modern technological systems. These systems are ubiquitous in our modern day life, ranging from computer networks, the internet, mobile networks, the power grid, and even implantable medical devices. This ubiquity highlights the essential need for a large research effort in order to strengthen the resiliency of these systems against attacks, intentional and unintentional misuse, and inadvertent failures.

The study of cyber-security problems in the existing literature can be divided into two main categories: *static* and *dynamic*.

Static problems concern settings where the agents, commonly considered to be an attacker and a defender, receive no new information during the time horizon in which decisions are made. Problems of this type in the security literature can largely be classified under the category of *resource allocation*, where both the defender and attacker make a single decision as to where to allocate their respective resources. The main bodies of work involve infrastructure protection [3, 7, 9] and mitigation of malware and virus spread in a network [5, 6, 8, 16]. Some of the above works consider settings where the agents are strategic [3, 9]. The presence of strategic agents results in a game between the attacker and defender. The strategic approaches in the above works are commonly referred to as *allocation games*. The survey by Roy et al. [18], as well as [20], provide useful outlines of some static game models in security.

R. Poovendran and W. Saad (Eds.): GameSec 2014, LNCS 8840, pp. 99–117, 2014.
© Springer International Publishing Switzerland 2014

Dynamic security problems are those that evolve over time, with the defender taking actions while observing some new information from the environment.[1] The formulation of a security problem as a dynamic problem, instead of a static one, offers numerous advantages. The first advantage is clear; since real-world security problems have an inherently dynamic aspect, dynamic models can more easily capture realistic security settings, compared to static models. Also, most attacks in cyber-security settings are *progressive*, meaning more recent attacks build upon previous attacks (such as denial-of-service attacks, brute-force attacks, and the replication of viruses, malware, and worms, to name a few). This progressive nature is more easily modeled in a dynamic setting than in a static setting.

The literature within the dynamic setting can be further subdivided into two areas: models based on control theory [10, 13, 14, 17, 19] and models based on game theory [11, 18, 21, 22].

The control theory based security models in the literature differ in the ways in which the dynamics are modeled. The work by Khouzani et al. [10] studies the problem of a malware attack in a mobile wireless network; the dynamics of the malware spread are modeled using differential equations. A large part of the literature on control theory based models focuses on problems where the dynamics are modeled by finite state automata. The works of [13, 14, 19] implement specific control policies (protocols) for security purposes. The work of Schneider [19] uses a finite state automaton to describe a setting where signals are sent to a computer. Given a set of initial possible states, the signals cause the state of the computer to evolve over time. An entity termed the *observer* monitors the evolution of the system and enforces security in real-time. Extensions of Schneider's model are centered around including additional actions for the observer. Ligatti et al. [13] extend Schneider's model by introducing a variety of abstract machines which can edit the actions of a program, at run-time, when deviation from a specified control policy is observed. More recent work [14] develops a formal framework for analyzing the enforcement of more general policies. Another category of dynamic defense concerns scenarios where the defender selects an *adaptive attack surface*[2] in order to change the possible attack and defense policies. A notion termed *moving target defense* (a term for dynamic system reconfiguration) is one class of such dynamic defense policies. The work of Rowe et al. [17] develops control theoretic mechanisms to determine maneuvers that modify the attack surface in order to mitigate attacks. The work involves first developing algorithms for estimation of the security state of the system, then formalizing a method for determining the cost of a given maneuver. The model uses a logical automaton to describe the evolution of the state of the system; however, it does not propose an analytical approach for *determining* an optimal defense policy.

---

[1] This new information could consist of the attacker's actions, events in nature, or the state of a some underlying system.

[2] For example, changing the network topology.

The next set of security models in the literature are based on the theory of dynamic games. The work in [15] considers a stochastic dynamic game to model the environment of conflict between an attacker and a defender. In this model, the state of the system evolves according to a Markov chain. This paper has many elements in common with our model; however, it assumes the attacker and defender have perfect observations of the system state. In our paper, we consider the problem from the defender's point of view and assume that the defender has imperfect information about the system state. The work by Khouzani [11] studies a zero-sum two-agent (malware agent and a network agent) dynamic game with perfect information. The malware agent is choosing a strategy which trades off malware spread and network damage while the network agent is choosing a counter-measure strategy. The authors illustrate that *saddle-point strategies* exhibit a threshold form. The work of Yin et al. [22] (dynamic game version of [3]) studies a Stackelberg game where the defender moves first and commits to a strategy. The work addresses how the defender should choose a strategy when it is uncertain whether the attacker will observe the first move. Van Dijk et al. [21] propose a two player dynamic game, termed *Flipit*, which models a general setting where a defender and an attacker fight (in continuous time) over control of a resource. The results concern the determination of scenarios where there exist dominant strategies for both players. We refer the reader to Roy et al. [18], and references therein, for a survey on the application of dynamic games to problems in security.

While models based on game theory have generated positive results in the static setting, there has been little progress in the dynamic setting. We believe this is for two reasons; first, dynamic security has not been fully investigated in a non-strategic context and second, the results in the theory of dynamic games are limited.

In this paper, we develop a (supervisory) control theory approach to a dynamic cyber-security problem and determine the optimal defense policy against progressive attacks. We consider a network of K computers, each of which can be in one of four security states, as seen in Figure 1. The state of the system is the K-tuple of the computer states and evolves in time with both defender and attacker actions. We use a finite state logical automaton to model the dynamics of the system. The defender adjusts to attacks based on the information available.

Our model takes a different approach than the existing papers in the literature. One fundamental difference of our work from the existing literature that make use of automata is the development of an *analytical* framework for *determining optimal defense policies* within a restricted set of policies. Other works involving automata propose methods for *enforcing* a predetermined policy, rather than determining an optimal policy. Also, our control theoretic approach considers imperfect information regarding attacker actions, which we feel is an aspect that is engrained into security problems.

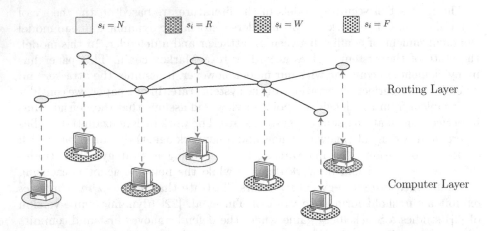

Fig. 1. *An instance of the problem that we consider. Computers are connected through a routing layer. Each computer can be in one of four security states:* normal *(N)*, compromised *(R)*, fully compromised *(W)*, or remote compromised *(F)*.

1.1   Contribution

The contribution of this paper is the development of a formal model for analyzing a dynamic cyber-security problem from the defender's point of view. Our approach has the following desirable features: (i) It captures the progressive nature of attacks; (ii) It captures the fact that the defender has imperfect knowledge regarding the state of the system; this uncertainty is a result of the fact that all attacks are uncontrollable and most are unobservable, by the defender; (iii) It allows us to quantify the cost incurred at every possible state of the system, as well as the cost due to every possible defender action; (iv) It allows us to quantify the performance of various defender policies and to determine the defender's optimal control policy, within a restricted set of policies, with respect to a min-max performance criterion.

1.2   Organization

The paper is organized as follows. In Section 2 we discuss our dynamic defense model. This is done by introducing the assumptions on the computer network and corresponding state, as well as the events which drive the evolution of the system state. In Section 3, we model the defender's problem of keeping the computer network as secure as possible while subjected to progressive attacks. We provide a simplified problem formulation that is tractable. In Section 4, we determine an optimal control policy for the defender based on dynamic programming. We discuss the nature of the optimal policy in Section 5. We offer conclusions and reflections in Section 6.

2 The Dynamic Defense Model

The key features of our model are characterized by assumptions **(A1)** – **(A6)**. We first describe the assumptions related to the *computer network*, discussed in assumption **(A1)**. In assumption **(A2)** we introduce the notion of the computer network *system state*. Next, in assumptions **(A3)** – **(A5)**, we discuss the *events* that can occur within the system. We describe how the events cause the system state to evolve, as well as specify which events are controllable and observable by the defender. In **(A6)** we discuss an assumption on the rules of interaction between the attacker and the defender. As mentioned in the introduction, we consider the cyber-security problem from the defender's viewpoint; the model we propose reflects this viewpoint.

Assumption 1 - Computer Network: *We assume a set of networked computers, $\mathcal{N} = \{1, 2, \ldots, K\}$. Each computer, $i \in \mathcal{N}$, can be at security level $z^i \in \mathcal{M} = \{N, R, W, F\}$ where $\mathcal{M}$ is the set of security states.*

Each computer, $i \in \mathcal{N}$, is assumed to have three security boundaries, denoted by $\mathcal{B} = \{B_1, B_2, B_3\}$, representative of a layered structure to its security. These security boundaries partition the set of security states $\mathcal{M}$. Throughout this paper, we assume that the set of security states $\mathcal{M} = \{N, R, W, F\}$ is defined as follows.

- *Normal* ($z^i = N$): Computer i is in the *normal* state if none of the security boundaries have been passed by the attacker.
- *Compromised* ($z^i = R$): Computer i is *compromised* when security boundary B_1 has been passed by the attacker. In this state, the attacker has exploited some vulnerability on the computer and has managed to obtain user-level access privilege to the computer.
- *Fully Compromised* ($z^i = W$): Computer i is *fully compromised* when both boundaries B_1 and B_2 have been passed by the attacker. The attacker has exploited some additional vulnerability on the computer and has managed to obtain root level or execute privilege to the computer.
- *Remote Compromised* ($z^i = F$): Computer i is *remote compromised* when all security boundaries B_1, B_2, and B_3 have been passed by the attacker. The attacker has managed to obtain enough privileges to attack another computer and obtain user-level access privilege on that computer.

Assumption 2 - System State: *We assume that the computer network operates over an infinite time horizon, $\mathcal{T} = \{0, 1, 2, \ldots\}$. The state of the computer network, Z_t, which evolves with time $t \in \mathcal{T}$, is the combination of the states of all the computers at time t. Each state Z_t has a corresponding cost.*

The state of the network, denoted $Z_t = (z_t^1, z_t^2, \ldots, z_t^K) \in \mathcal{Z}$, is a K-tuple of all of the computer states.[3] The set $\mathcal{Z}$ denotes the set of all possible states,

---

[3] For example, a three computer network could have a network state of $Z_t' = (N, R, W)$. Notice that state Z_t' is distinct from state $Z_t'' = (R, N, W)$.

$$\mathcal{Z} = \{Z^1, Z^2, \ldots, Z^{|\mathcal{M}|^K}\} = \{(N, N, \ldots, N), (N, N, \ldots, R), \ldots, (F, F, \ldots, F)\},$$

where $|\mathcal{M}|^K$ is the number of system states.

The cost of the network state Z_t is defined by the costs of the states of the computers. We assign a cost, $c(z_t^i)$, to each computer i depending upon its state $z_t^i \in \mathcal{M}$. This cost is defined as follows

$$c(z_t^i) = \begin{cases} c_N & \text{if } z_t^i = N \\ c_R & \text{if } z_t^i = R \\ c_W & \text{if } z_t^i = W \\ c_F & \text{if } z_t^i = F \end{cases} \tag{1}$$

with $0 \leq c_N < c_R < c_W < c_F < \infty$. The cost of state Z_t is then defined as

$$C_{Z_t} = \sum_{i \in \mathcal{N}} c(z_t^i) \tag{2}$$

The state of the network, Z_t, evolves in time due to events, which we discuss in the next set of assumptions.

Assumption 3 - Events: *There is a set of events, $\mathcal{E} = \mathcal{A} \cup \mathcal{D}$, where $\mathcal{A}$ are the attacker's actions and $\mathcal{D}$ are the defender's actions.*

We assume that the attacker has access to three types of actions. The set of attacker actions, $\mathcal{A} = \{N^a, \{P_n^i\}_{i \in \mathcal{N}, n \in \mathcal{B}}, \{H^{ij}\}_{i,j \in \mathcal{N}}\}$, is defined as follows.

- N^a, *null*: The attacker takes no action. The null action does not change the system state and is admissible at any state of a computer.
- P_n^i, *security boundary attack*: Attacking the nth security boundary of computer i causes the security state of computer i to transition across the nth security boundary. Specifically, $P_{B_1}^i$ causes computer i to transition from normal, $z^i = N$, to compromised, $z^i = R$; $P_{B_2}^i$ from $z^i = R$ to $z^i = W$; and $P_{B_3}^i$ from $z^i = W$ to $z^i = F$. Actions $P_{B_1}^i$, $P_{B_2}^i$, and $P_{B_3}^i$ are only admissible from states $z^i = R$, $z^i = W$, and $z^i = F$, respectively.
- H^{ij}, *network attack*: Using a computer i in state $z^i = F$ to attack any other normal or compromised computer j in the network that is in state $z^j = \{N, R\}$ to bring computer j to state $z^j = W$. The action H^{ij} is admissible at state $z^i = F$ for $z^j \in \{N, R, W\}$.

We assume that the defender knows the set $\mathcal{A}$ as well as the resulting state transitions due to each action in $\mathcal{A}$.

The defender has access to three types of costly actions. These actions are admissible at any computer state. The set of defender actions, denoted by $\mathcal{D} = \{N^d, \{E_i\}_{i \in \mathcal{N}}, \{R_i\}_{i \in \mathcal{N}}\}$, is defined as follows.

- N^d, *null*: The defender takes no action. The null action does not change the system state.
- E_i, *sense computer i*: The *sense* action, E_i, reveals the state of computer i to the defender. The sense action does not change the system state.

- R_i, *re-image computer i*: The *re-image* action, R_i, brings computer i back to the normal state from any state that it is currently in. For example, R_3 applied to state (N, R, F) results in (N, R, N).

The costs of the actions in $\mathcal{D}$ are defined by $\hat{C}(N^d)$, $\hat{C}(E_i)$, $\hat{C}(R_i)$, where $0 \le \hat{C}(N^d) < \hat{C}(E_i) < \hat{C}(R_i) < \infty$ for all $i \in \mathcal{N}$.

Assumption 4 - Defender's Controllability of Events: *The actions in $\mathcal{A}$ are uncontrollable whereas the actions in $\mathcal{D}$ are controllable.*

Since the problem is viewed from the perspective of the defender, all actions in $\mathcal{D}$ are controllable. For the same reason, the defender is unable to control any of the attacker's actions $\mathcal{A}$.

Assumption 5 - Defender's Observability of Events: *All actions in $\mathcal{D}$ and some actions in $\mathcal{A}$ are assumed to be observable.*

Again, due to taking the defender's viewpoint, all actions in $\mathcal{D}$ are observable. Although we assume that the defender knows the set $\mathcal{A}$, we assume that it cannot observe N^a or any P_n^i actions; it can only observe actions of the type H^{ij}. One justification for this is that the the network attack H^{ij} involves passing sensitive information of computer j through the routing layer of the system to computer i.[4] We assume that the routing layer is able to detect the transfer of sensitive data through the network, and thus the defender is aware when an action of the form H^{ij} occurs.

Assumption 6 - Defender's Decision Epochs: *The defender acts at regular, discrete time intervals. At these time intervals, the defender takes only one action in $\mathcal{D}$. The attacker takes one action in $\mathcal{A}$ between each defender action.*

We require that the defender should consider taking a single action in $\mathcal{D}$ at regular time instances. We assume that between any two such instances, the attacker can only take one action in $\mathcal{A}$. This order of events is illustrated in Figure 2 for a given time $t = \tau$. We introduce *intermediate states*, denoted by $\tilde{\mathcal{Z}} = (\tilde{Z}^1, \tilde{Z}^2, \ldots, \tilde{Z}^{|\mathcal{M}|^K})$, which represent the system states at which events from $\mathcal{A}$ are admissible (that is, the states in which the attacker takes an action). The *system states*, denoted by $\mathcal{Z} = (Z^1, Z^2, \ldots, Z^{|\mathcal{M}|^K})$, are the states at which actions from $\mathcal{D}$ are admissible.

Assumption **(A6)** is, in our opinion, reasonable within the security context. Since time has value in security problems,[5] the defender should take actions at regular time intervals (note that at these instances the defender may choose N^d, that is, choose to do nothing). In general, a finite number of events in $\mathcal{A}$ may occur between any two successive defender actions; however, to reduce the dimensionality of the problem, we assume that only one event in $\mathcal{A}$ can occur.

---

[4] This sensitive information could be the login credentials of computer j.

[5] A computer that is compromised by the attacker for two time steps is more costly to the defender than a computer that is compromised for one time step.

Fig. 2. *Order of events for a given time-step. At time $t = \tau$, the cost of the current state C_{Z_τ} is realized. At τ^+, the defender takes an action in $\mathcal{D}$ (the cost of which is realized immediately). The resulting system state due to the defender's action is denoted by $\tilde{Z}_{\tau^+} \in \tilde{\mathcal{Z}}$. At τ^{++} the attacker takes an action in $\mathcal{A}$. At $\tau + 1$, the resulting system state is denoted by $Z_{\tau+1} \in \mathcal{Z}$.*

One important implication of assumption (**A6**) is related to the defender's observability of events in $\mathcal{A}$. By (**A6**), the defender is aware *when* an event in $\mathcal{A}$ occurs. Since the event H^{ij} is observable, if the defender does not observe H^{ij} when an event in $\mathcal{A}$ is known to occur, then it knows that one of the unobservable events, N^a or one of $\{P^i_n\}_{i \in \mathcal{N}, n \in \mathcal{B}}$, has occurred. To incorporate this fact into the defender's knowledge about the system's evolution, we group the above mentioned unobservable events into one event, denoted $X = \{N^a, \{P^i_n\}_{i \in \mathcal{N}, n \in \mathcal{B}}\}$. This philosophy is used in constructing the system automaton from the defender's point of view, as well as in defining the defender's information state (discussed in Section 3). As a result of the above grouping, the set of events $\mathcal{A}' = \{X, \{H^{ij}\}_{i,j \in \mathcal{N}}\}$ is observable by the defender. Notice, however, that by performing this grouping, we have introduced non-determinism into the system; that is, the event X can take the system to many possible system states. All unobservable events in the problem have been eliminated due to Assumption (**A6**) and the grouping of unobservable events in $\mathcal{A}$.

As a result of assumptions (**A1**) – (**A6**), the evolution of the system state, Z_t, from the defender's viewpoint, can be modeled by a discrete event system represented by a finite state automaton, which we term the *system automaton*. Due to assumption (**A6**), we duplicate the system states by forming the set of intermediate states, denoted by $\tilde{\mathcal{Z}} = (\tilde{Z}^1, \tilde{Z}^2, \ldots, \tilde{Z}^{|\mathcal{M}|^K})$. The set of intermediate states represents the states at which an event from $\mathcal{A}$ can occur. The set of system states, denoted by $\mathcal{Z}$, are the states at which the defender takes an action $d \in \mathcal{D}$. The resulting automaton has $2|\mathcal{M}|^K$ states. The set of events that can occur is described by the set $\mathcal{E}' = \mathcal{A}' \cup \mathcal{D}$; the transitions due to these events follow the rules discussed in assumption (**A3**). The system automaton takes the form of a bipartite graph, as seen in Figure 3. Notice that, like the null action, the sense actions, E_i, for all $i \in \mathcal{N}$, do not change the underlying system state. The purpose of sense is to update the defender's information state, which will be defined and explained in the following section.

3   The Defender's Problem

We now formulate the defender's problem – protecting the computer network. The defender must decide which costly action to take, at each time step, in order

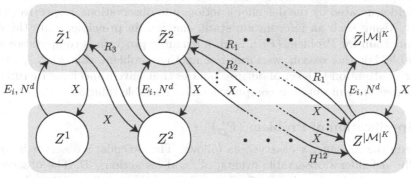

Fig. 3. *The system automaton represented as a bipartite graph of intermediate states,* $\tilde{\mathcal{Z}} = (\tilde{Z}^1, \tilde{Z}^2, \ldots, \tilde{Z}^{|\mathcal{M}|^K})$, *and system states,* $\mathcal{Z} = (Z^1, Z^2, \ldots, Z^{|\mathcal{M}|^K})$, *with events* $\mathcal{E}' = \mathcal{A}' \cup \mathcal{D}$. *Notice the non-determinism of the event* $X \in \mathcal{A}'$.

to keep the system as secure as possible given that it has imperfect knowledge of the network's state.

3.1 The Defender's Optimization Problem

Let $g := \{g_t, t \in \mathcal{T}\}$, denote a control policy of the defender, where

$$g_t : \mathcal{D}^{t-1} \times \mathcal{A}'^{t-1} \to \mathcal{D}, \tag{3}$$

and $\mathcal{D}^{t-1}$ and $\mathcal{A}'^{t-1}$ denote the space of the defender's actions and observations up to $t-1$, respectively. Let $\mathcal{G} := \{g \mid g_t : \mathcal{D}^{t-1} \times \mathcal{A}'^{t-1} \to \mathcal{D} \text{ for all } t \in \mathcal{T}\}$ denote the space of admissible control policies for the defender.

The defender's optimization problem is

$$\min_{g \in \mathcal{G}} \max_{\{Z_t^g \in \mathcal{Z}, t \in \mathcal{T}\}} \left\{ \sum_{t \in \mathcal{T}} \beta^t \left[C_{Z_t^g} + \hat{C}(d_t) \right] \right\} \tag{P_D}$$

subject to   Assumptions **(A1)** – **(A6)**

where $\{Z_t^g \in \mathcal{Z}, t \in \mathcal{T}\}$ denotes a sequence of states generated by control policy g and d_t is the defender's action at t generated according to Equation (3). Problem (P_D) is a supervisory control problem with imperfect observations.

3.2 Discussion of Problem (P_D)

The notion of an information state [12] is a key concept in supervisory (and general) control problems with imperfect information. Because of the nature of the performance criterion and the fact that the defender's information is imperfect, an appropriate information state for the defender at time t is $\sigma(\mathcal{D}^{t-1}, \mathcal{A}'^{t-1})$, the σ-field generated by the defender's actions and observations, respectively, up to $t - 1$. Using such an information state, one can, in principle, write the dynamic program for Problem (P_D). Such a dynamic program is computationally intractable. For this reason, we formulate another problem, called (P'_D), where we restrict attention to a set of defense policies that have a specific structure; in this problem we can obtain a computationally tractable solution.

3.3   Specification of Problem (P'_D)

We define the *defender's observer* as follows. The defender's observer is built using the defender's observable events, $\mathcal{A}'$, and its actions, $\mathcal{D}$. The observer's state at time t, denoted by $S_t \subseteq \mathcal{Z}$, consists of the possible states that the network can be in at time t from the defender's perspective. We denote by $\mathcal{S}$ the space to which S_t belongs, for any $t \in \mathcal{T}$.

The evolution of the observer's state is described by the function $f : \mathcal{S} \times \mathcal{D} \times \mathcal{A}' \to \mathcal{S}$. The observer's state S_t follows the update

$$S_{t+1} = f(S_t, d_t, a'_t)$$

where $d_t \in \mathcal{D}$ is the realization of the defender's action and its effect at time t^+, and $a'_t \in \mathcal{A}'$ is the realization of the defender's observation at t^{++}. The precise form of the function f is determined by the dynamic defense model of Section 2. Thus, the dynamics of the defender's observer are described by a finite state automaton with state space $\mathcal{S}$ and transitions that obey the dynamics defined by the function $f(S_t, d_t, a'_t)$.

Using the defender's observer we formulate Problem (P'_D) as follows.

$$\min_{g \in \mathcal{G}'} \max_{Z^g_t \in \mathcal{Z}, t \in \mathcal{T}} \left\{ \sum_{t \in \mathcal{T}} \beta^t \left[C_{Z^g_t} + \hat{C}(d_t) \right] \right\} \qquad (P'_D)$$

$$\text{subject to} \quad \text{Assumptions } \textbf{(A1)} - \textbf{(A6)},$$
$$d_t = g_t(S_t), \, t \in \mathcal{T},$$
$$Z^g_t \in S_t, \, t \in \mathcal{T},$$
$$S_{t+1} = f(S_t, d_t, a'_t), \, t \in \mathcal{T}.$$

where $\mathcal{G}' := \{g \, | \, g := \{g_t, t \in \mathcal{T}\}, g_t : \mathcal{S} \to \mathcal{D} \text{ for all } t \in \mathcal{T}\}$.

4   Dynamic Programming Solution for the Defender's Problem

4.1   The Dynamic Program

We solve Problem (P'_D) using dynamic programming. The dynamic program corresponding to Problem (P'_D) is

$$V(S) = \min_{d \in \mathcal{D}} \max_{Z \in S} \left[C_Z + \hat{C}(d) + \max_{S' \in \mathcal{Q}(S, d, Z)} \beta V(S') \right]. \qquad (4)$$

for every $S \in \mathcal{S}$ (see [2,12]), where $\mathcal{Q}(S,d,Z)$ is the set of observer states that can be reached by S when the defender's action is d and the true system state in S is Z. The set $\mathcal{Q}(S,d,Z)$ is determined as follows. If at time t the observer's state is S and the defender takes action d then, before the effect of d at time t^+ and the observation at time t^{++} are realized, there will be several potential candidate observer states at $t+1$. Only a subset of these possible observer states can occur when the true state of the system at time t is $Z \in S$. This subset is $\mathcal{Q}(S,d,Z)$. We illustrate the form of the set $\mathcal{Q}(S,d,Z)$ by the following example.

Example 1. Assume a network of three computers and a current observer state of

$$S_t = \{(F,N,N),(F,N,R),(F,R,N)\}.$$

If the defender takes action E_2 then, before the effect of E_2 and the observation $H^{1,2}$ at t^{++} are realized, the possible observer states S_{t+1} are

$$\{\{(F,W,N),(F,W,R)\},\{(F,W,N)\}\}.$$

If the true system state is $Z_t = (F,N,R)$ then

$$\mathcal{Q}(S_t,E_2,Z_t) = \{(F,W,N),(F,W,R)\}.$$

$\triangle$

4.2   Solution of the Dynamic Program

We obtain the solution of the dynamic program, Equation (4), via *value iteration* [2,12]. For that matter, we define the operator T by

$$TV(S) := \min_{d \in \mathcal{D}} \max_{Z \in S} \left[C_Z + \hat{C}(d) + \max_{S' \in \mathcal{Q}(S,d,Z)} \beta V(S') \right]. \tag{5}$$

We prove the following result.

Theorem 1. *The operator T, defined by Equation (5), is a contraction map.*

Proof. We use Blackwell's sufficiency theorem (Theorem 5, [4]) to show that T is a contraction mapping. We show:

i) *Bounded value functions*: First, note that $|\mathcal{S}|,|\mathcal{D}| < \infty$, and that we have bounded costs, $C_Z \leq M_1 < \infty, \forall S \in \mathcal{S}$; $\hat{C}(d) \leq M_2 < \infty, \forall d \in \mathcal{D}$. Starting from any bounded value function, $V(S) \leq M_3 < \infty$ with $M_3 > \frac{M_1+M_2}{1-\beta}$ we have

$$TV(S) = \min_{d \in \mathcal{D}} \max_{Z \in S} \left[C_Z + \hat{C}(d) + \max_{S' \in \mathcal{Q}(S,d,Z)} \beta V(S') \right]$$
$$\leq M_1 + M_2 + \beta M_3 < M_3 < \infty$$

for all $S \in \mathcal{S}$.

ii) *Monotonicity*: Assume $V_2(S) \geq V_1(S) \; \forall \, S \in \mathcal{S}, Z \in S$ and $d \in D$,

$$C_Z + \hat{C}(d) + \max_{S' \in \mathcal{Q}(S,d,Z)} \beta V_2(S') \geq C_Z + \hat{C}(d) + \max_{S' \in \mathcal{Q}(S,d,Z)} \beta V_1(S')$$

Therefore, for all $S \in \mathcal{S}$ and $d \in D$

$$\max_{Z \in S} \left[C_Z + \hat{C}(d) + \max_{S' \in \mathcal{Q}(S,d,Z)} \beta V_2(S') \right] \geq$$

$$\max_{Z \in S} \left[C_Z + \hat{C}(d) + \max_{S' \in \mathcal{Q}(S,d,Z)} \beta V_1(S') \right]$$

Hence,

$$TV_2(S) = \min_{d \in D} \max_{Z \in S} \left[C_Z + \hat{C}(d) + \max_{S' \in \mathcal{Q}(S,d,Z)} \beta V_2(S') \right]$$

$$\geq \min_{d \in D} \max_{Z \in S} \left[C_Z + \hat{C}(d) + \max_{S' \in \mathcal{Q}(S,d,Z)} \beta V_1(S') \right] = TV_1(S).$$

iii) *Discounting*: Assume $V_2(S) = V_1(S) + a$. Then, for all $S \in \mathcal{S}$

$$TV_2(S) = \min_{d \in D} \max_{Z \in S} \left[C_Z + \hat{C}(d) + \max_{S' \in \mathcal{Q}(S,d,Z)} \beta(V_1(S') + a) \right]$$

$$= \min_{d \in D} \max_{Z \in S} \left[C_Z + \hat{C}(d) + \max_{S' \in \mathcal{Q}(S,d,Z)} \beta V_1(S') \right] + \beta a$$

$$= TV_1(S) + \beta a.$$

By Blackwell's sufficiency theorem, the operator T is a contraction mapping. $\square$

Since T is a contraction mapping, we can use value iteration to obtain the solution to Equation (4), which we term the stationary value function, $V^*(S)$. From the stationary value function, we can obtain an optimal policy, g^*, as follows

$$g^*(S) = \arg\min_{d \in D} \max_{Z \in S} \left[C_Z + \hat{C}(d) + \max_{S' \in \mathcal{Q}(S,d,Z)} \beta V(S') \right]$$

The optimal policy, $g^*(S)$, is not always unique. That is, for a given observer state $S \in \mathcal{S}$, there could be multiple $d \in D$ which achieve the same minimum value of $\min_{d \in D} \max_{Z \in S} \left[C_Z + \hat{C}(d) + \max_{S' \in \mathcal{Q}(S,d,Z)} \beta V(S') \right]$. We denote by $\mathcal{D}^*(S)$ the set of optimal actions for a given observer state S. In the event that $\mathcal{D}^*(S)$ is not a singleton for a given state S, we choose a single action $d^*(S) \in \mathcal{D}^*(S)$ based on a quantity we define as the *confidentiality threat*. The confidentiality threat is a measure of the degree to which computer i is presumed (by the defender) to be compromised and is defined as follows

$$\tilde{T}_i = \sum_{Z \in S} c(z^i), \quad i \in \mathcal{N}$$

where $c(z^i)$, $z^i \in \mathcal{M}$, is the cost of the state, as defined in Equation (1), of the ith computer in the candidate system state $Z \in S$. Summing over all candidate system states in the observer state S for a given computer i, we obtain the confidentiality threat $\tilde{T}_i$. Next, we compare the confidentiality threat of each computer and choose the action $d^*(S) \in \mathcal{D}^*(S)$ that corresponds to the highest confidentiality threat. In the case of equal confidentiality threats (which arise when the observer state is symmetric), we choose the action in $\mathcal{D}^*(S)$ corresponding to the computer with the lower index $i \in \mathcal{N}$.[6]

5   Optimal Defender's Policy

We now discuss the characteristics of the optimal policy for Problem (P'_D), henceforth referred to as the *optimal policy*. We illustrate sensitivity analysis via numerical results for both a two computer and a three computer network. We also discuss some qualitative observations of the optimal policy.

First we note that determining the set of observer states and its associated dynamics is not a trivial computational task, even for moderately sized networks. Our calculations show for the case of a two computer network, the defender's observer automaton consists of 87 states and 1207 transitions. Extending the system to a three computer network results in 1423 states with 65602 transitions. To automate the procedure, we have developed a collection of programs which makes use of the UMDES-LIB software library [1]. The specific procedure is discussed in Appendix A.

The sensitivity analysis studies how the cost of re-imaging affects the optimal policy. For both the two computer and three computer networks, we increase the re-image cost, $\hat{C}(R_i) = r$, $\forall i \in \mathcal{N}$, and observe how the optimal policy behaves. Since the number of observer states in the two computer network, denoted $|\mathcal{S}_2|$, is modest, $|\mathcal{S}_2| = 87$, we are able to plot the behavior for each observer state $S \in \mathcal{S}_2$, as seen in Figure 4(a).[7] In the three computer network, the size of observer state space, $|\mathcal{S}_3| = 1423$, is much larger than that of the two computer network. As a result, we plot the percentage of observer states that have the optimal action d, for all $d \in \mathcal{D}$, and analyze how the percentage changes as we increase r, as seen in Figure 4(b).

The behavior of the optimal policy due to increasing re-image costs, r, is intuitive. As r increases, the optimal policy exhibits a threshold form,[8] switching from specifying more expensive actions to less expensive actions. For very low re-image costs, the optimal policy specifies R_i in the majority of the observer states. As r increases, observer states for which R_i was optimal, switch to either sense, E_i, or null, N^d. Once the optimal action is null, it remains null for all higher values of r. For the observer states where the action switched to sense, a further increase in r may result in a switch to null; however, there exist some

---

[6] This choice is arbitrary; we could randomize the choice as well.
[7] The "ordering" of these states is arbitrary.
[8] In the simulations that we have performed.

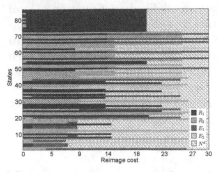
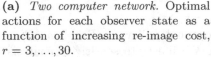
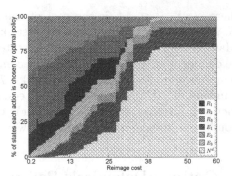

(a) *Two computer network.* Optimal actions for each observer state as a function of increasing re-image cost, $r = 3, \ldots, 30$.

(b) *Three computer network.* Percentage of observer states that have optimal actions $d \in \mathcal{D}$ as a function of increasing re-image cost, $r = 0.2, \ldots, 60$.

Fig. 4. *Sensitivity analysis for varying re-image cost r, where $r = \hat{C}(R_i)$ for all $i \in \mathcal{N}$. Other parameters are $\hat{C}(N_d) = 0$, $\hat{C}(E_i) = 0.1 \; \forall i \in \mathcal{N}$, $c_N = 0$, $c_R = 1$, $c_W = 2$, $c_F = 8$, and $\beta = 0.9$.*

observer states where the optimal action is sense for all higher values of r. This threshold behavior is clearly depicted in Figure 4(a).

As a result of the aforementioned threshold behavior, for high enough values of r, the optimal policy eventually specifies N^d or E_i for all states $S \in \mathcal{S}$. The argument to see why there is no re-image action for high values of r is straightforward; at these values of r the cost of re-imaging is prohibitively expensive and the defender would rather incur the cost of being in a poor system state (see Equation (2)).

An interesting (related) observation can be seen by analyzing the characteristics of the observer states and how these characteristics influence *when* the policy undergoes a switch as r increases. Consider Figure 4(a), and observe the behavior of the optimal policy around the re-image cost of $r = 20$. There is a collection of observer states (with indices 74 – 87) that contain the (F, F) element (both computers are in the remote compromised state) where the optimal policy specifies a switch from re-image to null. In these observer states, the defender believes that the true system state is so poor that, even if the a computer were to be re-imaged, the events in $\mathcal{A}$ would cause the system to transition back to a poor state in so few iterations that the defender would just be wasting its resources by re-imaging. That is, the number of time steps that it takes for the system to return to a poor state is not high enough to justify the cost that the defender must incur to keep the system in a secure operating mode. For this reason, in these observer states, the defender exhibits the passive behavior of *giving up* by choosing the cheapest action, N^d. An interesting related observation is that for other observer states in the system (the observer states that do not contain the element (F, F)) the optimal policy specifies a switch away from re-image at a higher re-image cost (around $r \in [25 \; 26]$). In these observer states the defender views the process of securing the system as economically efficient because it can be returned to a secure operating mode in a small enough number of iterations (compared to the observer states that contain the system state (F, F)). This observed behavior reflects the fact that attacks are progressive and that time has value in our model.

Another observation is that there are sets of parameters for which the sense action is useful (as seen starting in Figure 4(a) around $r = 2$ and peaking in Figure 4(b) around $r = 25$). In these cases the act of sensing a computer results in a split observer state that has a lower future cost than if the defender were to choose either null or re-image. Thus, paying the cost to sense can result in the defender having a better idea of the underlying system state and thus make a wiser decision on which future action to take. However, for low values of r, we can see that the defender prefers to re-image over obtaining a better estimate of the system (and similarly for high values of r, the defender prefers to take the null action). This behavior highlights the duality between estimation and control.

Interestingly, sensing remains an optimal action even for high values of r when there is no re-image action prescribed in the optimal defense policy. In these cases, even though sensing does not change the state of the network, it refines the defender's information which then results in a lower future cost for the defender. Even though the sense action is more expensive than the null action, this lower future cost causes the defender to choose sense over null.

The intent of determining an optimal policy is to offer a set of procedures for the defender such that the network is able to be kept as secure as possible. After the defender specifies its costs for actions and costs for states, the optimal policy specifies a procedure that the defender should follow. For each action the defender takes, $d \in \mathcal{D}$, and for each event it observes, $a' \in \mathcal{A}'$, the resulting observer state is known through the dynamics of the observer state. For each of these observer states resulting from the sequence of defender actions and observed events, the optimal policy specifies whether to sense or re-image a particular computer, or to wait and do nothing. The resulting defender behavior will keep the network as secure as possible under the min-max cost criterion.

6   Conclusion and Reflections

In this paper we have proposed a supervisory control approach to dynamic cyber-security. We have taken the viewpoint of the defender whose task is to defend a network of computers against progressive attacks. Some of the attacker actions are unobservable by the defender, thus the defender does not have perfect knowledge of the true system state. We define an observer state for the defender to capture this lack of perfect knowledge.

We have assumed that the defender takes a conservative approach to preserving the security of the system. We have used the min-max performance criterion to capture the defender's conservative approach.

Dynamic programming was used to obtain an optimal defender policy to Problem (P'_D). The numerical results show that the optimal policy exhibits a threshold behavior when the cost of actions are varied. We have also observed the duality of estimation and control in our optimal policy.

We believe that our approach is suitable for modeling interactions between an attacker and a defender in general security settings. In general, we can use our approach to study dynamic defense against attacks in a network of N resources each with $\mathcal{M}$ (orderable) security levels and $\mathcal{M} - 1$ security boundaries. The attack actions can penetrate through some of these boundaries to compromise a resource, or use a compromised resource to attack other resources in the network. Some of these actions can be unobservable to the defender. On the other hand, the defender can take actions to change the state of resources to a more secure operating mode or sense the system state to obtain more refined information about the system's status.

The model we have defined is rich enough to be extended to capture more complicated environments. Some examples of such environments can be heterogeneity of the network's computers[9] or the introduction of a dummy computer[10] into the system so as to increase the network's resiliency to attacks.

One bottleneck of our approach is that the number of states and transitions grows exponentially with the number of computers. One solution to this is to use a hierarchical decomposition for the system. For example an *Internet Service Provider* (ISP) can model a collection of nodes in their network as one region (resource). Once a non-secure region is observed in the system, the ISP can more carefully analyze the nodes within that region and take appropriate actions. Approximate dynamic programming methods could also be useful in dealing with systems with a large number of computers.

Acknowledgement. This work was supported in part by NSF grant CNS-1238962 and ARO MURI grant W911NF-13-1-0421. The authors are grateful to Eric Dallal for helpful discussions.

References

[1] Umdes-lib (August 2000), https://www.eecs.umich.edu/umdes/toolboxes.html
[2] Bertsekas, D.P.: Dynamic programming and optimal control, vol. 1. Athena Scientific, Belmont (1995)
[3] Bier, V., Oliveros, S., Samuelson, L.: Choosing what to protect: Strategic defensive allocation against an unknown attacker. Journal of Public Economic Theory 9(4), 563–587 (2007)
[4] Blackwell, D.: Discounted dynamic programming. The Annals of Mathematical Statistics, 226–235 (1965)
[5] Bloem, M., Alpcan, T., Başar, T.: Optimal and robust epidemic response for multiple networks. Control Engineering Practice 17(5), 525–533 (2009)

---

[9] Placing an *importance* weight on each computer.
[10] The dummy computer contains no sensitive information and is meant to mislead the attacker.

[6] Bloem, M., Alpcan, T., Schmidt, S., Basar, T.: Malware filtering for network security using weighted optimality measures. In: IEEE International Conference on Control Applications, CCA 2007, pp. 295–300. IEEE (2007)

[7] Böhme, R., Félegyházi, M.: Optimal information security investment with penetration testing. In: Alpcan, T., Buttyán, L., Baras, J.S. (eds.) GameSec 2010. LNCS, vol. 6442, pp. 21–37. Springer, Heidelberg (2010)

[8] Chen, T.M., Jamil, N.: Effectiveness of quarantine in worm epidemics. In: IEEE International Conference on Communications, ICC 2006, vol. 5, pp. 2142–2147. IEEE (2006)

[9] Hart, S.: Discrete colonel blotto and general lotto games. International Journal of Game Theory 36(3-4), 441–460 (2008)

[10] Khouzani, M., Sarkar, S., Altman, E.: Maximum damage malware attack in mobile wireless networks. IEEE/ACM Transactions on Networking 20(5), 1347–1360 (2012)

[11] Khouzani, M., Sarkar, S., Altman, E.: Saddle-point strategies in malware attack. IEEE Journal on Selected Areas in Communications 30(1), 31–43 (2012)

[12] Kumar, P.R., Varaiya, P.: Stochastic systems: Estimation, identification and adaptive control. Prentice-Hall, Inc. (1986)

[13] Ligatti, J., Bauer, L., Walker, D.: Edit automata: Enforcement mechanisms for run-time security policies. International Journal of Information Security 4(1-2), 2–16 (2005)

[14] Ligatti, J., Bauer, L., Walker, D.: Run-time enforcement of nonsafety policies. ACM Transactions on Information and System Security (TISSEC) 12(3), 19 (2009)

[15] Lye, K., Wing, J.M.: Game strategies in network security. International Journal of Information Security 4(1-2), 71–86 (2005)

[16] Mastroleon, L.: Scalable resource control in large-scale computing/networking infrastructures. ProQuest (2009)

[17] Rowe, J., Levitt, K.N., Demir, T., Erbacher, R.: Artificial diversity as maneuvers in a control theoretic moving target defense. In: National Symposium on Moving Target Research (2012)

[18] Roy, S., Ellis, C., Shiva, S., Dasgupta, D., Shandilya, V., Wu, Q.: A survey of game theory as applied to network security. In: 2010 43rd Hawaii International Conference on System Sciences (HICSS), pp. 1–10. IEEE (2010)

[19] Schneider, F.B.: Enforceable security policies. ACM Trans. Inf. Syst. Secur. 3(1), 30–50 (2000)

[20] Schwartz, G.: Blotto games for security, review and directions. Presented at NetEcon Meeting. University of California, Berkeley (Private communication) (2013)

[21] Van Dijk, M., Juels, A., Oprea, A., Rivest, R.L.: Flipit: The game of stealthy takeover. Journal of Cryptology 26(4), 655–713 (2013)

[22] Yin, Z., Korzhyk, D., Kiekintveld, C., Conitzer, V., Tambe, M.: Stackelberg vs. nash in security games: Interchangeability, equivalence, and uniqueness. In: Proceedings of the 9th International Conference on Autonomous Agents and Multiagent Systems, vol. 1, pp. 1139–1146, International Foundation for Autonomous Agents and Multiagent Systems (2010)

A   Appendix – UMDES-LIB

The UMDES-LIB library [1] is a collection of C-routines that was built to study discrete event systems that are modeled by finite state automata. Through specification of the states and events of a system automaton (along with the controllability and observability of events), the library can construct an entity termed the *observer automaton*. In our problem the observer automaton is the defender's observer automaton, since we take the viewpoint of the defender. Thus, the observer automaton consists of the defender's observer states.

In this appendix we describe an automated process[11] for extracting the defender's observer state from the system automaton that makes use of UMDES-LIB. This requires first *constructing the system automaton* in an acceptable format for the library while preserving all the features of our model. After running the library on the provided system automaton, we *extract the defender's observer state* from the observer automaton output. This method allows one to construct the defender's observer state for any number of computers.[12]

Constructing the System Automaton. The input that we provide to UMDES-LIB is the system automaton from the defender's viewpoint, as illustrated earlier in Figure 3.

In order to preserve all features of our model in the resulting observer automaton, we need to introduce additional sensing actions. Recall that the sense action, $\{E_i\}_{i \in \mathcal{N}}$, causes the system automaton to transition to the same state as the null action, N^d (see Figure 3). However, as stated in Section 2, the sense action updates the information state of the defender. In order to ensure that UMDES-LIB captures this functionality, we expand the sense action E_i for each computer i into $|\mathcal{M}|$ distinct actions, denoted by $E_i^{z^i}$, which represent sensing computer i when it is in state $z^i \in \mathcal{M}$. This results in a reduced level of uncertainty for the defender as it splits the observer state into, at most, $|\mathcal{M}|$ possible sets of observer states. The admissible actions from $\{E_i^{z^i}\}_{z^i \in \mathcal{M}}$, at a given system state, are the sense actions that correspond to the true system state. For example, from the system state $Z_t = (N, R, W)$, the admissible sense actions are E_1^N, E_2^R, and E_3^W. The above example of the expanded sense action is perhaps worrisome at first glance – if the only admissible sense actions from the current state are the ones that correspond to the current state of the computer, then the defender will know what the current state of each computer is, eliminating the need for a sense action. However, the observer state that is obtained from each expanded sense action is the same as the observer state that is obtained if the defender were to observe the true, unknown state of a computer.

Running UMDES-LIB on the system automaton with the expanded sense actions results in the observer automaton.

---

[11] Source code is available upon request.

[12] The only bottleneck being the (potentially large) dimensionality of the problem.

Extracting the Defender's Observer State. The output of UMDES-LIB is the observer automaton, from which we must extract the defender's observer state. First, since the defender does not have the ability to choose the expanded sense actions, $E_i^{z^i}$, we re-group them into a single, non-deterministic action, $E_i \in \mathcal{D}$, for each $i \in \mathcal{N}$. Next, we need to extract the function, $f : \mathcal{S} \times \mathcal{D} \times \mathcal{A}' \rightarrow \mathcal{S}$ from the observer automaton. The observer automaton, generated by UMDES-LIB, takes the form of a bipartite graph; one collection of states of the bipartite graph is observer states over system states $\mathcal{Z}$, denoted $\mathcal{S}$, whereas the other collection is observer states over intermediate states $\tilde{\mathcal{Z}}$, denoted $\tilde{\mathcal{S}}$. Defender actions, $\mathcal{D}$, are the only admissible actions from observer states $\mathcal{S}$. The defense action $d \in \mathcal{D}$ causes a transition[13] to an observer state in $\tilde{\mathcal{S}}$, where only events in $\mathcal{A}'$ are admissible. Each event $a' \in \mathcal{A}'$ causes a transition back to an observer state in $\mathcal{S}$. Repeating this process for all observer states in $\mathcal{S}$, actions $d \in \mathcal{D}$, and events $a' \in \mathcal{A}'$, the function $f : \mathcal{S} \times \mathcal{D} \times \mathcal{A}' \rightarrow \mathcal{S}$ is defined. To construct the set $\mathcal{Q}(S, d, Z)$ we follow the approach described in Section 4.1 and illustrated by Example 1.

---

[13] This transition may be non-deterministic due to the sense action.

Numerical Computation
of Multi-goal Security Strategies

Stefan Rass[1] and Benjamin Rainer[2]

[1] Universität Klagenfurt, Institute of Applied Informatics,
System Security Group, Klagenfurt, Austria
stefan.rass@aau.at
[2] Universität Klagenfurt, Institute of Information Technology,
Klagenfurt, Austria
benjamin.rainer@aau.at

Abstract. Security is often investigated in terms of a single goal (e.g., confidentiality), but in practical settings mostly a compound property comprising multiple and often interdependent aspects. *Security strategies* are behavior profiles that guarantee some performance *regardless* of how the adversary really behaves (provided that it stays within its action set). While security strategies towards a single goal are easy to compute via Nash-equilibria (or refinements thereof), playing safe towards multiple security goals induces the notion of Pareto-optimal security strategies. These were recently characterized via Nash-equilibria of multi-player games, for which solution algorithms are intricate and may fail for small instances already. Iterative techniques, however, exhibited good stability even for large games. In this work, we thus report on theoretical and practical results how security strategies for multiple (interdependent) goals can be computed via a set of simple transformations and a final application of humble fictitious play.

Keywords: Pareto-optimality, security strategies, game theory, equilibrium, fictitious play, security.

1 Introduction

Security strategies have been introduced in [18], as a mean of optimizing behavior under uncertainty of the opponent. That is, a security strategy gives the best payoff for player 1 under arbitrary, especially not equilibrium, behavior of player 2 in a two-person game. This models situations in which only the opponent's action space is known, but the player remains uncertain about the other's payoff structure(s). Information security is a natural incarnation of this, as we seek the optimal defense against arbitrary actions of an adversary, whose possible actions are known, but nothing about its particular behavior can be assumed reasonably. Treating a single security goal in that sense yields scalar two-person games in the style "honest-vs-adversary". However, most practical settings require simultaneous defense strategies against various different threats, such as violations

R. Poovendran and W. Saad (Eds.): GameSec 2014, LNCS 8840, pp. 118–133, 2014.

of confidentiality, integrity, availability and authenticity (CIA+ security). Security strategies accounting for simultaneously optimal payoffs in various perhaps strongly interdependent goals have not been studied very extensively so far, and are subject of this work.

Security strategies in the scalar case, i.e., when only a single security goal is of interest, are easily identified as Nash-equilibria of zero-sum games. In a multi-dimensional case, i.e., for security in multiple possibly interdependent aspects, Pareto-optimal security strategies are sought. Applications of these are manifold, e.g., creating high-security communication lines that are confidential, robust and authentic, can be achieved by multipath-transmission and multipath-authentication, which in turn leads to straightforward game-models (an example is given in section 6.2).

Searching for security strategies is interesting from a theoretical and practical point of view, as it can provide quantitative risk estimates. For example, setting up a transmission channel between to peers by virtue of multipath transmission, the game can be defined with the sender acting as player 1, who chooses the transmission configuration (in particular the paths over which information is conveyed). Player two is the adversary, who chooses nodes to attack. The game's payoff function is the fraction of correctly delivered messages, where "correctly" here covers confidentiality and integrity (at least). Given a particular network infrastructure (topology), what is the likelihood of achieving the two security goals upon a single transmission? The answer lets the sender utilize the network in a proper way so as to minimize the risk of security breaches, and can be used to enhance the network infrastructure (by additional protections at the most likely targets for the opponent (adversary) in the network infrastructure). So, the practical aspect of game-theory in network security is related to *topological vulnerability analysis*, where the competition between the (honest) network users and the adversary points out best practices to use the network, as well as neuralgic spots being indicated as the most likely attack strategies for the adversary (opponent player 2). We revisit this use case later.

Our focus here is, however, not on game-theoretic models of applied cryptography, but rather on covering a numerical problem in the computation of Pareto-optimal security strategies. These can be computed to support or enhance processes of topological vulnerability analysis and quantitative risk management. Especially the latter may call for efficient updates following changes (enhancements) to the system. Therefore, the efficiency of computing security strategies may be of interest besides its theoretical value.

In fact, relying on the characterization as obtained in prior literature (and cited below), "standard" algorithms to compute Nash-equilibria may be applied. Unfortunately, however, the whole armory of algorithms that ships with the GAMBIT software [10], rapidly failed to compute the sought results even for small examples (numerical instabilities occurred already in example instances with, e.g., three goals and eight strategies per player). On the bright side, fictitious play exhibited good numerical stability (though slow convergence) and has been proven capable of computing the sought security strategies even for large games like those arising in our example application of security risk assessment of multipath communication channels.

2   Preliminaries

Throughout this work, normal font denotes scalars and bold face font denotes vectors. Sets are written in upper-case latin letters like N. The cardinality of a set N is denoted by $|N|$.

A *game* is a triple $\Gamma = (N, S, H)$, where N is the – in our case always finite – set of $n = |N|$ players, $S = \{PS_1, \ldots, PS_n\}$ contains the individual action sets for each player, and $H = \{u_1, \ldots, u_n\}$ is the family of payoff functions $u_i : \prod_{i=1}^n PS_i \to \mathbb{R}$ for each $i \in N$.

As a standard shorthand notation, we write PS_{-i} for the cartesian product of all $PS_j \in S$, excluding PS_i. The vector $(s_1, \ldots, s_{i-1}, s_{i+1}, \ldots, s_n) \in PS_{-i}$ is abbreviated as s_{-i}.

Hereafter, we write s for pure strategies, but mostly consider mixed strategies, i.e., probability distributions over the action sets. For simplicity, we thus denote S_i as the set of all probability distributions supported on a set PS_i of actions, also called *pure strategies*. This is the set of *mixed* strategies. Such mixed strategies and general probability distributions are denoted by lowercase Greek letters, e.g., $\theta, \phi \in S_i$. We will hereafter drop the attribute "mixed", as we will not explicitly talk about pure strategies any more (and because pure strategies arise via degenerate mixed strategies anyway). Random variables are denoted by uppercase letters like X; their distribution θ is told by the symbol $X \sim \theta$.

A *Nash-equilibrium* in an n-person game is a set of strategies $(\theta_1^*, \ldots, \theta_n^*)$ so that all players $i \in N$ receive for all $\theta_i \in S_i$ an expected payoff $\mathsf{E}_{(\theta_i^*, \theta_{-i}^*)} u_i(X_i, X_{-i}) \geq \mathsf{E}_{(\theta_i, \theta_{-i}^*)} u_i(X_i, X_{-i})$, where the expectation is taken over the probability distributions noted in the subscripts of the expectation operator. By a slight abuse of notation for the sake of simplicity, we let $u_i(\theta_i, \phi_i)$ also denote the long-run average payoff (over an infinite number of repetitions of the game[1]), as we will exclusively speak about expected payoffs in in the context of mixed strategies. In that notation, the Nash-equilibrium condition in a two-person zero-sum game (expected payoff functions being u_1 and $-u_1$) can compactly be written as

$$u_1(\theta, \phi^*) \leq u_1(\theta^*, \phi^*) \leq u_1(\theta^*, \phi) \quad \forall \theta \in S_1, \phi \in S_2, \tag{1}$$

where the pair (θ^*, ϕ^*) denotes the equilibrium, and we call $v = u_1(\theta^*, \phi^*)$ its *(saddle-point) value*.

---

[1] Even if the game cannot be repeated, then using indicator variables for the payoffs turns the expected payoffs into probabilities. In this setting, the Nash-equilibrium is the likelihood to win (or loose) in a single round of the game, thus making the concept applicable even if the game is not repeatable.

3 Security Strategies

Towards an axiomatic characterization of security strategies in general games (finite or infinite), captured as definition 1, we take known results in the scalar case as the template for upcoming definitions.

3.1 The Scalar Case

The following is a well-known fact (cf. [2] among others).

Lemma 1. *Let* $\Gamma = (N, S, H = \{u_1, u_2\})$ *be a two-person game with continuous payoff functions. Define the zero-sum game* $\Gamma_0 = (N, S, H_0 = \{u_1, -u_1\})$, *with Nash-equilibrium* $v = u_1(\theta^*, \phi^*)$. *Then, player 1 always receives* $u_1(\theta^*, \phi) \geq v$ *in* Γ, *no matter how player 2 actually behaves. Moreover, there is a strategy* $\phi' \in S_2$ *so that* $u_1(\theta^*, \phi') = v$ *in* Γ.

The lower bound provided by the zero-sum equilibrium value is easily obtained by observing that player 2 due to a perhaps different payoff structure in Γ most likely deviates from the optimal zero-sum strategy ϕ^* in Γ, thus leaving player 1 with more than the zero-sum equilibrium payoff v. The existence of a strategy ϕ' achieving equality directly follows from the continuity of the payoff functions.

The ordering of $\mathbb{R}$ that lets us define the equilibrium condition is lost upon the transition to $\mathbb{R}^k$ for $k > 1$. This unfortunate fact renders the proof of lemma 1 non-transferable to $\mathbb{R}^k$, and calls for more sophisticated concepts.

3.2 The Multi-criteria Case

A *multi-objective game* (MOG) has vector-valued payoffs. That is, the i-th player receives r_i different payoffs, denoted by the function $u_i : \prod_{i=1}^{n} PS_i \to \mathbb{R}^{r_i}$, $(s_i, s_{-i}) \mapsto (u_i^{(1)}(s_i, s_{-i}), \dots, u_i^{(r_i)}(s_i, s_{-i}))$. For two vectors $a = (a_1, \dots, a_k)$, $b = (b_1, \dots, b_k) \in \mathbb{R}^k$, we write $a \leq b$, if $a_i \leq b_i$ for all $i = 1, 2, \dots, k$. The complement relation is $a >_1 b$ and holds iff an index $1 \leq j \leq k$ exists such that $a_j > b_j$, no matter what the other components do. The vector-relations $\geq, <_1, \leq_1$ and $\geq_1$ are defined accordingly.

The sibling of Nash-equilibrium in the scalar case is the Pareto-Nash equilibrium in the multivariate case: here, we require the inequalities in (1) to fail in at least one component upon a deviation from the optimum. That is, an n-player MOG $\Gamma = (N, S, H)$ admits a *Pareto-Nash equilibrium* $(\theta_1^*, \dots, \theta_n^*)$ if for every player $i \in N$, we have $u_i(\theta_i, \theta_{-i}^*) \leq_1 u_i(\theta_i^*, \theta_{-i}^*)$ for every θ_i. For two players, the resulting pair of inequalities resembles the equilibrium condition (1) by requiring that optimality fails in at least one goal by any deviation from the Pareto-Nash strategy profile $(\theta_i^*, \theta_{-i}^*)$.

In [13], a precursor definition towards an axiomatic characterization of *network provisioning* security strategies is given. We adapt this construction into our definition 1 here that is not confined to problems of secure data delivery.

Definition 1. *A strategy $\theta^* \in S_1$ in a two-person multi-criteria game Γ with continuous payoff $\boldsymbol{u}_1 : S_1 \times S_2 \to \mathbb{R}^k$ for player 1, is called a* multi-criteria security strategy *(MCSS) with assurance vector $\boldsymbol{v} = (v_1, \ldots, v_k)$, if the following two conditions hold:*

1. *The assurances are the component-wise guaranteed payoff for player 1, i.e. for all components i, we have*

$$v_i \leq u_1^{(i)}(\theta^*, \phi) \qquad \forall \phi \in S_2, \tag{2}$$

   with equality being achieved by at least one choice $\phi_i \in S_2$.
2. *At least one assurance becomes void if player 1 deviates from $\boldsymbol{x}^*$ by playing $\theta \neq \theta^*$. In that case, some $\phi \in S_2$ exists such that*

$$\boldsymbol{u}_1(\theta, \phi) \leq_1 \boldsymbol{v}. \tag{3}$$

Observe that the above definition transforms the assertions of lemma 1 in the scalar case into axioms in the multi-dimensional case. The existence of multi-dimensional security strategies has been studied in the literature, where the following characterization was established:

Theorem 1 ([13]). *Let Γ be a two-player MOG. The distribution θ^* constitutes a multi-criteria security strategy (MCSS) $\boldsymbol{v}$ for player 1 and k goals in the game Γ, if and only if it is a Pareto-Nash equilibrium strategy for player 0 in the following $(k+1)$-player multi-objective auxiliary game $\overline{\Gamma} = (N, S, H)$, where: $N = \{0, 1, \ldots, k\}$, $S = \{PS_1, PS_2, \ldots, PS_2\}$ (i.e. a multiset with $|S| = k + 1$) and the payoffs are $\overline{\boldsymbol{u}}_0(s_0, \ldots, s_k) := (u_1^{(1)}(s_0, s_1), \ldots, u_1^{(k)}(s_0, s_k))$ for player 0 (vector-valued), and $\overline{\boldsymbol{u}}_i(s_0, \ldots, s_k) := -u_1^{(i)}(s_0, s_i)$ (scalar-valued) for the opponents $i = 1, 2, \ldots, k$.*

From theorem 1, the existence of security strategies is not immediately evident, but can be concluded from results of [9] concerning the existence of Pareto-Nash equilibria in multiobjective games (MOG).

Theorem 2 ([9]). *Let $\Gamma = (N, S, H)$ be a MOG, where each $PS_i \in S$ is convex and compact, and each $\boldsymbol{u}_i \in H$ is continuous. Moreover, assume that for each player $i \in N$, every individual payoff $u_i^{(j)}(s_i, \boldsymbol{s}_{-i})$ for $1 \leq j \leq r_i$ is a concave function of s_i on PS_i, whenever the remaining values $\boldsymbol{s}_{-i}$ are fixed. Then, Γ has a Pareto-Nash equilibrium.*

From this we easily obtain the existence of MCSS under various conditions. For example, every finite game admits multi-criteria security strategies, which re-proves a known result of [1] by a humble application of theorems 1 and 2:

Corollary 1 (Existence of MCSS in matrix games). *Every finite MOG admits a multi-criteria security strategy.*

We will not go into further details about existence of MCSS, beyond stressing the fact that definition 1 is not limited to finite games or games with a finite number of players. In that sense, the characterization theorem 1 can be obtained with alternative results to theorem 2 to establish the existence of MCSS for various other classes of games.

For simplicity, e.g. security risk management in multipath communication networks, can work with corollary 1 to handle the arising matrix-games.

The proof of theorem 2 is "constructive" in the sense of equating the set of Pareto-Nash equilibria to the set of Nash-equilibria in a scalarized version of the MOG. Specifically, [9] prescribe the following steps to find a Pareto-Nash equilibrium in the n-player MOG Γ:

1. Fix an arbitrary set of real numbers $\alpha_{11}, \alpha_{12}, \ldots, \alpha_{1r_1}, \alpha_{21}, \ldots, \alpha_{2r_2}, \ldots, \alpha_{n1}$, $\ldots, \alpha_{nr_n}$ that satisfy condition (4):

$$\left.\begin{array}{ll} \sum_{\kappa=1}^{r_i} \alpha_{i\kappa} = 1 & \text{for } i = 1, 2, \ldots, n, \text{ and} \\ \alpha_{i\kappa} > 0 & \text{for } \kappa = 1, 2, \ldots, r_i \text{ and } i = 1, 2, \ldots, n. \end{array}\right\} \qquad (4)$$

2. Form a (scalar) game $\Gamma_s = (N, S, H')$ with $H' = \{f_1, \ldots, f_n\}$ and

$$f_i = \sum_{\kappa=1}^{r_i} \alpha_{i\kappa} u_i^{(\kappa)}. \qquad (5)$$

3. Find a Nash-equilibrium $\theta^* = (x_1^*, \ldots, x_n^*)$ in Γ_s, which is then a Pareto-Nash equilibrium in Γ.

Notice that the Nash-equilibria found by the above algorithm depend on the particular choice of weights. Indeed, the full set of equilibria is given as the union of all equilibria over all admissible choices of α's in (4) [9].

4 Numerical Computation of MCSS

Although there exist sophisticated algorithms and implementations to compute Nash-equilibria in multi-person games, an experimental implementation of our transformation using the GAMBIT software [10] showed that these algorithms fail on games with many players and strategies. It therefore appears advisable to prefer iterative numeric techniques over analytic ones for practical settings, in which we can expect a large number of strategies and security goals, the latter of which correspond to players. Our method of choice is fictitious play.

4.1 Fictitious Play in Multi-criteria Compound Games

Briefly speaking, fictitious play is the process of repeatedly playing the game while every player notes and learns the other player's moves, while at the same time optimizing his/her own behavior based on the so-far recorded behavior profiles. More concretely, let $t \in \mathbb{N}$ be the sequence of discrete time steps. Player i moves along a sequence of actions $(s_i(t))_{t \in \mathbb{N}} \in PS_i$ and maintains beliefs for each opponent $j \neq i$ that are discrete probability distributions for each $t \in \mathbb{N}$ of the form $\left(\beta_i(t) = \frac{1}{t}\sum_{\tau=1}^{t}\delta_i(\tau)\right)_{i=1}^{k}$. Here, δ_i is the Dirac probability distribution that assigns unit mass to action s_i (by this convention PS_i is included in S_i as extremal points). Player i's next move at time $t+1$ is then the optimal response to its recorded opponent behavior profile $(\beta_i^1(t), \ldots, \beta_i^{i-1}, \beta_i^{i+1}, \ldots, \beta_i^n(t))$ at time t. We say that a game has the *fictitious play property*, if this process approaches an equilibrium θ^* in the sense that for every $\varepsilon > 0$ there is some t_0 such that for every $t \geq t_0$, we have $\|(\beta_i^1(t), \ldots, \beta_1^n(t)) - \theta^*\| < \varepsilon$ in some norm. See [17] for a more comprehensive account.

4.2 Computing MCSS by Fictitious Play

In the terminology of [17], the auxiliary game $\overline{\Gamma}$ is a "one-against-all" multi-player game or *compound game*, which can be solved iteratively by fictitious play if it were zero-sum. Although theorem 1 specifies $\overline{\Gamma}$ not as zero-sum, this can be fixed easily without changing the set of equilibria. Indeed, it is the scalarization (5) that will become helpful in a twofold manner, as it lets us apply standard fictitious play and it lets us prioritize our security goals.

Given a two-player MOG Γ and its auxiliary game $\overline{\Gamma}$, we prepare the latter for fictitious play by making it zero-sum before the necessary scalarization. To this end, recall that player 1 in Γ, who is player 0 in $\overline{\Gamma}$, has k goals to optimize, each of which is represented as another opponent in the auxiliary game $\overline{\Gamma}$. We define the payoffs in a *compound game* ("one-against-all") from the payoffs in $\overline{\Gamma}$, while making the scalar payoffs vector-valued to achieve the zero-sum property:

- player 0:

$$\overline{u}_0 : PS_1 \times \prod_{i=1}^{k} PS_2 \to \mathbb{R}^k,$$

$$u_0(s_0, \ldots, s_k) = (u_1^{(1)}(s_0, s_1), u_1^{(2)}(s_0, s_2), \ldots, u_1^{(k)}(s_0, s_k))$$

- i-th opponent for $i = 1, 2, \ldots, k$:

$$\overline{u}_i = (0, 0, \ldots, 0, -u_1^{(i)}, 0, \ldots, 0). \tag{6}$$

Obviously, the "vectorization" of the opponents payoffs does not affect any equilibrium conditions, so the so-modified game comes with the same set of equilibria as $\overline{\Gamma}$. To numerically compute (one of) them, we scalarize as follows: to each of player 0's k goals, we assign a weight $\alpha_{01}, \ldots, \alpha_{0k}$. The scalarization in (5) is via

$$\alpha_{ji} := \alpha_{0i} \text{ for } i = 1, 2, \ldots, k \text{ and } j = 1, 2, \ldots, k.$$

With these weights, the payoffs in the scalarized compound game are:

- for player 0: $f_0 = \alpha_{01}\overline{u}_1 + \alpha_{02}\overline{u}_2 + \cdots + \alpha_{0k}\overline{u}_k$,
- for the i-th opponent, where $i = 1, 2, \ldots, k$

$$f_i = \alpha_{01} \cdot 0 + \alpha_{02} \cdot 0 + \cdots + \alpha_{0,i-1} \cdot 0 + \alpha_{0i} \cdot (-u_1^{(i)}) + \alpha_{0,i+1} \cdot 0 + \alpha_{0k} \cdot 0$$
$$= -\alpha_{0i} \cdot u_1^{(i)} \tag{7}$$

Concluding the transformation, we obtain a scalar compound game

$$\overline{\Gamma}_{sc} = (\{0, 1, \ldots, k\}, \{PS_1, \underbrace{PS_2, \ldots, PS_2}_{k \text{ times}}\}, \{f_0, \ldots, f_k\}) \qquad (8)$$

from the original two-person MOG Γ with payoffs $u_1^{(1)}, \ldots, u_1^{(k)}$ that can directly be be plugged into expressions (6) and (7).

Towards a numerical computation of equilibria in $\overline{\Gamma}_{sc}$, we need yet another transformation due to [17]: for the moment, let us consider a general compound game Γ_c as a collection of k two-person games $\Gamma_1, \ldots, \Gamma_k$, each of which is played independently between player 0 and one of its k opponents. With Γ_c, we associate a two-person game Γ_{cr} that we call the *reduced game*. The strategy sets and payoffs of player 0 in Γ_{cr} are the same as in Γ_c. Player 2's payoff in the reduced game is given as the *sum* of payoffs of all opponents of player 0 in the compound game.

Lemma 2 ([17]). *A fictitious play process approaches equilibrium in a compound game Γ_c, if and only if it approaches equilibrium in its reduced game Γ_{cr}.*

So, it suffices to consider the reduced game $\overline{\Gamma}_{scr}$ belonging to $\overline{\Gamma}_{sc}$. It is a trivial matter to verify the following fact (by substitution).

Lemma 3. *The reduced game $\overline{\Gamma}_{ocr}$ of the scalarized compound game $\overline{\Gamma}_{sc}$ defined by (8) is zero-sum.*

So by the famous result of [15] on the convergence of fictitious play in two person zero-sum games, we obtain the following final result:

Theorem 3. *The scalarized compound game $\overline{\Gamma}_{sc}$ defined by (8) has the fictitious play property.*

Theorem 3 induces the following procedure to compute multi-criteria security strategies according to definition 1:

Algorithm to compute MCSS: Given a two-player MOG Γ with k payoffs $u_1^{(1)}, \ldots, u_1^{(k)}$ for player 1 (and possibly unknown payoffs for player 2), we obtain a MCSS along the following steps:

1. Assign strictly positive weights $\alpha_{01}, \ldots, \alpha_{0k}$, satisfying $\sum_{i=1}^{k} \alpha_{0i} = 1$, to each goal, and set up the scalarized auxiliary compound game $\overline{\Gamma}_{sc}$ by virtue of expressions (8), (6) and (7).
   Observe that, as we can choose the weights arbitrarily, these give us a method to *prioritize* different goals. However, practical experiments indicated that different choices of priorities (α-values) have only a minor if not negligible effect on the particular result of the computation.
2. Run fictitious play in $\overline{\Gamma}_{sc}$, stopping when the desirable precision of the equilibrium approximation is reached. In our experiments, we stopped when the difference between the intermediate result vectors θ_{t-1}^* and θ_t^* at steps t and $t - 1$ has become less than an adjustable threshold $\delta > 0$ in the 1-norm.

3. The result vector θ^* is directly the sought multi-criteria security strategy, whose assurances are given by the respective expected payoffs of the opponents. In case of matrix games, where the i-th payoff is given by a matrix A_i, the sought assurances are $v_i = (\theta^*)^T A_i \phi_i^*$ for $i = 1, 2, \ldots, k$, where $\phi_1^*, \ldots, \phi_k^*$ are the other player's equilibrium strategy approximations obtained along the fictitious play.

5 Experimental Evaluation

We stress that theorem 3 asserts the fictitious play property for the games constructed, yet does not limit numerical solution techniques to a particular algorithm (not even to fictitious play). Our experimental implementation used the basic (and non-optimized) fictitious play procedure [15], but can easily be replaced by more sophisticated algorithms (e.g., [20]) to gain speed. Our tests were done on a 3 x AMD Opteron 6212 machine, having 2.6 GHz 24 cores (virtualized), 96 GB RAM, and 1 TB disk space.

Towards a (non-application-specific) performance evaluation, we created random payoff matrices to simulate arbitrary matrix game structures (matrices with independent and uniformly distributed Bernoulli random entries) ranging from 2 to 170 strategies (in steps of 2) for the honest player, seeking to secure its behavior in terms of two security goals. In each setting, we ran (at least) 50 trials, taking the average number of iterations until convergence as the empirical performance indicator. Convergence is said to be reached once the change in the payoff-values $v_1, \ldots, v_k$ (per security goal) between two iterations has become less than a threshold $\delta = 0.01$ in the 1-norm.[2] Figure 1a plots the results.

Fictitious play has shown to be numerically stable, yet suffers from slow convergence (without optimizations) and memory shortage in case of games with many goals (each of which corresponds to a player with its own payoff structure). In the latter cases, the computation may be parallelized towards a speed-up by assigning each player its own processor and memory. The temporal speed under parallelization is then mostly determined by the communication overhead, which in a multi-processor CPU is not too much of a problem.

As expected, the maximal number of iterations grows with the size of the strategy sets and the number of security goals. Towards an empirical estimate of asymptotic complexity in terms of the game's *size* (number of strategies), we fitted a linear model to the plot of $N(n)$ (Figure 1a). Here, n is the number n of strategies, and the model took the form $N = a \cdot n + b + \varepsilon$ with an error term ε being normally distributed. The parameter estimates came up to $a \approx 71.5657, b \approx -507.7625$. The normality hypothesis on the residual term ε was accepted by a Shapiro-Wilks test with a p-value of ≈ 0.8918 at a confidence level of $\alpha = 0.95$. Hence, we may – on empirical evidence – assume a growth of the iteration count N that is proportional to the number n of strategies, giving

---

[2] Notice that convergence in the fictitious play process as defined above implies convergence under our modified criterion by the continuity of the payoff functions.

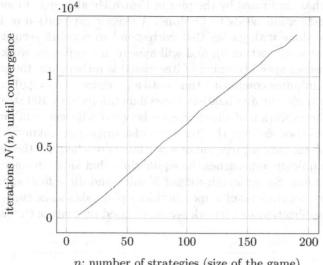

(a) Convergence, depending on the strategy set sizes

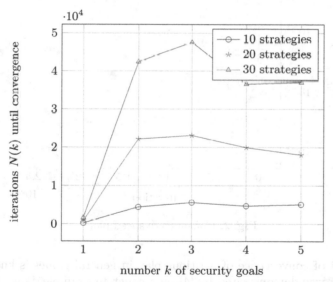

(b) Convergence, depending on the number of security goals

Fig. 1. Complexity of computing two-criteria security strategies

linear asymptotic average-case complexity $N \in O(n)$. The same linear relationship was also confirmed for trials in 3 and 4 dimensions (using smaller games in terms of strategy counts, though). Interestingly, the constants within the big-O were roughly equal between 2, 3 and 4 dimensions, indicating that convergence rates are only mildly affected by the number of security goals (dimensions).

This is somewhat confirmed by the plot in Figure 1b, although a more thorough empirical investigation needs to be done. A deeper exploration of both observations will be done with games that correspond to network security protocols (see the related work section 6), and will appear in companion work to this.

The convergence speed (number of iterations) is rather slow: the computation took about 15 minutes computing time until a precision of $\delta = 0.01$, and another 15 minutes to undercut $\delta = 0.001$ in three dimensions with 100 strategies. Figure 2 shows the evolution of the difference between adjacent equilibrium profile approximations (beliefs) over the iterations of a single run, taking 200 strategies in two dimensions until a precision of $\delta = 0.001$ is reached. As the figure shows, the algorithm quickly approaches the equilibrium, but slows down substantially near the optimum. So, although we get a quick-and-dirty first approximation, retrieving more accurate results upon fictitious play takes some time. Section 6.2 describes an application to network security, based on multipath transmission.

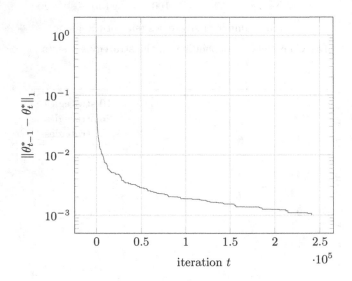

Fig. 2. Convergence speed plot

The speed of convergence of fictitious play in general games is known to be very slow, as was demonstrated by [4] on a concrete example game, where the FP process takes exponentially many rounds until the equilibrium is reached. Alternatively, convergence may be measured by considering the difference in the *payoffs*, rather than the behavior profiles (beliefs), such as we did in our experimental implementation. These may converge even though the distributions themselves may oscillate.

However, the slow convergence of regular fictitious play may – in large games – become unhandy, thus calling for replacements by more refined and sophisticated learning techniques. Inspecting the applicability of such alternatives is an interesting direction of future research.

6 Related Work

The idea of Pareto-optimal security strategies (POSS) is not new and has previously been introduced in [7,6,18]. This prior work appears as a special case of definition 1 when the games are finite. Infinity of action spaces, which arise when continuous parameters (such as timing) were not covered by this preliminary work. Treating communication as a game is a well-researched field, with a comprehensive account given by [2], and much precursor work (such as [21]). Game-theory has in the past as well been used to negotiate optimal service and operational level agreements (see [11,8] among others) and to quantitatively analyze security in ad hoc networks [22] under several optimality concepts (among which is Pareto-optimality). Our work aids and further substantiates this direction of research. An interesting yet unexplored relation to our work also exists in the results of [16], who consider a "non-static" gameplay. This direction is one of future considerations.

6.1 Multipath Transmission

A fruitful application is a game-theoretic model of multipath transmission. Roughly speaking, the game is about an honest sender attempting to communicate over a network that is partially under the attacker's control. The attacker is not constrained in its computational power, but limited to control a fixed maximal number of nodes, by which it can read and insert network traffic at its own will. The honest player's goal is to deliver a message to a designated receiver, while the payload remaining *confidential* and *authentic*, and with the maximum probability of delivery (*availability*). The gameplay is by the honest party (player 1) randomly choosing transmission paths, while the attacker (player 2) randomly chooses nodes to sniff, which – in its simplest form just described – makes the scenario almost a diagonal game. An illustration is given in Figure 3.

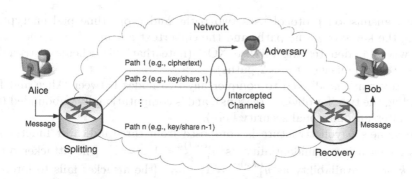

Fig. 3. Illustration of multipath transmission

We leave the protocol-, game- and cryptography-details aside here (referring the interested reader to [5,19,3,12,14] to fill these gaps), and confine ourselves

to stating that experimental evaluations on real life enterprise network topologies lead to small games (after eliminating redundant and dominated strategies)
that are easy to handle. This is mostly due to low connectivity (many enterprise network *backbones* have a graph vertex connectivity of two, for reasons
of redundancy). Realistic wide area topologies would follow an Erdös-Rényi or
scale-free topology, which we simulate in the course of a research project (see
the acknowledgement) on which we will report in subsequent work. Here, for the
sake of generality, this example shall merely substantiate the applicability of the
theoretical concept of Pareto-optimal security strategies, while our evaluation
will be on matrix games with randomly chosen payoff structures.

6.2   Example: Security of Multipath Transmission

Nevertheless, the method appears viable to compute quantitative security of
multipath transmission on a given network topology. As an example, consider
the network topology depicted in Figure 4, where Alice wishes to securely send
a message to Bob over the network. Hereby, a message m is called *secure*, if its
transmission is confidential, the payload is authentic and the delivery does not
fail (availability). Hence, we have three goals, i.e., three dimensions.

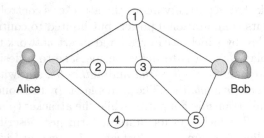

Fig. 4. Example network

The transmission protocol uses two paths and a one-time pad encryption,
sending the key k over one path, and the ciphertext $c = m \oplus k$ over the second
path, where $\oplus$ denotes the bitwise XOR (note that this scheme is trivial to
generalize to the usage of $n > 2$ paths).

The adversary is allowed to conquer any two nodes between Alice and Bob
(excluding the two, for obvious reasons), and is computationally unbounded (i.e.,
we are after unconditional security here).

The game's payoff structure is composed from three indicator functions of
success, measuring confidentiality as $u_1^{(conf)} = 1 : \iff$ [the attacker misses
either k or c], availability as $u_1^{(avail)} = 1 : \iff$ [the attacker fails to intercept
k or c], and authenticity. This is achieved by the protocol in [14], and yields
$u_1^{(auth)} = 1 : \iff$ [the attacker fails to conquer at least one of the chosen paths].
The strategy set for player 1 is the set of pairs of disjoint transmission paths (a
total of $|PS_1| = 3$ strategies). The strategy set for player 2 is the set of two-
element subsets of $\{1, 2, 3, 4, 5\}$, giving a total of $|PS_2| = \binom{5}{2} = 10$ strategies. The

payoff for player 1 is the vector $\boldsymbol{u}_1 = (u_1^{(\text{conf})}, u_1^{(\text{avail})}, u_1^{(\text{auth})})$. The importance weights are $(\alpha_{0,\text{conf}}, \alpha_{0,\text{avail}}, \alpha_{0,\text{auth}}) = (1/3, 1/3, 1/3)$.

The fictitious play process converged within 6 iterations until an accuracy of $\delta < 10^{-3}$, giving the final multicriteria security strategy $\theta^* = (\frac{1}{3}, \frac{1}{3}, \frac{1}{3})$, with assurance $\boldsymbol{v} = (\frac{2}{3}, 0, \frac{2}{3})$. This is indeed what we expect, since if the attacker intercepts one of the paths, the message remains confidential and cannot be forged unnoticeably ($u_1^{(\text{conf})} = 1 = u_1^{(\text{auth})}$), but it can become destroyed ($u_1^{(\text{avail})} = 0$). The assurance vectors thus give the conditional probability $\Pr[m$ *is authentic and has not been disclosed* $\mid m$ *was correctly delivered*$] \geq 2/3$, but the unconditional likelihood $\Pr[\text{delivery of } m \text{ can be disrupted}] = 1$. By the properties of MCSS, this is the best that the attacker can do. The protocol is as such insecure, as it is vulnerable to denial-of-service, although it can be made arbitrarily and unconditionally secure against eavesdropping (under the given adversary model) by repeating the process on a sequence of packets $m_1, m_2, \ldots, m_\ell$ whose bitweise XOR recovers $m = m_1 \oplus m_2 \oplus \cdots \oplus m_\ell$. Then, the likelihood to disclose m is $2^{-O(\ell)}$, if all ℓ messages are delivered according to the security strategy θ^*.

It is straightforward to apply the technique to other more efficient protocols like [19,5], and to take further probabilistic security in the network into account, by replacing the payoff functions accordingly.

7   Conclusion

Fictitious play has been demonstrated as a working method to numerically compute security strategies towards playing safe in multiple regards (security goals). The axiomatic characterization of multi-criteria security strategies as Pareto-Nash equilibria, which in turn can be computed as Nash-equilibria of multi-player games, induces a sequence of simple and straightforward transformations that culminate in a game enjoying the fictitious play property. In addition, we gain a degree of freedom to assign importance weights to different security goals, although these seem to have only minor (if not negligible) influence on the actual outcome (equilibrium) that is computed. Nevertheless, it adds an interesting aspect to practical applications by showing that a "prioritization" between security goals is not necessarily useful in general.

Aspects of future work are non-static game-plays, improved variants of fictitious play and examining complexities to more detail. As a showcase application, we will apply our algorithms to problems of establishing confidential, authentic and reliable communication in large scale computer networks by means of multipath transmission. Given the available cryptographic fundament, quantifying security in terms of Pareto-optimal security strategies then boils down to a straightforward application of our numerical method presented here.

Acknowledgment. This work was supported by the Austrian Research Promotion Agency (FFG) under project grants no. 836287. Furthermore, we thank the anonymous reviewers for valuable suggestions and for drawing our attention to interesting aspects of future research.

References

1. Acosta Ortega, F., Rafels, C.: Security strategies and equilibria in multiobjective matrix games. Working Papers in Economics 128, Universitat de Barcelona. Espai de Recerca en Economia (2005), http://ideas.repec.org/p/bar/bedcje/2005128.html
2. Alpcan, T., Başar, T.: Network Security: A Decision and Game Theoretic Approach. Cambridge University Press (2010)
3. Ashwin Kumar, M., Goundan, P.R., Srinathan, K., Pandu Rangan, C.: On perfectly secure communication over arbitrary networks. In: PODC 2002: Proceedings of the Twenty-First Annual Symposium on Principles of Distributed Computing, pp. 193–202. ACM, New York (2002)
4. Brandt, F., Fischer, F., Harrenstein, P.: On the rate of convergence of fictitious play. In: Kontogiannis, S., Koutsoupias, E., Spirakis, P.G. (eds.) SAGT 2010. LNCS, vol. 6386, pp. 102–113. Springer, Heidelberg (2010)
5. Fitzi, M., Franklin, M.K., Garay, J.A., Vardhan, S.H.: Towards optimal and efficient perfectly secure message transmission. In: Vadhan, S.P. (ed.) TCC 2007. LNCS, vol. 4392, pp. 311–322. Springer, Heidelberg (2007)
6. Ghose, D.: A necessary and sufficient condition for pareto-optimal security strategies in multicriteria matrix games. Journal of Optimization Theory and Applications 68(3), 463–481 (1991)
7. Ghose, D., Prasad, U.R.: Solution concepts in two-person multicriteria games. Journal of Optimization Theory and Applications 63(2), 167–189 (1989)
8. Kaminski, H., Perry, M.: A framework for automatic SLA creation. Tech. rep. The University of Western Ontario, Computer Science Publications (2008)
9. Lozovanu, D., Solomon, D., Zelikovsky, A.: Multiobjective games and determining pareto-nash equilibria. Buletinul Academiei de Stiinte a Republicii Moldova Matematica 3(49), 115–122 (2005), ISSN 1024-7696
10. McKelvey, R.D., McLennan, A.M., Turocy, T.L.: Gambit: Software tools for game theory, version 0.2007.12.04 (2007), http://gambit.sourceforge.net
11. Moroni, S., Figueroa, N., Jofre, A., Sahai, A., Chen, Y., Iyer, S.: A game-theoretic framework for creating optimal SLA/contract. Tech. Rep. HPL-2007-126, HP Laboratories Palo Alto (2007)
12. Rass, S., Schartner, P.: A unified framework for the analysis of availability, reliability and security, with applications to quantum networks. IEEE Transactions on Systems, Man, and Cybernetics – Part C: Applications and Reviews 41(1), 107–119 (2011)
13. Rass, S.: On game-theoretic network security provisioning. Springer Journal of Network and Systems Management 21(1), 47–64 (2013)
14. Rass, S., Schartner, P.: Multipath authentication without shared secrets and with applications in quantum networks. In: Proceedings of the International Conference on Security and Management (SAM), July 12-15, vol. 1, pp. 111–115. CSREA Press (2010)
15. Robinson, J.: An iterative method for solving a game. Annals of Mathematics 54, 296–301 (1951)
16. Ryu, C., Sharman, R., Rao, H., Upadhyaya, S.: Security protection design for deception and real system regimes: A model and analysis. European Journal of Operational Research 201(2), 545–556 (2010), http://www.sciencedirect.com/science/article/B6VCT-4VXTSK1-2/2/9ffe61e9aa467ce2271adfa338f27842

17. Sela, A.: Fictitious play in 'one-against-all' multi-player games. Economic Theory 14, 635–651 (1999), http://dx.doi.org/10.1007/s001990050345
18. Voorneveld, M.: Pareto-optimal security strategies as minimax strategies of a standard matrix game. Journal of Optimization Theory and Applications 102(1), 203–210 (1999)
19. Wang, Y., Desmedt, Y.: Perfectly secure message transmission revisited. IEEE Transactions on Information Theory 54(6), 2582–2595 (2008)
20. Washburn, A.: A new kind of fictitious play. Tech. rep., Operations Research Department, Naval Postgraduate School, Monterey, California 93943, copyright by John Wiley & Sons, Inc. (2001)
21. Ying, Z., Hanping, H., Wenxuan, G.: Network security transmission based on bimatrix game theory. Wuhan University Journal of Natural Sciences 11(3), 617–620 (2006)
22. Yu, W., Liu, K.J.R.: Game theoretic analysis of cooperation stimulation and security in autonomous mobile ad hoc networks. IEEE Transactions on Mobile Computing 6(5), 507–521 (2007)

Realizable Rational Multiparty Cryptographic Protocols*

John Ross Wallrabenstein and Chris Clifton**

Dept. of Computer Science, Purdue University, USA
{jwallrab,clifton}@cs.purdue.edu

Abstract. In this work, we describe how to realize rational cryptographic protocols in practice from abstract game specifications. Existing work requires strong assumptions about communication resources in order to preserve equilibria between game descriptions and realized protocols. We argue that for real world protocols, it must be assumed that players have access to point-to-point communication channels. Thus, allowing signaling and strategy correlation becomes unavoidable. We argue that ideal world game descriptions of realizable protocols should include such communication resources as well, in order to facilitate the design of protocols in the real world. Our results specify a modified ideal and real world model that account for the presence of point-to-point communication channels between players, where security is achieved through the simulation paradigm.

Keywords: Rational Multiparty Computation, Game Theory, Non-Cooperative Computation.

1 Introduction

The field of *rational cryptography* departs from modeling players as either honest or malicious, and instead models all players as rational utility-maximizing agents: each player chooses those actions that maximize their utility function $\mu(\cdot)$, which expresses their preferences over outcomes. All players may arbitrarily depart from the protocol specification if doing so is a utility-maximizing strategy. This approach to modeling removes the strong assumption of the semi-honest model: that honest players follow the protocol specification, regardless of whether or not it is in their best interest. By considering all players as rational agents, the standard properties of cryptographic protocols (e.g. privacy, correctness and fairness) are modeled through the utility functions of the players. Security of the protocol is then deduced from whether or not the stable equilibrium of the original game specification is reachable given the players' utility functions.

In secure multiparty computation (SMPC), the security of protocols are demonstrated through the *simulation paradigm*. Define an *ideal* protocol for

---

* The rights of this work are transferred to the extent transferable according to title 17 §105 U.S.C.
** This material is based upon work performed while Dr. Clifton was serving at the National Science Foundation.

R. Poovendran and W. Saad (Eds.): GameSec 2014, LNCS 8840, pp. 134–154, 2014.

computing a functionality f that invokes an incorruptible and universally trusted third party (TTP). Similarly, define a *real* protocol π for computing f where no TTP exists. Security is established if an adversary $\mathcal{A}$ in the real model has no advantage over a simulator $\mathcal{S}$ in the ideal model [1].

A major obstacle when defining security for rational multiparty protocols is the potential for players to form *coalitions*, colluding to undermine the security of the protocol. The strongest result, by Izmalkov et al. [2], allows any function to be computed securely by rational players using the approach of Goldreich et al. [1]. Although a universal result, it relies on strong assumptions including forced actions and physical primitives. A weaker notion, referred to as *collusion-free* computation [3–5], removes the ability of players to communicate additional information subliminally through the protocol communication resources. The result relies on a trusted mediator at the center of a star network topology, where all messages pass through the mediator and are re-randomized in order to prevent steganographic communication between the players. This result relies on adversarial independence, where simulators and adversaries are disallowed communication in the protocol. However, a collusion-free protocol may still cause issues when executed as part of a larger protocol. For example, the collusion-free protocols of Izmalkov et al. [2, 5] provide no guarantees when all players are malicious. This observation led to the work of Alwen et al. [6], where communication restrictions are further weakened to achieve *collusion-preserving* computation, which preserves any potential for collusion present in the original game specification. Although this result removes the requirement of a trusted mediator, it rules out a large class of communication resources (e.g. point-to-point and broadcast channels). Kamara et al. [7] consider a setting where adversaries have the capability to communicate additional information during protocol execution, yet choose to be *non-colluding*. Fuchsbauer et al. [8] give constructions under standard communication channels by forcing parties to send only unique messages as part of the protocol. Thus, collusion within the protocol is avoided, but communication outside of the protocol execution still facilitates collusion.

From this collection of work, addressing the issue of collusion appears to require strong limitations on the type of communication resources granted to players. As the general goal of rational cryptography is to provide a more realistic view of how players behave in cryptographic protocols, we consider what can be achieved when players have access to point-to-point communication channels - an unavoidable aspect in real world applications. Thus, in this work we define a security model where players may communicate information over point-to-point channels both inside and outside the protocol execution.

Our work proposes a new security framework for rational agents that models player access to point-to-point communication channels in the ideal world model. From this, we describe how to demonstrate the security of protocols in a real world model that implements games specified in our modified ideal world model. We note that imposing restrictions on the ideal world to capture unavoidable behavior exists currently in the cryptographic literature: it is a core feature of the malicious model, which extends the semi-honest model to consider more powerful adversaries. In the malicious setting, the ideal world must capture the ability of an adversary to coordinate the actions and inputs of players it corrupts, and force aborts during protocol execution; these actions are unavoidable in the presence of a monolithic malicious adversary. Our model necessarily limits the class of games that may be modeled in the ideal world formulation of our framework, as point-to-point communication channels must exist in the original game. Our work differs from existing formulations, which attempt to realize all games at the expense of restricting the communication interface available to players.

Throughout the remainder of the introduction, we argue that when point-to-point communication channels are unavoidable, it is meaningful to consider what games are realizable in their presence. We demonstrate that a non-trivial class of games constructed in our modified ideal world model have realizable implementations in the real world model through the Signaling game in Section 1.2, and the classic prisoner's dilemma in Section 2. We give our technical contribution, a security model for realizing protocols from game specifications in the presence of point-to-point communication channels, in Section 3. We demonstrate the power of our model relative to others through a full proof of security for the rational secret sharing protocol of Halpern and Teague [9] in Section 4, which is inadmissible under existing frameworks due to the presence of point-to-point communication channels. These examples demonstrate the key contribution of our model, which is less restrictive than prior work yet is able to correctly model the games' equilibria when played in the real world.

1.1   Local Adversaries

Translating the standard simulation paradigm to the game theoretic setting of rational cryptography requires addressing how adversaries should be modeled. In the original formulation, a centralized semi-honest or malicious adversary corrupts a subset of the players. However, rational cryptography makes no such distinction[1] between honest and corrupted players, and assumes all players are rational and acting to maximize their local utility function. Thus, translating the concept of an adversary is not immediate. Alwen et al. [6] give a collusion preserving framework where each player has an associated *local* adversary. Thus, the monolithic adversary of the standard model is shattered into an adversary for each individual player. Canetti et al. [11] argue that a local adversary should be defined for each ordered pair of players, as this provides a more granular model of the flow of information. Canetti et al. then demonstrate that the *local universal composition* (LUC) model can preserve the incentive structure in games.

We follow this modeling trend of shattering the monolithic adversary $\mathcal{A}$ into a *set* of local adversaries $\mathcal{A} = \{\mathcal{A}_i\}_{i \in [1...n]}$ such that each player $P_i \in \mathcal{P}$ is associated with adversary $\mathcal{A}_i$. Rather than considering local adversaries that "corrupt" their associated player P_i, we simply require that the adversary selects

---

[1] A mixed model has been proposed by Lysyanskaya et al. [10] where one subset of players are arbitrarily malicious, and the other subset are utility-maximizing rational agents.

the actions of P_i to maximize their local utility function μ_i. Thus, we preserve the assumption in rational cryptographic protocols that all players are purely rational and bound to a utility function, rather than remaining honest unless corrupted by a monolithic adversary.

1.2 Communication Resources

A core issue with existing work is how communication resources are modeled in game descriptions. In order to prevent players from signaling information or coordinating their actions, available communication resources are tightly restricted. For example, Izmalkov et al. [2] propose *rational secure computation* where only those equilibria in the game description exist in the realized protocol. However, this result comes at the cost of requiring forced actions and physical primitives such as opaque envelopes and ballot boxes[2]. Although not impossible to realize, in practice it has limited applicability.

In the ideal world model of secure multiparty computation, a protocol is viewed as an interaction between a set of players and a universally trusted third party (TTP). An ideal computation of a function has each player send their private input to the TTP, who computes the function and returns the results to each player. Restricting communication resources is not necessary, as players are assumed to be mutually distrustful. Further, any collusion between players is modeled through a monolithic adversary $\mathcal{A}$ that coordinates the actions of the players it corrupts.

In order to implement *arbitrary* games as protocols, strict notions of privacy preservation and the prevention of signaling and correlation must be satisfied. Arbitrary game specifications may impose restrictions on the communication resources available to players. Thus, the corresponding protocol implementation must not allow players to communicate more information than is possible in the ideal game specification. We briefly review the characteristics a model for implementing arbitrary games must satisfy[3]. We make the argument that even if a protocol satisfies all of these characteristics, it is likely to fall short of satisfying the ideal world model: communication between players outside of the protocol is unavoidable in real world settings. Thus, the model we present is not bound to satisfy these restrictions, and is a more accurate representation of what is achievable for protocols executed in the real world.

Privacy. A protocol π implementing an arbitrary game Γ must preserve both *pre-game privacy* and *post-game privacy* in addition to preserving the equilibrium of Γ. The notion of pre-game privacy ensures that the private input of each party is not revealed, as this will affect the actions of other parties. However, protocols implementing arbitrary games must also preserve the notion of post-game privacy, where nothing beyond the intended result (and what can be

---

[2] This result is a direct application of the GMW protocol [1].

[3] The ECRYPT summary report [12] on rational cryptographic protocols provides background on modeling techniques used to address privacy, signaling and correlated actions.

inferred from this) is revealed. This notion is necessary so that the equilibria of future games are not perturbed by information revealed in previous games.

Signaling. Similar to the notions of pre- and post-game privacy are the notions of *pre-game signaling* and *post-game signaling*. The ability to signal other players allows protocol participants to coordinate their actions to achieve a higher payoff. For example, consider two players A and B with inputs a and b. The payoff function is defined as $\Pi(\Gamma) := a \oplus b$, and described in Table 1:

Table 1. Signaling Game

	A sets $a = 1$	A sets $a = 0$
B sets $b = 1$	(0,0)	(1,1)
B sets $b = 0$	(1,1)	(0,0)

If A or B can signal even a single bit to the other, each will receive a payoff of 1 as opposed to an expected payoff of $\frac{1}{2}$. Thus, similar to the restriction on privacy, preventing pre- and post-game signaling is necessary to preserve the equilibria of individual and future games when constructing protocols for *arbitrary* games.

The signaling game specification can be formulated under existing frameworks as a protocol, and demonstrated to preserve the mixed equilibrium of the original game. Yet by ignoring the ability of players to communicate outside of the protocol, the protocol formulation is invalidated in real world settings: players will collude to achieve a payoff of 1, rather than the expected payoff of $\frac{1}{2}$ of the original game specification.

We only consider those game specifications that allow point-to-point communication, as these channels are unavoidable in the real world. Thus, our model correctly predicts a payoff of 1 for players in the signaling game, as point-to-point communication channels allow signaling.

Correlated Actions. Correlated actions are similar to signaling, but allow parties to coordinate actions without exchanging information. This is usually accomplished through a shared value, such as a *common reference string* (CRS). The parties need not distribute information, but rather rely on the shared CRS to coordinate their actions. As with signaling, protocol constructions for arbitrary games must prevent pre- and post-game correlation to preserve equilibria in local as well as future games.

2 Prisoner's Dilemma

As a classic example, we consider the Prisoner's Dilemma[4]: a game between two suspects A and B that have been accused of committing both a principal and

---
[4] The concept was originally proposed by Flood and Dresher while working at the RAND corporation, and is described in detail by Poundstone [13].

lesser crime. The Authority has sufficient evidence to convict both A and B on the lesser crime, punishable by 1 year in prison. However, there is insufficient evidence to convict A or B on the principal crime. The Authority separates A and B, and offers the following proposal: confess and serve no time while your partner serves 3 years in prison. Players A and B are then subject to the following dilemma:

1. If both A and B remain silent, they will each be convicted on the lesser crime and serve 1 year in prison.
2. If one confesses while the other remains silent, the confessor is set free while the other serves 3 years in prison.
3. If both A and B confess, each will serve 2 years in prison.

Table 2. Prisoner's Dilemma Game

	A Remains Silent	A Confesses
B **Remains Silent**	(-1,-1)	(0,-3)
B **Confesses**	(-3,0)	(-2,-2)

From the player payoffs listed in Table 2, note that each player maximizes their utility by confessing to the principal crime regardless of the strategy of their partner. We use this example to illustrate the necessity of removing monolithic adversaries, as well as how communication assumptions should be formulated in the ideal game description. Note that the original ideal game specification of the prisoner's dilemma requires that the suspects A and B are physically separated: thus unable to communicate or otherwise coordinate their actions. However, we will construct a modified formulation in the presence of point-to-point communication channels with an *equivalent equilibrium* to the original formulation under our proposed model.

2.1 Monolithic Adversaries

Traditionally, cryptographic protocols are analyzed with respect to their resilience to a monolithic adversary $\mathcal{A}$ corrupting some subset of the players. Protocol resilience to adversarial corruption is quantified by the fraction of players that may be corrupted before the protocol security is violated.

In the game theoretic setting of rational cryptography, this model has been called into question by Alwen et al. [6] and Canetti et al. [11]. The goal of rational cryptography is to model each player as bound to their local utility function, rather than controlled by a monolithic adversary with a global utility function. The monolithic adversary in both of their models is shattered into a set of *local* adversaries unique to each player. Removing the monolithic adversary in favor of a set of local adversaries is critical to preserving game theoretic equilibria. In the running example of the Prisoner's Dilemma, consider the case where $\mathcal{A}$ corrupts both A and B. As $\mathcal{A}$ controls both players, A and B may be forced to remain silent and achieve payoff $(-1, -1)$. However, consider the case where A (resp. B) has a *local* adversary $\mathcal{A}_A$ (resp. $\mathcal{A}_B$): as $\mathcal{A}_A$ is bound to the utility function $\mu_A(\cdot)$ of A, $\mathcal{A}_A$ maximizes $\mu_A(\cdot)$ by confessing as in the ideal specification of the game. An identical argument holds for $\mathcal{A}_B$ as well. Thus, a monolithic adversary is capable of introducing a stable collusion equilibrium that does not exist in the ideal game specification, whereas the local adversary model preserves the original incentive structure.

2.2 Realistic Communication Model

To prevent pre- and post-game signaling and strategy correlation, many rational cryptographic frameworks impose strong restrictions on the communication resources available to players. This issue is most pronounced in the multiparty setting, where communication resources may enable collusion. To prevent communication resources from perturbing the equilibria of the ideal world game, existing constructions require forced player action and physical primitives [2], trusted mediators and forced broadcast channels [4], as well as the cooperation of adversarial players to deliver messages [6].

While these results provide strong guarantees under restrictive communication resource assumptions, the security guarantees are with respect to the protocol only. That is, assuming players may only interact through the protocol and its communication resources, the equilibria of the ideal world game is preserved. However, we argue that this results in a false sense of security for protocols realized in the real world, where players typically have access to point-to-point communication channels - undermining the strict communication assumptions of the protocol.

Our example of the prisoner's dilemma illustrates a salient point: the necessary and sufficient condition for preserving the equilibrium of the original formulation is the ability of A and B to *privately communicate* with the Authority. The original game specification requires the two players A and B to be physically separated, and thus unable to communicate. However, the key to preserving the equilibrium (confess, confess) of the original game Γ only requires preventing A and B from observing their interaction with the Authority. Consider a modified game $\bar{\Gamma}$ where all players $\{A, B, \text{Authority}\} \in \mathcal{P}$ have access to a point-to-point communication resource $\mathcal{R}$. As long as the communication links $\mathcal{R}_{A,\text{Authority}}, \mathcal{R}_{B,\text{Authority}}$ are private, the original equilibrium is preserved despite the presence of point-to-point communication channels. In game theoretic terms, communication between A and B through $\mathcal{R}_{A,B}$ is considered *cheap talk*, as both A and B will claim to play silent, yet as utility maximizing agents they choose to confess, which strictly dominates silent. As neither A nor B can observe the message sent by the other to Authority, the coalition is unstable and disintegrates despite the presence of point-to-point communication channels.

3 Our Contribution

We argue that ideal world protocols should assume that players have the ability to communicate over point-to-point channels. As in the standard SMPC ideal world model, players may not wish to communicate due to mutual distrust. However, the option to do so should be part of the model, as this is unavoidable in the real world. Thus, we present a modified ideal world model capturing the presence of point-to-point communication channels between all players. Specifically, we answer the following questions:

1. How is security formalized when all players are rational and have access to point-to-point communication channels?
2. What benefits result from weakening the security guarantees of the standard malicious model by considering rational players with local adversaries?

3.1 Unstable Coalitions

A powerful aspect of the rational cryptographic setting with local adversaries is the ability to design protocols where coalitions are unstable. As each player has a local adversary that selects their actions in order to maximize a utility function, protocols may be designed to incentivize players to *leave* coalitions [14]. This benefit of modeling each player as an independently rational agent is frequently overlooked, and allows game equilibria to be preserved despite the presence of point-to-point communication channels. We have illustrated the power of unstable coalitions through our example of the prisoner's dilemma. We now consider coalition stability in the setting of *rational secret sharing*, as it is the most familiar example of a rational cryptographic protocol.

Rational Secret Sharing Candidate definitions for achieving security against rational agents should accurately model well-studied problems in rational cryptography. The most familiar rational cryptographic protocol is *rational secret sharing* [8, 15–19]. The goal of *threshold* secret sharing is to split a secret among n parties such that any k shares are sufficient to recover the secret value, using a scheme such as the polynomial interpolation approach proposed by Shamir [20]. Rational secret sharing, introduced by Halpern and Teague [9], is particularly concerned with the process of recovering the secret from the shares[5]. As noted by Halpern et al. [9], rational players' utility functions are assumed to value *exclusivity*, where preference is given to learning the output of the function while preventing other players from doing so. Under this assumption, no party has any incentive to distribute their share to the other parties, which destabilizes coalition formation. The equilibrium is to wait for other players to distribute their shares, as this is the only action that increases a player's utility function.

---

[5] Maleka et al. [21] consider rational secret sharing in the context of repeated games, and Nojoumian et al. [22] consider the repeated game setting from a socio-rational perspective where player reputation is important.

Thus, a player that does not distribute their share has the potential to be the exclusive player to recover the secret.

The authors demonstrate that this implies no deterministic protocol exists where rational parties are willing to disseminate their shares to other players. Their randomized protocol is a modified game where players are distributed a *set* of shares, where only one share is correct. In each round k, players distribute their shares which evaluate to either the secret or a default value $\perp$. The solution relies on the fact that parties are unaware whether the current round k is terminal (k^*, allowing the secret to be recovered), or merely a "test" round $k \neq k^*$ (where the secret cannot be recovered, but players who do not distribute shares are caught as cheaters). By choosing k^* from a geometric distribution, as in Groce et al. [18], cheating players that choose strategy $\sigma = \perp$ when $k \neq k^*$ are caught and the game may be terminated. Thus, players now have an incentive to distribute their share, as playing $\perp$ only yields positive utility when $k = k^*$.

A candidate security definition should accept this probabilistic protocol for rational secret sharing as secure against rational agents. However, the strong restrictions on communication channels imposed by existing work preclude the above protocol from satisfying their security definitions, despite refinements considering the problem under standard communication models [8, 23–25]. That is, the rational secret sharing protocol of Halpern and Teague [9] assumes players have access to a non-rushing broadcast channel. This clearly violates the assumptions of models assuming physical primitives [2], and even fails to satisfy the weakest security definition that has been proposed: collusion-preserving computation [6]. Ideally, the original rational secret sharing protocol of Halpern and Teague should be demonstrably secure against rational agents under a general security framework. Our framework allows point-to-point communication in the ideal model, and thus is able to accurately model the original solution to rational secret sharing, which we demonstrate in Section 4.

3.2 Adversarial Model

Traditionally, an adversary $\mathcal{A}$ is viewed as a monolithic entity with a specified computational complexity and ability to "corrupt" players in a static or dynamic fashion. In our model, we consider all players to have the ability to act in an adversarial manner. Thus, rather than considering a monolithic adversary $\mathcal{A}$, we endow each player $P \in \mathcal{P}$ with a local adversary $\mathcal{A}_P$. The adversary is bound to the player's utility function $\mu_P(\cdot)$ and selects actions for P in order to maximize $\mu_P(\cdot)$. Note that as we bind player actions to a local adversary seeking to maximize a utility function, we cannot bound the number of players that deviate from the protocol. This is an unavoidable consequence of modeling players as rational agents; they select strategies to maximize a local utility function and follow the protocol only when doing so is advantageous. As cryptographic protocols typically require a number of rounds of interaction, we allow the rational players to update their strategy based on observations throughout the game Γ. Thus, we assume each local adversary is *mobile* [26], and may choose to deviate or follow the protocol at each round in a dynamic fashion. Additionally, players may choose probabilistic strategies[6], so we must introduce a random tape r_P for each player P. Thus, each local adversary is adaptive, mobile, probabilistic, malicious, runs in *probabilistic polynomial-time* (PPT) and is presumed rational: bound to the player's local utility function.

Given the above definition of adversaries, the following actions are unavoidable:

- **Refusal to Participate:** Players may refuse to participate in the protocol. Constructions satisfying our definition thus assume that it is advantageous for players to engage in the protocol, and that this constitutes a utility maximization strategy with respect to their local utility function.
- **Input Substitution:** Players may supply an input to the protocol different from their true input.
- **Premature Abort:** Players may abort the protocol prior to completion.
- **Collusion:** Players may privately communicate over point-to-point communication channels, and collude to influence the protocol execution.

3.3   Ideal World Model

We now formalize the *ideal world* model, under which an ideal game specification Γ is constructed. We assume familiarity with standard game theoretic concepts in our exposition[7]. We first define the game specification of Γ under the *extensive form game* representation. In the game theoretic literature, *normal form game* representation is generally used for single round games where actions are played simultaneously. As cryptographic protocols typically proceed in a series of rounds where actions are played asynchronously, we prefer *extensive form game* representation, where the ideal game specification Γ is represented as a tree. At each node in the game tree, a subset $P \subset \mathcal{P}$ of the players select and simultaneously play an action.

Definition 1. *An **extensive form game** Γ consists of:*

1. *A finite set $\mathcal{P} = \{P_i\}_{i=1}^n$ of players.*
2. *A (finite) set of sequences $\mathcal{H}$ called the history. The empty sequence $\emptyset$ is a member of $\mathcal{H}$. We let k denote the current decision node. If $(a^k)_{k=1,\ldots,K} \in \mathcal{H}$ and $L < K$ then $(a^k)_{k=1,\ldots,L} \in \mathcal{H}$. If an infinite sequence $(a^k)_{k=1}^\infty$ satisfies $(a^k)_{k=1,\ldots,L} \in \mathcal{H}$ for every positive integer L then $(a^k)_{k=1}^\infty \in \mathcal{H}$. A history $(a^k)_{k=1,\ldots,K} \in \mathcal{H}$ is a terminal history if it is infinite or if there is no a^{K+1} such that $(a^k)_{k=1,\ldots,K+1} \in \mathcal{H}$. The set of actions available after the nonterminal history h is denoted $A(h) = \{a : (h, a) \in \mathcal{H}\}$ and the set of terminal histories is denoted $\mathcal{Z}$. We let $\mathcal{H}^k$ denote the history through round k.*

---

[6] In a game theoretic setting, such strategies are referred to as *mixed*.

[7] For a proper introduction to the subject, Katz [27] describes the current effort to combine game theoretic and cryptographic concepts, while Osborne et al. [28] and Fudenberg et al. [29] give a complete introduction to game theory.

3. *A player function P that assigns to each nonterminal history (each member of $\mathcal{H}/\mathcal{Z}$) a member of $\mathcal{P} \cup \{nature\}$. When $P(h) = nature$, then nature determines the action taken after history h.*
4. *For each player $P_i \in \mathcal{P}$ a partition $\mathcal{I}_i$ of $\{h \in \mathcal{H} : P(h) = i\}$ with the property that $A(h) = A(h')$ whenever h and h' are in the same member of the partition. For $I_i \in \mathcal{I}_i$ we denote by $A(I_i)$ the set $A(h)$ and by $P(I_i)$ the player $P(h)$ for any $h \in I_i$. Thus, $\mathcal{I}_i$ is the information partition of player i, while the set $I_i \in \mathcal{I}_i$ is an information set of player i.*
5. *For each player $P_i \in \mathcal{P}$ a preference relation $\precsim_i$ on lotteries[8] over $\mathcal{Z}$ that can be represented as the expected value of a payoff function defined on $\mathcal{Z}$.*

Throughout, we replace the preference relation $\precsim_i$ by a *utility function* μ_i : $A \to \mathbb{R}$, such that $\mu_i(a) \geq \mu_i(b)$ when $b \precsim_i a$.

We make the following modeling choices:

- **Extensive Form Games:** The ideal game specification Γ is described by a game tree in extensive form representation.
- **Imperfect Information:** A game specification is said to have *imperfect information* if players may have non-singleton information sets $I_i \in \mathcal{I}_i$. That is, at a given round in the game, players may be unaware of the move selected by the previous player(s). Thus, their information set may contain more than one node in the game tree at any given round.
- **Local Simulators:** Each player $P_i \in \mathcal{P}$ in the ideal model has a *local* simulator $\mathcal{S}_i$ that forces P to play those actions that maximize $\mu_i(\cdot)$, the utility function of player P_i. Each simulator $\mathcal{S}_i$ has an associated adversary $\mathcal{A}_i$ in the real world execution model, denoted $\mathcal{S}_i = \mathsf{Sim}(\mathcal{A}_i)$.
- **Point-to-Point Communication Resources:** Each player pair $(P_i, P_j)_{i \neq j} \in \mathcal{P}$ has a secure point-to-point communication resource $\mathcal{R}_{ij}$.

As we consider all players to be rational agents, we model the ideal world protocol as a game specification Γ that aims to achieve an equilibrium. The ideal game specification is an interaction between a set of n players $\mathcal{P} = \{P_i\}_{i=1}^n$, their local utility functions $\boldsymbol{\mu} = \{\mu_i\}_{i=1}^n$ and action sets A_i, which contains those actions playable by player P_i. Frequently, a deterministic choice of an action $a \in A_i$ will not yield a Nash equilibrium. Thus, we allow players to choose a *strategy* σ_i: a probability distribution over A_i. The standard equilibrium concept in the rational cryptographic literature is a *computational* Nash equilibrium [24, 25, 30–32], given by Definition 2.

Definition 2. *A **computational Nash equilibrium** of a two-party extensive-form game Γ is an independent strategy profile $\boldsymbol{\sigma}^* = \{\sigma_i^*\}_{i=1}^n$, such that*

1. *$\forall \sigma_i^* \in \boldsymbol{\sigma}^*, \sigma_i^*$ is PPT computable.*

---

[8] Even if all actions are deterministic, moves by *nature* can induce a probability distribution over the set of terminal histories.

2. *for each player* P_i, *any other PPT computable strategy* $\sigma_i' \neq \sigma_i^*$, *we have*
   $$\mu_i(\sigma_i', \boldsymbol{\sigma}_{-i}^*) \leq \mu_i(\boldsymbol{\sigma}^*) + negl(\lambda)$$

where $\sigma_{-i} \stackrel{def}{=} (\sigma_j)_{j \in [1\ldots n]/\{i\}}$.

Intuitively, no player P_i has an incentive to deviate from strategy σ_i given that every other player P_j selects their equilibrium strategy σ_j. The definition of a computational Nash equilibria adds a negligible term $negl(\lambda)$ with respect to a security parameter λ. This is necessary in the computational setting, as security rests on the premise that breaking cryptographic primitives occurs with only negligible probability. Thus, this notion must be incorporated into the equilibrium definition. Although computational Nash equilibria are the weakest of the equilibrium concepts described in the rational cryptographic literature, preserving only computational Nash equilibria in our framework is sufficient for extensions to more powerful equilibrium concepts.

The standard ideal world model has players interact with an incorruptible trusted third party (TTP) that accepts player inputs, computes the ideal functionality f, and distributes the output to players. In the setting of rational cryptography, we will consider a **Mediator** that enforces the ideal game specification.

Input Distribution:	Each player $P_i \in \mathcal{P}$ receives its input x_i, random coins r_i and auxiliary input[a] z_i. Each player has the option of inputting a different input $x_i \neq x_i$, as this is unavoidable.
Game Execution:	The **Mediator** allows the subset of players $P \subseteq \mathcal{P}$ specified at each node of the game specification Γ to simultaneously play their actions. Note that games where only a single player moves at each node (asynchronous play) are fully supported, as this is modeled by setting the subset $P = \{P_i\}$.
Payoff Assignment:	If the current node k is terminal (i.e. $k \in \mathcal{Z}$), then **Mediator** distributes the payoffs associated with k to all players $P_i \in \mathcal{P}$.

---
[a] An auxiliary input is provided to all players to model additional information available to them [33].

Protocol 3.1. Ideal World Game Execution

Definition 3. *Let* Γ *represent the ideal game specification in extensive form representation,* $\mathcal{R}$ *a point-to-point communication resource available between all pairs of players in* $\mathcal{P}$, $\mathcal{S}$ *the set of local simulators,* $\boldsymbol{\mu}$ *the set of player utility functions and* z *any auxiliary information provided to a player. We denote by* $\bar{x}$ *the set of inputs for players (which may differ from the set of their true inputs* x*) and by* r *the random coins provided to a player. We then define the* i^{th} *output of* an ideal world execution *for players* $\mathcal{P}$ *in the presence of local simulators* $\mathcal{S}$ *as:*

$$\left\{\text{IDEAL}_{\Gamma,\mathcal{R},\mathcal{P},\mathcal{S},\mu,z}^{(i\in[1...n])}(\lambda,\bar{\boldsymbol{x}};r)\right\}_{\lambda\in\mathbb{N},\bar{\boldsymbol{x}},r\in\{0,1\}^*} \triangleq \{\boldsymbol{\sigma}^*,\mathcal{I}\}$$

where $\boldsymbol{\sigma}^*$ *is the equilibrium in the ideal game specification* Γ, $\mathcal{S} = \{\mathcal{S}_i\}_{i\in[1...n]}$ *is the set of simulators such that* $\mathcal{S}_i = \text{Sim}(\mathcal{A}_i)$, $\mathcal{I}$ *is the information partition set for* $\mathcal{P}$, $|\bar{x}_i| = |\bar{x}_j| \forall i \neq j$ *and* $|z| = \text{poly}(|\bar{x}_i|)$.

This ideal world model necessarily limits the class of games that may be realized, as any game specification that disallows point-to-point communication channels between all parties cannot be modeled in the presence of $\mathcal{R}$. However, we will demonstrate that a broad class of games that initially appear inadmissible under our model are realizable through minor modifications to the game specification, and which preserve the equilibria of the original game.

3.4   Real World Model

We now introduce the real world model protocol Π that implements the ideal game specification Γ. In order to translate ideal game specifications into realizable protocols, we assume the existence of a public key infrastructure (PKI) in the real world model. That is, we must translate the ideal world point-to-point communication resource $\mathcal{R}$ into an implementation allowing point-to-point private communication between all players $P_i, P_j \in \mathcal{P}$ during the execution of Π. We denote the real world PKI communication resource by $\mathcal{C}$, where $\forall(P_i, P_j)_{i\neq j} \in \mathcal{P}, \exists \mathcal{C}_{ij} \in \mathcal{C}$.

In the real world execution, each player P_i has an associated local adversary $\mathcal{A}_i$, rather than a simulator $\mathcal{S}_i$ as in the ideal world game. The local adversary $\mathcal{A}_i$ selects the actions of P_i to maximize the player's local utility function μ_i. Similarly, in the real world execution there is no Mediator, as the goal is to remove reliance on trusted third parties.

Input Distribution:	Each player $P_i \in \mathcal{P}$ receives its input x_i, random coins r_i and auxiliary input z_i. Each player has the option of inputting a different input $\bar{x}_i \neq x_i$, as this is unavoidable.
Protocol Execution:	The execution of Π proceeds in a series of rounds, where at each round a subset of players $P \subseteq \mathcal{P}$ specified at each node play their actions. Each player pair $(P_i, P_j)_{i\neq j} \in \mathcal{P}$ is connected by a private authenticated point-to-point communication channel $\mathcal{C}_{ij}$, and may exchange messages throughout the protocol execution.
Payoff Assignment:	If the current node k is terminal (i.e. $k \in \mathcal{Z}$), then each player $P_i \in \mathcal{P}$ receives its associated payoff.

Protocol 3.2. Real World Protocol Execution

Definition 4. *Let Π represent the real world protocol implementing Π, $\mathcal{C}$ a point-to-point authenticated and private PKI communication resource available between all pairs of players in $\mathcal{P}$, $\mathcal{A}$ the set of local adversaries, $\boldsymbol{\mu}$ the set of player utility functions and z any auxiliary information provided to a player. We denote by $\bar{\boldsymbol{x}}$ the set of inputs for players (which may differ from the set of their true inputs $\boldsymbol{x}$) and by r the random coins provided to a player. We then define the ith output of a* real world execution *for players $\mathcal{P}$ in the presence of local adversaries $\mathcal{A}$ as:*

$$\left\{ \mathrm{REAL}_{\Pi,\mathcal{C},\mathcal{P},\mathcal{A},\mu,z}^{(i\in[1...n])}(\lambda,\bar{\boldsymbol{x}};r) \right\}_{\lambda\in\mathbb{N},\bar{\boldsymbol{x}},r\in\{0,1\}^*} \triangleq \{\boldsymbol{\sigma}^*,\mathcal{I}\}$$

where $\boldsymbol{\sigma}^*$ is the equilibrium in the real world protocol Π, $\mathcal{I}$ is the information partition set for $\mathcal{P}$, $|\bar{x}_i| = |\bar{x}_j| \forall i \neq j$ and $|z| - \mathrm{poly}(|\bar{x}_i|)$.

3.5   Establishing the Security of Realized Protocols

The security of protocols is established by demonstrating that the real and ideal world distribution ensembles are computationally indistinguishable[9]. This guarantees that any attack available to an adversary $\mathcal{A}$ in the real model is also available to a simulator $\mathcal{S}$ in the ideal model.

Definition 5. *(Security against Rational Adversaries) Let Γ be an n-player ideal game specification and Π be an n-party real world protocol. We say that Π* securely realizes Γ *if there exists a set $\{Sim_i\}_{i\in[1...n]}$ of PPT transformations admissible in the ideal model such that for all PPT rational adversaries $\mathcal{A} = \{\mathcal{A}_i\}_{i\in[1...n]}$ admissible in the real model, for all $x \in (\{0,1\}^*)^n$ and $z \in (\{0,1\}^*)^n$, and for all $i \in [1...n]$,*

$$\left\{ \mathrm{IDEAL}_{\Gamma,\mathcal{R},\mathcal{P},\mathcal{S},\mu,z}^{(i\in[1...n])}(\lambda,\bar{\boldsymbol{x}};r) \right\}_{\lambda\in\mathbb{N},\bar{\boldsymbol{x}},r\in\{0,1\}^*} \stackrel{c}{\equiv} \left\{ \mathrm{REAL}_{\Pi,\mathcal{C},\mathcal{P},\mathcal{A},\mu,z}^{(i\in[1...n])}(\lambda,\bar{\boldsymbol{x}};r) \right\}_{\lambda\in\mathbb{N},\bar{\boldsymbol{x}},r\in\{0,1\}^*}$$

where $\mathcal{S} = \{\mathcal{S}_i\}_{i\in[1...n]}$ is the set of simulators such that $\mathcal{S}_i = Sim(\mathcal{A}_i)$, $\mathcal{I}$ is the information partition set for $\mathcal{P}$ and r is chosen uniformly at random.

Thus, to establish the security of a realized protocol Π, we must construct a simulator $\mathcal{S}_i$ for all players $P_i \in \mathcal{P}$ such that for all probabilistic polynomial-time distinguishers $\mathcal{D}$, the distributions of $\mathcal{S}$ in the ideal world and $\mathcal{A}$ in the real world can only be differentiated with probability negligibly greater than $\frac{1}{2}$.

4   Demonstrating the Model on Rational Secret Sharing

To illustrate the power of our model, we return to the example of rational secret sharing. We demonstrate that, despite the presence of point-to-point communication channels, the original game specification is admissible in our ideal world

---

[9] That is, any probabilistic polynomial-time (PPT) distinguisher $\mathcal{D}$ cannot distinguish between an execution of Γ in the ideal world model and an execution of Π in the real world model with probability non-negligibly greater than $\frac{1}{2}$.

model, and realizable in the real world model. This violates the assumptions of existing security frameworks, which disallow point-to-point communication either within the protocol execution, outside of the protocol execution, or both.

4.1 Ideal World Game Specification

The ideal world game Γ is an interaction between a set of players $\mathcal{P} = \{P_i\}_{i \in [1...n]}$, where P_i has access to a point-to-point communication resource $\mathcal{R}_{P_i,P_j} \forall j \neq i$. That is, P_i may privately communicate with any other player P_j. We now demonstrate that Γ is admissible in our ideal world definition.

Input Distribution:	Each player $P_i \in \mathcal{P}$ receives its input share x_i, random coins r_i and auxiliary input z_i. Each player has the option of inputting a different share $\bar{x}_i \neq x_i$ or aborting the protocol at any time, as this is unavoidable.
"Cheap Talk":	Player P_i is free to collaborate with all players $P_j \in \hat{\mathcal{P}}$ over $\mathcal{R}_{P_i,P_j}$, where $\hat{\mathcal{P}}$ is the set of colluding players. Proposition 1 demonstrates that communication over $\mathcal{R}$ is considered "cheap talk" (it *does not affect* the strategy selection of the player), and that the local simulator $\mathcal{S}_i$ for each player will select $a_i = \mathsf{reveal}$, as this maximizes μ_i.
Game Execution:	The $\mathtt{Mediator}$ instructs $P_i, \forall i \in n$ to play their action a_i at each round k, where $a_i \in \{\mathsf{silent}^a, \mathsf{reveal}\}$.
Payoff Assignment:	At the terminal round k^* where the shares yield the secret, $\mathtt{Mediator}$ distributes the payoffs to $P_i \in \mathcal{P}$.

---

^{a} Note that selecting $a_i = \mathsf{silent}$ is equivalent to aborting.

Protocol 4.1. Ideal World Game Γ Execution

Let Γ be the ideal game specification for rational secret sharing, with player set $\mathcal{P} = \{P_i\}_{i \in [1...n]}$ and associated set of local simulators $\mathcal{S} = \{\mathcal{S}_i\}_{i \in [1...n]}$ that select actions for players to maximize their local utility functions, resource set $\mathcal{R} = \{\mathcal{R}_{P_i,P_j}\}_{\forall i, i \neq j}$, and all players $P_i \in \mathcal{P}$ have utility functions defined as

$$\mu_i(\sigma_i) \mapsto \begin{cases} (\mu^{++})(p) : \sigma_i = \mathsf{silent}, k = k^* \\ (\mu^-)(1-p) : \sigma_i = \mathsf{silent}, k \neq k^* \\ (\mu^+) : \qquad \sigma_i = \mathsf{reveal} \end{cases} \tag{1}$$

where μ^+ represents positive utility, μ^- represents negative utility, and $\mu^{++} > \mu^+$ as players value exclusivity.

Proposition 1. *For all players $P_i \in \mathcal{P}$ in Γ with utility function defined as $\mu_i(\sigma_i)$ in Equation 1, strategy $\{\sigma_{P_i}^* = \mathsf{reveal}\}_{\forall i \in n} > \{\sigma_{P_i} = \mathsf{silent}\}_{\forall i \in n}$ when $p < \frac{\mu^+}{\mu^{++}}$.*

Proof. In the original rational secret sharing protocol, the strategy $\sigma^* = \{\sigma^*_{P_i} = \text{reveal}\}_{\forall i \in n}$ is the only Nash equilibrium, as the true final round k^* (where combining shares reveals the shared secret) is chosen from a geometric distribution. As the probability of correctly guessing the final round k^* is the parameter p, the expected utility for $\sigma_{P_i} = \text{silent}$ is at most $(\mu^{++})(p)$. We set $\mu^{++} > \mu^+$, as players are assumed to value exclusivity (recovering the secret while preventing other players from doing so). If a player remains silent in any round $k < k^*$, they are caught by the other players as a cheater and excluded from future rounds (receiving negative utility μ^-). By choosing p such that $p < \frac{\mu^+}{\mu^{++}}$, we have $(\mu^{++})(p) < \mu^+$ which implies $\mu_{P_i}(\text{silent}) < \mu_{P_i}(\text{reveal})$. Thus revealing the share for each round strictly dominates remaining silent. Players in our ideal model Γ may communicate over $\mathcal{R}$ and attempt to convince other players that they will select silent. This provides a greater degree of exclusivity, as only those colluding players in $\hat{\mathcal{P}} \subseteq \mathcal{P}$ will recover the secret. However, this communication is considered cheap talk, as each player maximizes μ_i by selecting $\sigma_i = \text{silent}$ regardless of the messages sent over $\mathcal{R}$ when $p < \frac{\mu^+}{\mu^{++}}$.

4.2   Real World Protocol Construction

We now translate the ideal game specification Γ to a real world protocol Π, and demonstrate that there exist simulators such that the distribution of the ideal world game is computationally indistinguishable from the distribution of the real world protocol execution.

Input Distribution:	Each player $P_i \in \mathcal{P}$ receives its input share x_i, random coins r_i and auxiliary input z_i. Each player has the option of inputting a different share $\bar{x}_i \neq x_i$ or aborting the protocol at any time, as this is unavoidable.
"Cheap Talk":	Player P_i is free to collaborate with all players $P_j \in \hat{\mathcal{P}}$ over $\mathcal{C}_{P_i, P_j}$, where $\hat{\mathcal{P}}$ is the set of colluding players. Proposition 1 demonstrates that communication over $\mathcal{C}$ is considered "cheap talk" (it *does not affect* the strategy selection of the player), and that the local adversary $\mathcal{A}_i$ for each player selects $a_i = \text{reveal}$, as this maximizes μ_i.
Game Execution:	Each player $P_i \in \mathcal{P}$ selects and plays their action a_i at each round k, where $a_i \in \{\text{silent}^a, \text{reveal}\}$.
Payoff Assignment:	At the terminal round k^* where the shares yield the secret, each player $P_i \in \mathcal{P}$ receives its associated payoff.

---

[a] Note that selecting $a_i = \text{silent}$ is equivalent to aborting.

Protocol 4.2. Real World Protocol Π Execution

In the real world model, the communication resource $\mathcal{R}$ is replaced with a public key infrastructure $\mathcal{C}$. Each pair of players $(P_i, P_j) \in \mathcal{P}$ has access to a private and authenticated point-to-point communication channel C_{ij}. Let Π be a real world protocol, with player set $\mathcal{P} = \{P_i\}_{i \in [1...n]}$ and associated set of local adversaries $\mathcal{A} = \{\mathcal{A}_i\}_{i \in [1...n]}$ that select actions for players to maximize their local utility functions, communication channel set $\mathcal{C} = \{C_{ij}\}_{\forall i \neq j}$, and all players have identical utility functions defined as in Equation 1.

Clearly Π is admissible under the real world model, as the PKI infrastructure $\mathcal{C}$ facilitates the point-to-point communication channels between all players. The real world protocol Π for rational secret sharing proceeds as in Protocol 4.2. Again, the original equilibrium of $\sigma^* = \{\sigma_{P_i} = \mathsf{reveal}\}$ is preserved despite the presence of the communication channel $\mathcal{C}$.

4.3   Demonstrating Protocol Π Security

We use the simulation paradigm [33] to demonstrate the security of the construction by proving the distribution of the real world protocol is computationally indistinguishable from the ideal world distribution.

Theorem 1. *(Security of Π against Rational Adversaries) Let Γ be the n-party ideal world game specification of Protocol 4.1 and let Π be the n-party real world execution of Protocol 4.2. There exists a set $\{Sim_i\}_{i \in [1...n]}$ of PPT transformations admissible in the ideal model such that for all PPT rational adversaries $\mathcal{A} = \{\mathcal{A}_i\}_{i \in [1...n]}$ admissible in the real model, for all $x \in (\{0,1\}^*)^n$ and $z \in (\{0,1\}^*)^n$, and for all $i \in [1...n]$,*

$$\left\{ \mathrm{IDEAL}_{\Gamma,\mathcal{R},\mathcal{P},\mathcal{S},\mu,z}^{(i \in [1...n])}(\lambda, \bar{x}; r) \right\}_{\lambda \in \mathbb{N}, \bar{x}, r \in \{0,1\}^*} \stackrel{c}{\equiv} \left\{ \mathrm{REAL}_{\Pi,\mathcal{C},\mathcal{P},\mathcal{A},\mu,z}^{(i \in [1...n])}(\lambda, \bar{x}; r) \right\}_{\lambda \in \mathbb{N}, \bar{x}, r \in \{0,1\}^*}$$

establishing that Π securely realizes Γ.

Proof. To prove the security of Π against rational adversaries $\mathcal{A} = \{\mathcal{A}_i\}_{i \in [1...n]}$ we must construct a set of simulators $\mathcal{S} = \{\mathcal{S}_i\}_{i \in [1...n]}$ whose output in the ideal game specification Γ is indistinguishable from the output of $\mathcal{A}$ in the real world execution.

To achieve this, we construct simulators $\mathcal{S}_i = \mathsf{Sim}(\mathcal{A}_i)$ that simulate all messages and the output of $\mathcal{A}_i$ in the real world execution of Π, and is thus able to return these as its own. The simulated messages and output returned by $\mathcal{S}_i$ must be computationally indistinguishable such that, for all probabilistic polynomial-time distinguishers $\mathcal{D}$, the probability of differentiating the ideal world and real world distributions is at most negligibly greater than $\frac{1}{2}$.

Each simulator $\mathcal{S}_i$ will rely on the private communication resource $\mathcal{R}$ to simulate the messages exchanged and final output produced by $\mathcal{A}_i$ acting to maximize the utility function μ_i for player P_i. The simulator $\mathcal{S}_i$ given in Construction 4.1 holds for all players $\mathcal{P} = \{P_i\}_{i \in [1...n]}$.

The construction relies on the computational indistinguishability of the real world communication channel $\mathcal{C}$ from the ideal world private and authenticated communication resource $\mathcal{R}$. All messages sent by simulators $\mathcal{S}_i \in \mathcal{S}$ in the ideal world model are passed over $\mathcal{R}$. In the real world execution, messages are encrypted between players using the PKI communication resource $\mathcal{C}$. Thus, all probabilistic polynomial-time distinguishers $\mathcal{D}$ are able to distinguish the view of the ideal world execution from the real world execution with at most probability negligibly greater than $\frac{1}{2}$ by the security of the PKI communication resource $\mathcal{C}$.

Input Distribution:	The simulator $\mathcal{S}_i \in \mathcal{S}$ is given input share x_i, random coins r_i and auxiliary input z_i
"Cheap Talk":	The simulator $\mathcal{S}_i$ is free to communicate over $\mathcal{R}_{\mathcal{S}_i,\mathcal{S}_j}$ where $i \neq j$. $\mathcal{S}_i, \forall i \neq j$ must simulate the "cheap talk" between the other player's adversary $\mathcal{A}_j$. $\mathcal{S}_i$ uses its random coins r_i to construct a random message m, and sends m over resource $\mathcal{R}_{\mathcal{S}_i,\mathcal{S}_j}$. By definition, $\mathcal{R}$ is a private and authenticated point-to-point communication resource. Thus, the messages sent by the simulator are computationally indistinguishable from those sent in the real world execution, which are encrypted under the public key infrastructure communication resource $\mathcal{C}$. The local simulator $\mathcal{S}_i$ for each player selects $m_i = \mathsf{reveal}$, as this maximizes μ_i regardless of the messages exchanged during this phase.
Game Execution:	The simulator $\mathcal{S}_i$ sends a message m to $\mathcal{S}_j, \forall j \neq i$ over $\mathcal{R}_{\mathcal{S}_i,\mathcal{S}_j}$ with their decision, where $m \in \{\mathsf{silent}, \mathsf{reveal}\}$. By definition, $\mathcal{R}$ is a private and authenticated point-to-point communication resource. Thus, the messages sent by the simulator to $\mathcal{S}_j$ are computationally indistinguishable from those sent in the real world execution, which are encrypted under the public key infrastructure communication resource $\mathcal{C}$.
Payoff Assignment:	After $\mathsf{P}_j \in \mathcal{P}, \forall j \neq i$ has received $m_{\mathcal{S}_i}$, each simulator receives the payoff associated with the outcome.

Construction 4.1. Construction of Simulator $\mathcal{S}_i$

5 Conclusion

In this work, we have proposed a security definition capturing rational cryptographic protocols in the presence of standard point-to-point communication resources. Rather than limit the communication resources available to players, we answer the question of how game specifications admissible in an ideal model allowing point-to-point communication channels may be realized in practice. Thus, the ideal world model necessarily limits the class of games that are admissible and is not a general result. However, we have argued that point-to-point communication channels are unavoidable in real-world settings, and consequently must be incorporated into the definition of security. Further, we have demonstrated that not all game specifications forbidding point-to-point communication are inadmissible under our model. We presented the transformation for the classic prisoner's dilemma, which disallows point-to-point communication through physical assumptions, into a modified game that is admissible under our model and preserves the original equilibrium. Similarly, we have demonstrated that the signaling game has an expected payoff of 1 when executed in the presence of point-to-point channels, rather than an expected payoff of $\frac{1}{2}$: a distinction not captured by models that disallow communication outside of the protocol execution. Finally, we have presented a full security proof for rational secret sharing under our proposed framework. Although our results are not universal, we have demonstrated a powerful benefit of our model: assigning local adversaries may aid mechanism design in destabilizing the formation of coalitions. Thus, there are tangible benefits from adopting our definition of security against local rational adversaries in the presence of point-to-point communication resources.

References

1. Goldreich, O., Micali, S., Wigderson, A.: How to play any mental game. In: STOC 1987: Proceedings of the Nineteenth Annual ACM Symposium on Theory of Computing, pp. 218–229. ACM, New York (1987)
2. Izmalkov, S., Micali, S., Lepinski, M.: Rational secure computation and ideal mechanism design. In: 46th Annual IEEE Symposium on Foundations of Computer Science, FOCS 2005, pp. 585–594 (2005)
3. Alwen, J., Shelat, A., Visconti, I.: Collusion-free protocols in the mediated model. In: Wagner, D. (ed.) CRYPTO 2008. LNCS, vol. 5157, pp. 497–514. Springer, Heidelberg (2008)
4. Alwen, J., Katz, J., Lindell, Y., Persiano, G., Shelat, A., Visconti, I.: Collusion-free multiparty computation in the mediated model. In: Halevi, S. (ed.) CRYPTO 2009. LNCS, vol. 5677, pp. 524–540. Springer, Heidelberg (2009)
5. Lepinksi, M., Micali, S., Shelat, A.: Collusion-free protocols. In: Proceedings of the Thirty-Seventh Annual ACM Symposium on Theory of Computing, STOC 2005, pp. 543–552. ACM, New York (2005)
6. Alwen, J., Katz, J., Maurer, U., Zikas, V.: Collusion-preserving computation. In: Safavi-Naini, R., Canetti, R. (eds.) CRYPTO 2012. LNCS, vol. 7417, pp. 124–143. Springer, Heidelberg (2012)
7. Kamara, S., Mohassel, P., Raykova, M.: Outsourcing multi-party computation. Cryptology ePrint Archive, Report 2011/272 (2011), http://eprint.iacr.org/
8. Fuchsbauer, G., Katz, J., Naccache, D.: Efficient rational secret sharing in standard communication networks. In: Micciancio, D. (ed.) TCC 2010. LNCS, vol. 5978, pp. 419–436. Springer, Heidelberg (2010)
9. Halpern, J., Teague, V.: Rational secret sharing and multiparty computation: Extended abstract. In: Proceedings of the Thirty-Sixth Annual ACM Symposium on Theory of Computing, STOC 2004, pp. 623–632. ACM, New York (2004)
10. Lysyanskaya, A., Triandopoulos, N.: Rationality and adversarial behavior in multiparty computation. In: Dwork, C. (ed.) CRYPTO 2006. LNCS, vol. 4117, pp. 180–197. Springer, Heidelberg (2006)

11. Canetti, R., Vald, M.: Universally composable security with local adversaries. In: Visconti, I., De Prisco, R. (eds.) SCN 2012. LNCS, vol. 7485, pp. 281–301. Springer, Heidelberg (2012)
12. Nielsen, J.B., Alwen, J., Cachin, C., Nielsen, J.B., Pereira, O.: Summary report on rational cryptographic protocols (2007)
13. Poundstone, W.: Prisoner's Dilemma: John Von Neumann, Game Theory and the Puzzle of the Bomb, 1st edn. Doubleday, New York (1992)
14. Wallrabenstein, J.R., Clifton, C.: Privacy preserving tatonnement: A cryptographic construction of an incentive compatible market. In: Financial Cryptography and Data Security. LNCS. Springer, Heidelberg (2014)
15. Gordon, S.D., Katz, J.: Rational secret sharing, revisited. Cryptology ePrint Archive, Report 2006/142 (2006), http://eprint.iacr.org/
16. Micali, S., Shelat, A.: Purely rational secret sharing (extended abstract). In: Reingold, O. (ed.) TCC 2009. LNCS, vol. 5444, pp. 54–71. Springer, Heidelberg (2009)
17. Groce, A., Katz, J., Thiruvengadam, A., Zikas, V.: Byzantine agreement with a rational adversary. In: Czumaj, A., Mehlhorn, K., Pitts, A., Wattenhofer, R. (eds.) ICALP 2012, Part II. LNCS, vol. 7392, pp. 561–572. Springer, Heidelberg (2012)
18. Groce, A., Katz, J.: Fair computation with rational players. In: Pointcheval, D., Johansson, T. (eds.) EUROCRYPT 2012. LNCS, vol. 7237, pp. 81–98. Springer, Heidelberg (2012)
19. Zhang, Z., Liu, M.: Rational secret sharing as extensive games. Science China Information Sciences 56, 1–13 (2013)
20. Shamir, A.: How to share a secret. Commun. ACM 22(11), 612–613 (1979)
21. Maleka, S., Shareef, A., Rangan, C.P.: Rational secret sharing with repeated games. In: Chen, L., Mu, Y., Susilo, W. (eds.) ISPEC 2008. LNCS, vol. 4991, pp. 334–346. Springer, Heidelberg (2008)
22. Nojoumian, M., Stinson, D.R.: Socio-rational secret sharing as a new direction in rational cryptography. In: Grossklags, J., Walrand, J. (eds.) GameSec 2012. LNCS, vol. 7638, pp. 18–37. Springer, Heidelberg (2012)
23. Kol, G., Naor, M.: Games for exchanging information. In: Proceedings of the 40th Annual ACM Symposium on Theory of Computing, STOC 2008, pp. 423–432. ACM, New York (2008)
24. Kol, G., Naor, M.: Cryptography and game theory: designing protocols for exchanging information. In: Canetti, R. (ed.) TCC 2008. LNCS, vol. 4948, pp. 320–339. Springer, Heidelberg (2008)
25. Zhang, Z., Liu, M.: Unconditionally secure rational secret sharing in standard communication networks. In: Rhee, K.-H., Nyang, D. (eds.) ICISC 2010. LNCS, vol. 6829, pp. 355–369. Springer, Heidelberg (2011)
26. Ostrovsky, R., Yung, M.: How to withstand mobile virus attacks (extended abstract). In: Proceedings of the Tenth Annual ACM Symposium on Principles of Distributed Computing, PODC 1991, pp. 51–59. ACM, New York (1991)
27. Katz, J.: Bridging game theory and cryptography: Recent results and future directions. In: Canetti, R. (ed.) TCC 2008. LNCS, vol. 4948, pp. 251–272. Springer, Heidelberg (2008)
28. Osborne, M.J., Rubinstein, A.: A Course in Game Theory. MIT Press Books, vol. 1. MIT Press (1994)
29. Fudenberg, D., Tirole, J.: Game Theory. MIT Press (August 991)

30. Asharov, G., Canetti, R., Hazay, C.: Towards a game theoretic view of secure computation. In: Paterson, K.G. (ed.) EUROCRYPT 2011. LNCS, vol. 6632, pp. 426–445. Springer, Heidelberg (2011)
31. Dodis, Y., Halevi, S., Rabin, T.: A cryptographic solution to a game theoretic problem. In: Bellare, M. (ed.) CRYPTO 2000. LNCS, vol. 1880, pp. 112–130. Springer, Heidelberg (2000)
32. Miltersen, P.B., Nielsen, J.B., Triandopoulos, N.: Privacy-enhancing auctions using rational cryptography. In: Halevi, S. (ed.) CRYPTO 2009. LNCS, vol. 5677, pp. 541–558. Springer, Heidelberg (2009)
33. Goldreich, O.: Foundations of Cryptography, vol. 2. Cambridge University Press (2004)

Limiting Adversarial Budget
in Quantitative Security Assessment

Aleksandr Lenin[1,2] and Ahto Buldas[1,2,3,*]

[1] Cybernetica AS, Mäealuse 2/1, Tallinn, Estonia
[2] Guardtime AS, Tammsaare tee 60, Tallinn, Estonia
[3] Tallinn University of Technology, Ehitajate tee 5, Tallinn, Estonia

Abstract. We present the results of research of limiting adversarial budget in attack games, and, in particular, in the failure-free attack tree models presented by Buldas-Stepanenko in 2012 and improved in 2013 by Buldas and Lenin. In the previously presented models attacker's budget was assumed to be unlimited. It is natural to assume that the adversarial budget is limited and such an assumption would allow us to model the adversarial decision making more close to the one that might happen in real life. We analyze three atomic cases – the single atomic case, the atomic AND, and the atomic OR. Even these elementary cases become quite complex, at the same time, limiting adversarial budget does not seem to provide any better or more precise results compared to the failure-free models. For the limited model analysis results to be reliable, it is required that the adversarial reward is estimated with high precision, probably not achievable by providing expert estimations for the quantitative annotations on the attack steps, such as the cost or the success probability. It is doubtful that it is reasonable to face this complexity, as the failure-free model provides reliable upper bounds, being at the same time computationally less complex.

1 Introduction

The failure-free models [2,3] provide reliable utility upper bounds, however this results in systems that might be over-secured. It has not been studied how much extra cost the upper-bound oriented methods cause. We present the intermediate results of researching the model assuming that the adversarial budget is limited and compare the results of analysis using adaptive strategies with limited budget to the analysis results of the failure-free model, in which the adversary is not limited in any way. The adversarial limitation is the only limitation applied to

* The research leading to these results has received funding from the European Regional Development Fund through Centre of Excellence in Computer Science (EXCS), the Estonian Research Council under Institutional Research Grant IUT27-1, and the the European Union Seventh Framework Programme (FP7/2007-2013) under grant agreement ICT-318003 (TREsPASS). This publication reflects only the authors' views and the Union is not liable for any use that may be made of the information contained herein.

R. Poovendran and W. Saad (Eds.): GameSec 2014, LNCS 8840, pp. 155–174, 2014.
© Springer International Publishing Switzerland 2014

the adversary, all other assumptions and concepts are identical to the failure-free model.

The assumption that the adversarial budget is limited is natural, as this is what happens in reality. Limited budget models the adversarial strategic decision making in a better way, which is more close to the one likely to be observed in real life and the research on the adaptive strategies with limited budget is an important research area in quantitative security analysis based on attack trees.

We analyze three cases: the atomic attack case, the atomic AND, and the atomic OR analyzing the effect of limiting adversarial budget in fully-adaptive strategies [2,3]. We show that the atomic attack case and the atomic AND case do not provide whatsoever better or more reliable results, compared to the existing failure-free models. The atomic AND case might provide more precise result, but in this case analysts must estimate the adversarial reward with the required precision, which in real-life scenarios might be less than €1. If they fail to do that, the results of such an analysis are unreliable. In practice, it is doubtful that analysts would be able to come up with such precise estimations. Even if such precise estimations existed, the model would not provide reliable results, as there is still margin for human mistake and in case analysts might overlook the estimations provided to such parameters as cost of the attack step, or the adversarial reward, the results of the analysis would not be reliable. On the contrary, the existing failure-free models with unlimited adversarial budget provide reliable utility upper bounds, despite the fact that this may result in over-secured systems.

It seems that limited budget makes the model much more complex compared to the unlimited budget approach. For example, optimal strategies that were shown to be non-adaptive in the failure-free models [2,3] can be adaptive and more complex to analyze in the limited budget model. The best move to undertake in certain states of the game changes bouncing between the attack steps.

Even the elementary cases studied in this paper become quite complex considering limited budget assumption compared to the corresponding cases in the failure-free models [2,3]. It is doubtful that the more general case will have a graceful easy solution to derive optimal strategies. Considering the requirement to be able to estimate the adversarial reward very precisely it is doubtful that it is reasonable to face the complexity of the calculations on the limited adaptive strategies.

The outline of the paper is the following: Section 1 provides a high-level overview of the problem and briefly outlines the results obtained so far. Section 2 describes the work related to the presented approach, Section 3 provides definitions of terms used throughout the paper. Section 4 describes the effect of limited budget assumption on the fully-adaptive strategies and the strategic decision making undertaken by the adversary. Finally, Section 5 summarizes the obtained results, outlines questions still left open, and describes interesting problems for future research.

2   Related Work

In this section, we outline the work that has lead to and influenced the development of the presented model.

2.1   Schneier Attack Trees' Concept

The idea of analyzing security using the so-called attack trees was popularized by Schneier in [7]. The author suggested to use attack trees as a convenient hierarchical representation of an attack scenario. The analysis implied that the analysts had to estimate one single parameter they would like to reason about, for each of the leaves in the attack tree. Then the bottom-up parameter propagation approach was applied to propagate the results of calculations towards the root node of the tree, the result of the root node was considered the result of such an analysis. The suggested bottom-up parameter propagation method allowed to reason about such parameters like minimal/average/maximal cost of the attack scenario, likelihood of its success, etc. The analysis relied on an assumption that the analyzed parameters are mutually independent, which allowed to analyze them independently of each other and to derive some meaningful conclusions about the security of the systems based on the obtained results.

2.2   Buldas-Priisalu Model

The model of Buldas et al. [1] is remarkable for introducing the multi-parameter approach to the quantitative security risk analysis. The model is based on the assumption of a rational adversary who is always trying to maximize his average outcome. The authors state that in order to assess security it is sufficient to assess adversarial utility. If the utility is negative or zero, the system is reasonably secure, as attacking it is not profitable. If the utility is positive, the adversary has an incentive to attack and attacking is profitable for him. The adversary undertakes strategic decision-making in accordance with the rationality assumption – the adversary will start attacking iff it is profitable. Additionally, authors state that malicious actions are, as a rule, related to criminal behavior and for this reason they applied economic reasoning in their model which considers the risk of detection and potential penalties of the adversary. Their model introduced a novel way to think about security and gave start to multi-parameter quantitative security analysis. Jürgenson et al. have shown that Buldas et al. model is inconsistent with Mauw-Oostijk foundations [6] and introduced the so-called parallel model [4] and the serial model [5] which provided more reliable results, however in both models the adversary did not behave in a fully adaptive way.

2.3   Buldas-Stepanenko Fully Adaptive Model

In the Buldas-Stepanenko fully adaptive model [3] the adversaries behave in a fully adaptive way launching atomic attack steps in an arbitrary order, depending on the results of the previous trials. However, the model had force-failure states, when the adversary could not continue playing and thus adversarial fully adaptive behavior was limited. In their model optimal strategies are non-adaptive and in some cases, like atomic OR or atomic AND, may be easily derived by calculating certain invariants. In their failure-free model the adversary was expected to launch attack steps until success, thus the failure-free model is similar to the fully adaptive model with the difference that in the failure-free model success probabilities of the attack steps are equal to 1. The most significant contribution of the paper [3] is the upper bounds ideology by which the models should estimate adversarial utility from above, trying to avoid false-positive security results.

2.4   Improved Failure-Free Model

The improved failure-free model [2] improves the Buldas-Stepanenko failure-free model [3] by eliminating the force-failure states. In the improved model the adversarial behavior more fully conforms to the upper bounds ideology introduced in [3] – the adversary may repeat failed attack steps and play on when caught. It turned out that the elimination of the force failure states has made the model computationally easier. The authors show that in the new model optimal strategies always exist. Optimal strategies are single-branched BDD-s where the order of attack steps is irrelevant. Additionally, authors show that finding an optimal strategy in the new model is NP-complete. Two computational methods were introduced – the one allowing to compute the precise adversarial utility value, and the one which allowed to derive the approximated estimation of adversarial utility upper bound.

3   Definitions

Definition 1 (Derived function). *If $F(x_1, \ldots, x_m)$ is a Boolean function and $v \in \{0,1\}$, then by the derived Boolean function $F|_{x_j=v}$ we mean the function $F(x_1, \ldots, x_{j-1}, v, x_{j+1}, \ldots, x_m)$ derived from F by the assignment $x_j := v$.*

Definition 2 (Constant functions). *By $\mathbf{1}$ we mean a Boolean function that is identically true and by $\mathbf{0}$ we mean a Boolean function that is identically false.*

Definition 3 (Satisfiability game). *By a satisfiability game we mean a single-player game in which the player's goal is to satisfy a monotone Boolean function $F(x_1, x_2, \ldots, x_k)$ by picking variables x_i one at a time and assigning $x_i = 1$. Each time the player picks the variable x_i he pays some amount of expenses $\mathcal{E}_i \in \mathfrak{R}$, sometimes also modelled as a random variable. With a certain probability p_i the move x_i succeeds. Function F representing the current game instance is transformed to its derived form $F|_{x_i=1}$ and the next game iteration starts. The game ends when the condition $F \equiv \mathbf{1}$ is satisfied and the player wins the prize $\mathcal{P} \in \mathfrak{R}$, or when the player stops playing. With probability $1 - p_i$ the move x_i fails. The player may end up in a different game instance represented by the derived Boolean function $F|_{x_i \equiv 0}$ in the case of a game without move repetitions,* *and may end up in the very same instance of the game $\mathcal{F}$ in the case of a game with repetitions. Under certain conditions with a certain probability the game may end up in a forced failure state, i.e. if the player is caught and this implies that he cannot continue playing, i.e. according to the Buldas-Stepanenko model [3]. The rules of the game are model-specific and may vary from model to model. Thus we can define three common types of games:*

1. *SAT Game Without Repetitions - the type of a game where an adversary can perform a move only once.*
2. *SAT Game With Repetitions - the type of a game where an adversary can re-run failed moves again an arbitrary number of times.*
3. *Failure-Free SAT Game - the type of a game in which all success probabilities are equal to 1. It is shown in [2] that any game with repetitions is equivalent to a failure-free game (Thm. 5).*

Definition 4 (Satisfiability game with limited budget). *By a satisfiability game with limited budget we mean the SAT game with move repetitions in which the current state of the game is described by the Boolean function $\mathcal{F}(x_1, \ldots, x_k)$ and the budget λ – $\langle \mathcal{F}, \lambda \rangle$. Every move x_i made by the player changes the state of the game. If x_i succeeded, the game moves into the state $\langle \mathcal{F}|_{x_i=1}, \lambda - \mathcal{C}_i \rangle$ and if x_i has failed, the new state of the game is $\langle \mathcal{F}|_{x_i=0}, \lambda - \mathcal{C}_i \rangle$, where $\mathcal{C}_i$ is the cost of x_i. The game ends if the player has satisfied the Boolean function $\mathcal{F} \equiv 1$ and reached the state $\langle 1, \lambda \rangle$ thus winning the game, or when the player has reached the state $\langle \mathcal{F}, \lambda \rangle$ in the case of which the expenses of every possible move $\mathcal{E}_i > \lambda$ and $\mathcal{F}$ has not been satisfied, meaning the loss of the game.*

Definition 5 (Line of a game). *By a line of a satisfiability game we mean a sequence of assignments $\gamma = \langle x_{j_1} = v_1, \ldots, x_{j_k} = v_k \rangle$ (where $v_j \in \{0,1\}$) that represent the player's moves, and possibly some auxiliary information. We say that γ is a **winning line** if the Boolean formula $x_{i_1} \wedge \ldots \wedge x_{i_k} \Rightarrow \mathcal{F}(x_1, \ldots, x_n)$ is a tautology, where $\mathcal{F}$ is a Boolean function of the satisfiability game.*

Definition 6 (Strategy). *By a strategy $\mathcal{S}$ for a game $\mathcal{G}$ we mean a rule that for any line γ of $\mathcal{G}$ either suggests the next move $x_{j_{k+1}}$ or decides to give up.*

Strategies can be represented graphically as binary decision diagrams (BDDs).

Definition 7 (Line of a strategy). *A line of a strategy $\mathcal{S}$ for a game $\mathcal{G}$ is the smallest set $\mathcal{L}$ of lines of $\mathcal{G}$ such that (1) $\langle \rangle \in \mathcal{L}$ and (2) if $\gamma \in \mathcal{L}$, and $\mathcal{S}$ suggests x_j as the next move to try, then $\langle \gamma, x_j = 0 \rangle \in \mathcal{L}$ and $\langle \gamma, x_j = 1 \rangle \in \mathcal{L}$.*

Definition 8 (Branch). *A branch β of a strategy $\mathcal{S}$ for a game $\mathcal{G}$ is a line γ of $\mathcal{S}$ for which $\mathcal{S}$ does not suggest the next move. By $\mathcal{B}_{\mathcal{S}}$ we denote the set of all branches of $\mathcal{S}$.*

For example, all winning lines of $\mathcal{S}$ are branches.

Definition 9 (Expenses of a branch). *If $\beta = \langle x_{i_1=v_1}, \ldots, x_{i_k=v_k} \rangle$ is a branch of a strategy $\mathcal{S}$ for $\mathcal{G}$, then by expenses $\epsilon_{\mathcal{G}}(\mathcal{S}, \beta)$ of β we mean the sum $\overline{\mathcal{E}}_{i_1} + \ldots + \overline{\mathcal{E}}_{i_k}$ where by $\overline{\mathcal{E}}_{i_j}$ we mean the mathematical expectation of $\mathcal{E}_{i_j}$.*

Definition 10 (Prize of a branch). *The prize $\mathcal{P}_{\mathcal{G}}(\mathcal{S}, \beta)$ of a branch β of a strategy $\mathcal{S}$ is $\mathcal{P}$ if β is a winning branch, and 0 otherwise.*

Definition 11 (Utility of a strategy). *By the utility of a strategy $\mathcal{S}$ in a game $\mathcal{G}$ we mean the sum:* $\mathcal{U}(\mathcal{G}, \mathcal{S}) = \sum_{\beta \in \mathcal{B}_{\mathcal{S}}} Pr(\beta) \cdot [\mathcal{P}_{\mathcal{G}}(\mathcal{S}, \beta) - \epsilon_{\mathcal{G}}(\mathcal{S}, \beta)]$. *For the empty strategy* $\mathcal{U}(\mathcal{G}, \emptyset) = 0$.

Definition 12 (Prize and Expenses of a strategy). *By the expenses $\mathcal{E}(\mathcal{G}, \mathcal{S})$ of a strategy $\mathcal{S}$ we mean the sum* $\sum_{\beta \in \mathcal{B}_{\mathcal{S}}} Pr(\beta) \cdot \epsilon_{\mathcal{G}}(\mathcal{S}, \beta)$. *The prize $\mathcal{P}(\mathcal{G}, \mathcal{S})$ of $\mathcal{S}$ is* $\sum_{\beta \in \mathcal{B}_{\mathcal{S}}} Pr(\beta) \cdot \mathcal{P}_{\mathcal{G}}(\mathcal{S}, \beta)$.

It is easy to see that $\mathcal{U}(\mathcal{G}, \mathcal{S}) = \mathcal{P}(\mathcal{G}, \mathcal{S}) - \mathcal{E}(\mathcal{G}, \mathcal{S})$.

Definition 13 (Utility of a satisfiability game). *The utility of a SAT game $\mathcal{G}$ is the limit* $\mathcal{U}(\mathcal{G}) = \sup_{\mathcal{S}} \mathcal{U}(\mathcal{G}, \mathcal{S})$ *that exists due to the bound* $\mathcal{U}(\mathcal{G}, \mathcal{S}) \leqslant \mathcal{P}$.

Definition 14 (Optimal strategy). *By an optimal strategy for a game $\mathcal{G}$ we mean a strategy $\mathcal{S}$ for which* $\mathcal{U}(\mathcal{G}) = \mathcal{U}(\mathcal{G}, \mathcal{S})$.

It has been shown that for satisfiability games optimal strategies always exist [2].

4   Limiting Adversarial Budget in the Improved Failure-Free Model

In this paper we focus on the fully adaptive adversarial strategies assuming that the adversarial budget is limited. Budget limitation is the only limitation used, compared to the improved failure-free model [2]. Adversaries still behave in a fully adaptive way and are allowed to launch failed attack steps again in any order, until the budget gets so small that no attack steps can be launched. When the budget decreases by a considerable amount, monetary limitation starts effecting possible strategic choices of the attacker – possible set of choices reduces (the adversary may launch only some subset of the attack steps) and eventually, this subset becomes an empty set. It turns out that the optimal strategy depends on the amount of the monetary resource available to the adversary.

In the improved failure-free model the state of the game is represented by the Boolean function $\mathcal{F}$. If the attack step has failed, the adversary finds himself in the very same state of the game $\mathcal{F}$. Due to this non-adaptive strategies always exist in the set of optimal strategies of the game.

This is not always the case when we consider budget limitations – in general, optimal strategies are adaptive, except for some certain sets of parameters in case of which optimal strategies are non-adaptive. When we consider budget limitations the state of the game is represented by the Boolean function $\mathcal{F}$ and the budget λ. We denote the utility in a certain game state $\langle \mathcal{F}, \lambda \rangle$ with $\mathcal{U}^{\lambda}(\mathcal{F})$. When an attack step fails, the adversary finds himself in an another state of the game represented by $\mathcal{U}^{\lambda-C}(\mathcal{F})$, where C is the cost of the failed attack step. The relation between the utility upper bound $\mathcal{U}^{\infty}(\mathcal{F})$ in [2] and the utility $\mathcal{U}^{\lambda}(\mathcal{F})$ given budget λ is the following:

$$\mathcal{U}^{\infty}(\mathcal{F}) = \lim_{\lambda \to \infty} \mathcal{U}^{\lambda}(\mathcal{F}) \ .$$

In some certain cases optimal strategies are non-adaptive, but in general they are not. This makes computations reasonably complex. When the adversarial budget increases, his utility increases as well and approaches the adversarial utility upper bound in the improved failure-free model [2]. It turns out that in the case of a reasonably big budget the complexity added by the budget limitation does not add any value nor give any additional benefits, as the difference between the utility in the model with budget limitations and the utility upper bound becomes negligible.

In this paper we focus on the three elementary games – the single attack case, the atomic AND and the atomic OR game and show the effect of budget limitations in these games. Even these elementary cases become quite complex when taking budget limitations into account. It becomes doubtful if the practical application of the model with budget limitations is efficient and reliable. Using complex computational procedures we face the risk to make the model inapplicable for the practical cases, while the negligible deviation between the results of the model with budget limitations and the one without them in case of a reasonably big budget (which is the expected case in real-life scenarios) and much less complex and more efficient computations induces us to give preference to the model without budget limitations, despite the fact that it overestimates adversarial power and capabilities for the cases when the adversarial budget is reasonably small.

4.1 Single Atomic Attack Case

In case the adversary may choose from a single available choice, he will continue launching the attack step until it succeeds, or as long as the budget allows it. Such a strategy may be represented in the form of a single-branched BDD as in Fig. 1:

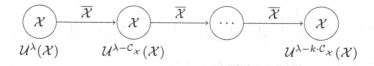

Fig. 1. An adaptive strategy suggesting to iterate attack step $\mathcal{X}$ until it succeeds or as long as the adversarial budget allows to launch the attack step

In accordance with the strategy, the adversary launches an attack step $\mathcal{X}$ with cost C and success probability p. If it succeeds, the adversary has accomplished the attack and has won the game. If $\mathcal{X}$ fails, the adversary finds himself in another state of the game $\langle \mathcal{X}, \lambda - \mathcal{C} \rangle$. Thus, adversarial utility may be expressed in the form of the relation (1):

$$\mathcal{U}^\lambda(\mathcal{X}) = \max \left\{ 0,\ \mathcal{U}(\mathcal{X}) + (1-p) \cdot \mathcal{U}^{\lambda-\mathcal{C}}(\mathcal{X}) \right\} . \tag{1}$$

It can be seen (see Fig. 2) that the adversarial utility changes in the points where the budget is multiples of the cost of the attack step. In case the adversarial budget is less than the cost of the attack step, the adversary cannot launch a single attack step and thus his utility is 0. The optimal strategy in this case is an empty strategy – the attacker will be better off not trying to attack. In case the budget exceeds the cost of the attack step, the utility grows with each subsequent trial to launch an attack step, as every subsequent trial increases the likelihood of success that the attack step will succeed. Thus, adversarial utility asymptotically approaches the utility upper bound in the model without budget limitations.

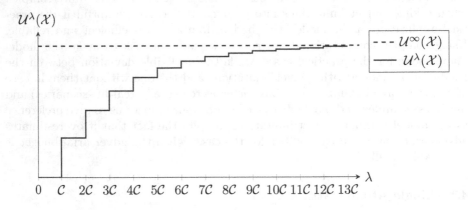

Fig. 2. Single atomic attack case

The utility value that the adversary may achieve, given budget λ, may be expressed in the form of equation (2):

$$\mathcal{U}^\lambda(\mathcal{X}) = \left[\mathcal{P} - \frac{\mathcal{C}}{p} \right] \cdot \left[1 - (1-p)^{\lfloor \frac{\lambda}{\mathcal{C}} \rfloor} \right] = \mathcal{U}^\infty(\mathcal{X}) \left[1 - (1-p)^{\lfloor \frac{\lambda}{\mathcal{C}} \rfloor} \right] , \tag{2}$$

where $\mathcal{U}^\infty(\mathcal{X})$ is the utility upper bound [2].

Comparison with the Improved Failure-Free Model

We will investigate the case when the improved failure-free model analysis result states that the system is *insecure*, while the budgeted model result states that the system is *secure*.

According to the improved failure-free model the adversarial utility $\mathcal{U}^\infty(\mathcal{X}) = \mathcal{P} - \frac{C}{p}$. The system is secure in case $\mathcal{P} \leqslant \frac{C}{p}$ and insecure in case $\mathcal{P} > \frac{C}{p}$.

$$\begin{cases} \mathcal{U}^\infty(\mathcal{X}) = \mathcal{P} - \frac{C}{p} > 0 \\ \mathcal{U}^\lambda(\mathcal{X}) = \left[\mathcal{P} - \frac{C}{p}\right]\left[1 - (1-p)^{\lfloor\frac{\lambda}{C}\rfloor}\right] \leqslant 0 \end{cases} \quad (3)$$

It can be seen that the condition (3) can be reached only when the adversary has no resources to attack ($\lambda < C$). Thus limiting adversarial budget does not provide more trustworthy nor more reliable results compared to the improved failure-free model in case of single atomic attack games. If in the case of some positive budget λ the adversarial utility is positive, it will be less or equal to zero in the model with budget limitations only if $\lambda < C$. In other words, if the system is insecure in the improved failure-free model, it will also be insecure in the model with budget limitations for any adversarial budget, sufficient to launch the attack step at least once.

4.2   Two Attack Steps

In the case of atomic games of 2 possible attack steps $\mathcal{X}_i$ and $\mathcal{X}_j$ and corresponding costs $\mathcal{C}_{x_i}$ and $\mathcal{C}_{x_j}$, the adversarial utility changes in the so-called *lattice points* which are the projections of points $(n\mathcal{C}_{x_i}, m\mathcal{C}_{x_j})$ in two-dimensional Euclidean space into one-dimensional space using the formula $\mathcal{L}_i = n\mathcal{C}_{x_i} + m\mathcal{C}_{x_j}$, where

$n \in \left\{1, 2, \ldots, \left\lfloor\frac{\lambda}{\mathcal{C}_{x_i}}\right\rfloor\right\}$, $m \in \left\{1, 2, \ldots, \left\lfloor\frac{\lambda}{\mathcal{C}_{x_j}}\right\rfloor\right\}$, $\forall i : \mathcal{L}_i \leqslant \lambda$ (see Fig. 3). In the case of three attack steps the utility changes in the projections of points in three-dimensional space into one-dimensional space. Thus with the increase in the amount of possible attack steps the lattice argument space becomes more complex.

It can be shown that the distance between the two adjacent lattice points has a lower bound.

Theorem 1. *If the relation of attack step costs may be expressed in terms of a rational fraction (a fraction of two rational numbers, corresponding cost values may be irrational)* $\frac{\mathcal{C}_{x_i}}{\mathcal{C}_{x_j}} = \frac{p}{q}$, *then the distance between two adjacent lattice points* $\mathcal{L}_i$ *and* $\mathcal{L}_{i+1}$ *will be not less than* $\frac{\mathcal{C}_{x_j}}{q}$.

Proof. The distance δ between the two adjacent lattice points $\mathcal{L}_i$ and $\mathcal{L}_{i+1}$ may be expressed as

$$\delta = |(n-n')\mathcal{C}_{x_i} + (m-m')\mathcal{C}_{x_j}| = |\underbrace{(n-n')p + (m-m')q}_{\alpha \in \mathbb{Z}}| \cdot \frac{\mathcal{C}_{x_j}}{q}$$

$$= \begin{cases} 0 & \text{, if } \alpha = 0, \\ \geqslant \frac{\mathcal{C}_{x_j}}{q} & \text{, if } \alpha \neq 0. \end{cases}$$

$\square$

If the ratio of the attack step costs is irrational, lattice points appear with increasing frequency eventually positioning infinitely close to each other. In real life we can expect the costs to be rational (it would be difficult to estimate an irrational value for the cost parameter) and for this reason the above mentioned bound exists in the practical cases.

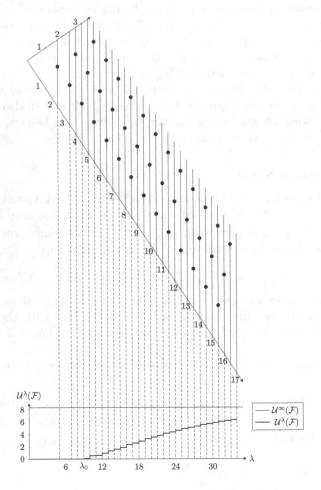

Fig. 3. Projections of the lattice points in the two-dimensional space into the one-dimensional space

Atomic OR Case

In the case of an atomic OR game in order to win it is sufficient that any of the two attack steps, $\mathcal{X} = \{\mathcal{X}_i, \mathcal{X}_j\}$ succeeds. The initial state of the game is $\langle \mathcal{X}_i \vee \mathcal{X}_j, \lambda \rangle$ and the subset of available attack steps to launch is $\{\mathcal{X}_i, \mathcal{X}_j\}$. In each state of the game the player may choose to launch any attack step from the subset of available attack steps, or to discontinue playing. The attacker launches an attack step $\mathcal{X}_k$ from this set. If $\mathcal{X}_k$ succeeded the game moves into the state $\langle 1, \lambda - C_k \rangle$, where C_k is the cost of the launched attack step, and the player has won the game. If the attack has failed, the game moves into the state $\langle \mathcal{X}_i \vee \mathcal{X}_j, \lambda - C_k \rangle$ and the game goes on while $\mathcal{E}_k \leqslant \lambda$. At some point the current λ will reduce the set of available attacks to one (cheapest) attack, and eventually, the set of possible attacks becomes an empty set. Upon reaching the state in which $\mathcal{E}_k > \lambda$ and the Boolean function of the game has not been satisfied – the player has lost the game.

Adversarial utility may be expressed in the form of the relation (4):

$$\mathcal{U}^\lambda(\mathcal{X}_i \vee \mathcal{X}_j) = \max \begin{cases} 0 \, , \\ \mathcal{U}(\mathcal{X}_i) + (1 - p_{x_i}) \, \mathcal{U}^{\lambda - C_{x_i}}(\mathcal{X}_i \vee \mathcal{X}_j) \, , \\ \mathcal{U}(\mathcal{X}_j) + (1 - p_{x_j}) \, \mathcal{U}^{\lambda - C_{x_j}}(\mathcal{X}_i \vee \mathcal{X}_j) \, . \end{cases} \qquad (4)$$

In certain cases under certain conditions the optimal strategy in the atomic OR case is non-adaptive and suggests to repeat one of the attacks independently of the current state of the game. We will bring an example of such a case.

Theorem 2. *If the costs of the attacks are equal, the attack having greater success probability will be best to try in every state of the game.*

Proof. Assume that $C_{x_i} = C_{x_j} = C$. The utility of the game may be expressed in the form of

$$\mathcal{U}^\lambda(\mathcal{X}_i \vee \mathcal{X}_j) = \max \begin{cases} 0 \, , \\ \mathcal{U}(\mathcal{X}_i) + (1 - p_{x_i}) \cdot \mathcal{U}^{\lambda - C}(\mathcal{X}_i \vee \mathcal{X}_j) \, , \\ \mathcal{U}(\mathcal{X}_j) + (1 - p_{x_j}) \cdot \mathcal{U}^{\lambda - C}(\mathcal{X}_i \vee \mathcal{X}_j) \, . \end{cases}$$

Optimal strategy will suggest to try attack $\mathcal{X}_i$ if

$$\mathcal{U}(\mathcal{X}_i) + (1 - p_{x_i}) \cdot \mathcal{U}^{\lambda - C}(\mathcal{X}_i \vee \mathcal{X}_j) > \mathcal{U}(\mathcal{X}_j) + (1 - p_{x_j}) \cdot \mathcal{U}^{\lambda - C}(\mathcal{X}_i \vee \mathcal{X}_j) \qquad (5)$$

Solving inequality (5) we reach condition $p_{x_i} > p_{x_j}$. $\qquad\square$

Algorithm 4.1 outlines the recursive procedure to calculate maximal adversarial utility in the atomic OR game given budget λ according to (4).

We show how the best move changes in the atomic OR game, depending on the current budget λ demonstrating it by several examples:

The first example (Fig. 4) shows that the best move bounces between the two attack steps when the budget is rather small, and sticks to one attack step later on. By $\emptyset$ we mean that the best move is not to start attacking at all.

The second example (Fig. 5) demonstrates the case when both of the attack steps are equally good when the budget is rather small and thus there is no difference for the attacker whether to launch attack step $\mathcal{X}_i$ or to launch attack step $\mathcal{X}_j$. But when the budget increases, the adversary has a clear preference for

---

Algorithm 4.1. Maximal utility of the atomic OR case with the given budget

---

Input: Attack step $\mathcal{X}_i$ cost i_cost
Input: Attack step $\mathcal{X}_i$ probability i_pr
Input: Attack step $\mathcal{X}_j$ cost j_cost
Input: Attack step $\mathcal{X}_j$ probability j_pr
Input: Prize of the game *prize*
Input: Budget *budget*
Output: Maximal adversarial utility value (a real number)

```
1   Procedure AtomicOr (i_cost, i_pr, j_cost, j_pr, prize, budget)
2   if budget is less than i_cost and j_cost then
3   |   return (0)
4   i_utility := -i_cost + i_pr · prize
5   j_utility := -j_cost + j_pr · prize
6   if budget is greater than i_cost then
7   |   u_i = i_utility + (1-i_pr) · AtomicOr (i_cost, i_pr, j_cost, J_pr, prize,
        budget-i_cost)
8   |   if u_i is negative then
9   |   |   u_i := 0
10  if budget is greater than j_cost then
11  |   u_j = j_utility + (1-j_pr) · AtomicOr (i_cost, i_pr, j_cost, j_pr, prize,
        budget-j_cost)
12  |   if u_j is negative then
13  |   |   u_j := 0
14  if u_i is not less than u_j then
15  |   maximal_utility := u_i
16  else
17  |   maximal_utility := u_j
18  return (maximal_utility)
```

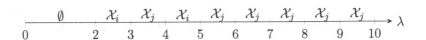

Fig. 4. Atomic OR case with parameters $\mathcal{C}_{\mathcal{X}_i} = 2, p_{\mathcal{X}_i} = 0.3, \mathcal{C}_{\mathcal{X}_j} = 3, p_{\mathcal{X}_j} = 0.48, Prize = 30$

one attack over the other one. By $=$ we mean that launching attack step $\mathcal{X}_i$ is as good as launching attack step $\mathcal{X}_j$.

The third example (Fig. 6) demonstrates the case when the costs of the attacks are irrational, but their relation may be expressed in terms of a fraction of rational numbers. It can be seen that the best move to undertake in a certain state of the game between attack steps $\mathcal{X}_i$ and $\mathcal{X}_j$.

The next example (Fig. 7) demonstrates that there are cases where the optimal strategy is non-adaptive and iterates one single attack step $\mathcal{X}_j$.

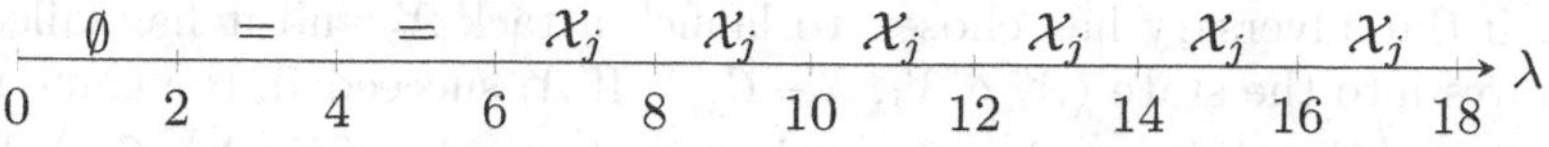

Fig. 5. Atomic OR case with parameters $C_{X_i} = 2, p_{X_i} = 0.05, C_{X_j} = 6, p_{X_j} = 0.9, Prize = 30$

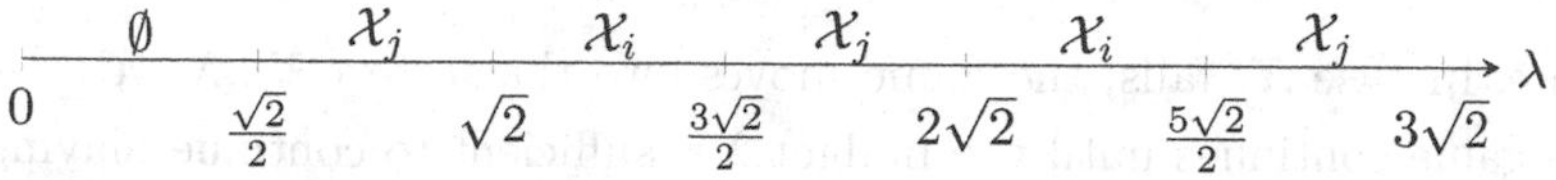

Fig. 6. Atomic OR case with parameters $C_{X_i} = \sqrt{2}, p_{X_i} = 0.8, C_{X_j} = \frac{\sqrt{2}}{2}, p_{X_j} = 0.45, Prize = 30$

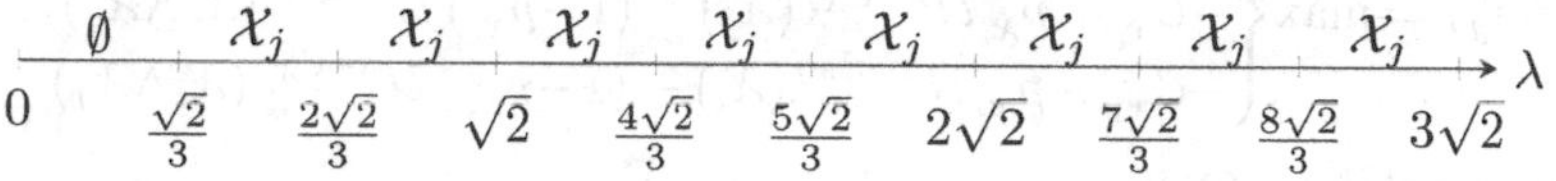

Fig. 7. Atomic OR case with parameters $C_{X_i} = \sqrt{2}, p_{X_i} = 0.1, C_{X_j} = \frac{\sqrt{2}}{3}, p_{X_j} = 0.38, Prize = 30$

Comparison with the Improved Failure-Free Model

We will show that the case when the improved failure-free model analysis result states that the system is *insecure*, while the budgeted model result states that the system is *secure* is impossible. Lets consider adversarial budget $\mathcal{I}$ for which the following inequalities hold:

$$\mathcal{U}^{\mathcal{I}}(X_i \vee X_j) > 0 ,\tag{6}$$
$$\mathcal{U}^{\mathcal{I}-\mathcal{C}}(X_i \vee X_j) \leqslant 0 ,\tag{7}$$

where $\mathcal{C}$ is the cost of any of the atomic attacks. Assuming $\mathcal{I}$ is greater than the costs of attacks X_i and X_j:

$$\mathcal{U}^{\mathcal{C}}(X_i \vee X_j) \leqslant 0 .\tag{8}$$

Let X_k with cost $\mathcal{C}$ and probability p be the optimal move in the considered state of the game. In this case:

$$\mathcal{U}^{\mathcal{I}}(X_i \vee X_j) = \mathcal{U}^{\mathcal{C}}(X_i \vee X_j) + (1 - p) \cdot \mathcal{U}^{\mathcal{I}-\mathcal{C}}(X_i \vee X_j) .\tag{9}$$

As $\mathcal{U}^{\mathcal{I}-\mathcal{C}}(X_i \vee X_j) \leqslant 0$ by (7) and $\mathcal{U}^{\mathcal{C}}(X_i \vee X_j) \leqslant 0$ by (8), it contradicts with the initial assumption $\mathcal{U}^{\mathcal{I}}(X_i \vee X_j) > 0$. Thus it seems that there is no point in limiting adversarial budget in the elementary OR case.

Atomic AND Case

In the case of atomic AND game in the initial state of the game the adversary has to choose either to launch the attack step X_i, or to launch X_j or not to start playing. If the adversary has chosen to launch attack $\mathcal{X}_i$ and it has failed, the game moves into the state $\langle \mathcal{X}_i \wedge \mathcal{X}_j, \lambda - C_{\mathcal{X}_i} \rangle$. If $\mathcal{X}_i$ succeeded, the game moves into the state $\langle \mathcal{X}_i \wedge \mathcal{X}_j |_{\mathcal{X}_i=1}, \lambda - C_{\mathcal{X}_i} \rangle$ which is identical to $\langle \mathcal{X}_j, \lambda - C_{\mathcal{X}_i} \rangle$. In this case, the attacker has the following choices: either to launch the remaining attack $\mathcal{X}_j$ (if λ is sufficient for it), or to discontinue playing the game. If $\mathcal{X}_j$ succeeds, the game moves into the state $\langle 1, \lambda - C_{\mathcal{X}_i} - C_{\mathcal{X}_j} \rangle$ and the adversary has won the game. In case $\mathcal{X}_j$ fails, the game moves into the state $\langle \mathcal{X}_j, \lambda - C_{\mathcal{X}_i} - C_{\mathcal{X}_j} \rangle$ and the game continues until the budget λ is sufficient to continue playing. Adversarial utility may be expressed in the form of the relation (10).

$$\mathcal{U}^\lambda(\mathcal{X}_i \wedge \mathcal{X}_j) = \max \begin{cases} 0 \\ -C_{\mathcal{X}_i} + p_{\mathcal{X}_i} \mathcal{U}^{\lambda - C_{\mathcal{X}_i}}(\mathcal{X}_j) + (1 - p_{\mathcal{X}_i}) \mathcal{U}^{\lambda - C_{\mathcal{X}_i}}(\mathcal{X}_i \wedge \mathcal{X}_j) \\ -C_{\mathcal{X}_j} + p_{\mathcal{X}_j} \mathcal{U}^{\lambda - C_{\mathcal{X}_j}}(\mathcal{X}_i) + (1 - p_{\mathcal{X}_j}) \mathcal{U}^{\lambda - C_{\mathcal{X}_j}}(\mathcal{X}_i \wedge \mathcal{X}_j) \end{cases} \quad (10)$$

where (according to (2)):

$$\mathcal{U}^{\lambda - C_{\mathcal{X}_i}}(\mathcal{X}_j) = \mathcal{U}^\infty(\mathcal{X}_j) \left[1 - (1 - p_{\mathcal{X}_j})^{\left\lfloor \frac{\lambda - C_{\mathcal{X}_i}}{C_{\mathcal{X}_j}} \right\rfloor} \right] ,$$

$$\mathcal{U}^{\lambda - C_{\mathcal{X}_j}}(\mathcal{X}_i) = \mathcal{U}^\infty(\mathcal{X}_i) \left[1 - (1 - p_{\mathcal{X}_i})^{\left\lfloor \frac{\lambda - C_{\mathcal{X}_j}}{C_{\mathcal{X}_i}} \right\rfloor} \right] .$$

In the atomic AND game the positive utility may not be achieved immediately by the adversary. We call the minimal value of the adversarial budget, sufficient to achieve positive utility the adversarial utility *budget lower bound*, which can be computed as:

$$\lambda_0 = \min \begin{cases} 0 , \\ \left[\log_{(1 - p_{\mathcal{X}_j})} \left[1 - \frac{C_{\mathcal{X}_i}}{p_{\mathcal{X}_i} \mathcal{U}^\infty(\mathcal{X}_j)} \right] \right] \cdot C_{\mathcal{X}_j} + C_{\mathcal{X}_i} , \\ \left[\log_{(1 - p_{\mathcal{X}_i})} \left[1 - \frac{C_{\mathcal{X}_j}}{p_{\mathcal{X}_j} \mathcal{U}^\infty(\mathcal{X}_i)} \right] \right] \cdot C_{\mathcal{X}_i} + C_{\mathcal{X}_j} . \end{cases} \quad (11)$$

Algorithm 4.2 outlines the recursive procedure to calculate maximal adversarial utility in the atomic AND game given budget λ according to (10).

We show how the best move changes in the atomic AND game, depending on the current budget λ demonstrating it by several examples

The first example (Fig. 8) shows that there are certain sets of parameters which make the adversary indifferent in whether to launch attack step $\mathcal{X}_i$ or attack step $\mathcal{X}_j$ in every state of the game.

The second example (Fig. 9) demonstrates the case when the best move bounces between attack step $\mathcal{X}_i$ and attack step $\mathcal{X}_j$. In some states of the game both of the attack steps are equally optimal to launch.

Algorithm 4.2. Maximal utility of the atomic AND case with the given budget

---

Input: Attack step $\mathcal{X}_i$ cost i_cost
Input: Attack step $\mathcal{X}_i$ probability i_pr
Input: Attack step $\mathcal{X}_j$ cost j_cost
Input: Attack step $\mathcal{X}_j$ probability j_pr
Input: Prize of the game $prize$
Input: Budget $budget$
Output: Maximal adversarial utility value (a real number)

1   Procedure **AtomicAnd** (i_cost, i_pr, j_cost, j_pr, prize, budget)
2   **if** *budget is less than the sum of i_cost and j_cost* **then**
3 | **return** (0)
4   i_inf := prize $- \frac{i\_cost}{i\_pr}$
5   j_inf := prize $- \frac{j\_cost}{j\_pr}$
6   i_rep := i_inf $\cdot \left[1 - (1 - j\_pr)^{\left\lfloor \frac{budget - i\_cost}{j\_cost} \right\rfloor}\right]$
7   j_rep := j_inf $\cdot \left[1 - (1 - i\_pr)^{\left\lfloor \frac{budget - j\_cost}{i\_cost} \right\rfloor}\right]$
8   ui = -i_cost + i_pr $\cdot$ j_rep + (1-i_pr) $\cdot$ **AtomicAnd** (i_cost, i_pr, j_cost, j_pr, prize, budget-i_cost)
9   **if** *ui is negative* **then**
10 | ui := 0
11 uj = -j_cost + j_pr $\cdot$ i_rep + (1-j_pr) $\cdot$ **AtomicAnd** (i_cost, i_pr, j_cost, j_pr, prize, budget-j_cost)
12 **if** *uj is negative* **then**
13 | uj := 0
14 **if** *ui is not less than uj* **then**
15 | maximal_utility := ui
16 **else**
17 | maximal_utility := uj
18 **return** (maximal_utility)

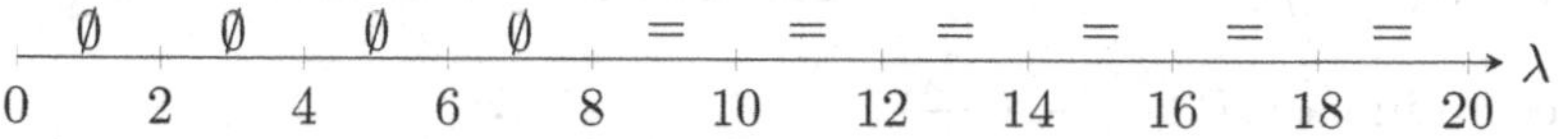

Fig. 8. Atomic AND case with parameters $\mathcal{C}_{\mathcal{X}_i} = 2, p_{\mathcal{X}_i} = 0.05, \mathcal{C}_{\mathcal{X}_j} = 6, p_{\mathcal{X}_j} = 0.9, Prize = 30$

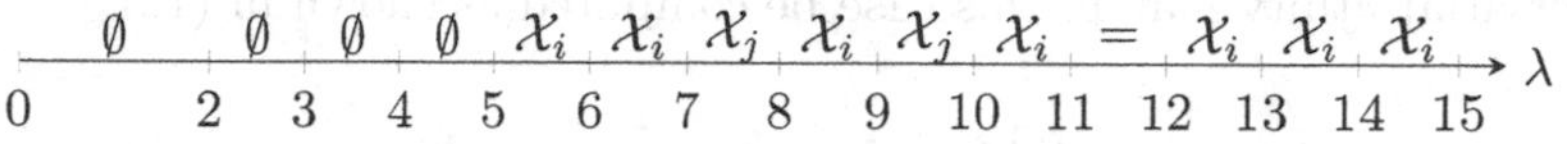

Fig. 9. Atomic AND case with parameters $\mathcal{C}_{\mathcal{X}_i} = 2, p_{\mathcal{X}_i} = 0.3, \mathcal{C}_{\mathcal{X}_j} = 3, p_{\mathcal{X}_j} = 0.48, Prize = 30$

The third example (Fig. 10) demonstrates the case when the costs of the attacks are irrational, but their relation may be expressed in terms of a fraction of rational numbers. It can be seen that with the given parameters optimal strategy will suggest to iterate attack step $\mathcal{X}_j$ and thus the optimal strategy is non-adaptive.

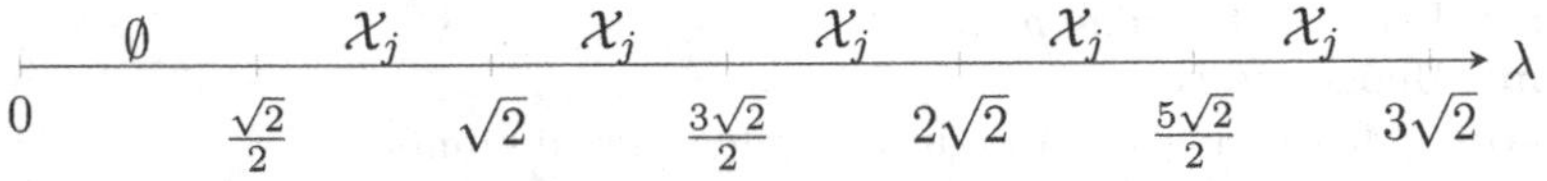

$$\begin{array}{c|ccccccc} & \mathcal{X}_j & \mathcal{X}_j & \mathcal{X}_j & \mathcal{X}_j & \mathcal{X}_j & \\ \hline 0 & \frac{\sqrt{2}}{2} & \sqrt{2} & \frac{3\sqrt{2}}{2} & 2\sqrt{2} & \frac{5\sqrt{2}}{2} & 3\sqrt{2} \end{array} \longrightarrow \lambda$$

Fig. 10. Atomic AND case with parameters $C_{\mathcal{X}_i} = \sqrt{2}, p_{\mathcal{X}_i} = 0.8, C_{\mathcal{X}_j} = \frac{\sqrt{2}}{2}, p_{\mathcal{X}_j} = 0.45, Prize = 30$.

The next example (Fig. 11) demonstrates the case when the optimal strategy is adaptive and the best move to undertake in a certain state of the game alternates between attack step $\mathcal{X}_i$ and attack step $\mathcal{X}_j$.

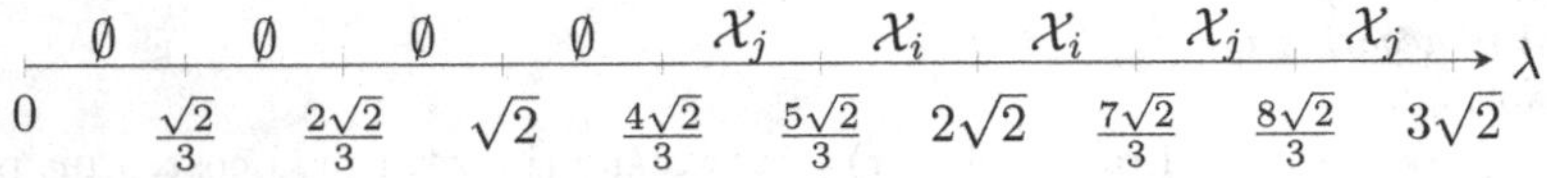

$$\begin{array}{c|ccccccccc} & \emptyset & \emptyset & \emptyset & \emptyset & \mathcal{X}_j & \mathcal{X}_i & \mathcal{X}_i & \mathcal{X}_j & \mathcal{X}_j \\ \hline 0 & \frac{\sqrt{2}}{3} & \frac{2\sqrt{2}}{3} & \sqrt{2} & \frac{4\sqrt{2}}{3} & \frac{5\sqrt{2}}{3} & 2\sqrt{2} & \frac{7\sqrt{2}}{3} & \frac{8\sqrt{2}}{3} & 3\sqrt{2} \end{array} \longrightarrow \lambda$$

Fig. 11. Atomic AND case with parameters $C_{\mathcal{X}_i} = \sqrt{2}, p_{\mathcal{X}_i} = 0.1, C_{\mathcal{X}_j} = \frac{\sqrt{2}}{3}, p_{\mathcal{X}_j} = 0.38, Prize = 30$

Comparison with the Improved Failure-Free Model

We will investigate the case when the improved failure-free model analysis result states that the system is *insecure*, while the budgeted model result states that the system is *secure*. According to the improved failure-free model the adversarial utility $\mathcal{U}^\infty (\mathcal{X}_i \wedge \mathcal{X}_j) = \mathcal{P} - \frac{C_{\mathcal{X}_i}}{p_{\mathcal{X}_i}} - \frac{C_{\mathcal{X}_j}}{p_{\mathcal{X}_j}}$. The system is secure in case $\mathcal{P} \leqslant \frac{C_{\mathcal{X}_i}}{p_{\mathcal{X}_i}} + \frac{C_{\mathcal{X}_j}}{p_{\mathcal{X}_j}}$ and insecure in case $\mathcal{P} > \frac{C_{\mathcal{X}_i}}{p_{\mathcal{X}_i}} + \frac{C_{\mathcal{X}_j}}{p_{\mathcal{X}_j}}$.

Let the adversarial budget λ suffice to launch m attack steps in total and the adversarial strategy may be the one as shown in Fig. 12 and for the sake of simplicity lets assume that $C_{\mathcal{X}_i} = C_{\mathcal{X}_j} = C$ and $p_{\mathcal{X}_i} = p_{\mathcal{X}_j} = p$.

Adversarial utility may in this case be computed as shown in (12).

$$\mathcal{U}^{m \times C}(\mathcal{X}_i \wedge \mathcal{X}_j) = \left[\mathcal{U}^\infty(\mathcal{X}_j) - \frac{C}{p} \right] \left[1 - (1-p)^{m-1} \right] - (m-1)(1-p)^{m-1} \left[p\, \mathcal{U}^\infty(\mathcal{X}_j) \right] \tag{12}$$

$$= \left[\mathcal{P} - \frac{2C}{p} \right] \left[1 - (1-p)^{m-1} \right] - (m-1)(1-p)^{m-1} \left[p\, \mathcal{P} - C \right]$$

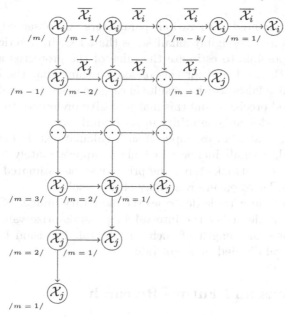

Fig. 12. An adaptive strategy consisting of two attack steps $\mathcal{X}_i$ and $\mathcal{X}_j$, with adversarial budget λ

According to the budgeted model the strategy is not profitable for an attacker, while the improved failure-free model states that the strategy is profitable if:

$$\frac{2C}{p} < P \leqslant \frac{2C}{p} \cdot \frac{1 - [1 + C(m-1)](1-p)^{m-1}}{1 - [1 + p(m-1)](1-p)^{m-1}} \,. \tag{13}$$

Inequality (13) shows the interval for the value of prize within which the result of the limited budget model and result of the improved failure-free models differ. We will show what happens to the results of the analysis of both models in the broader view.

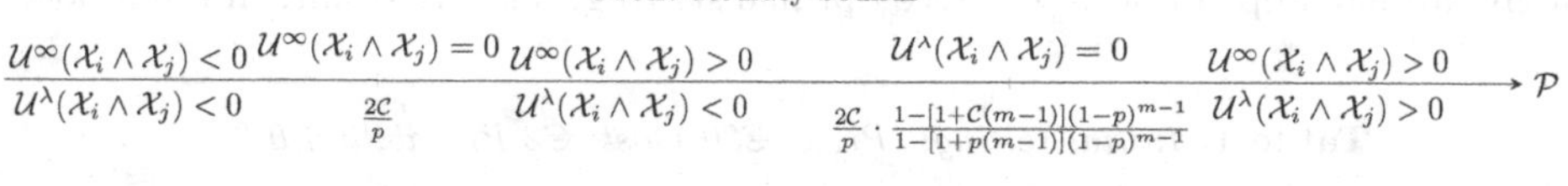

Profit accuracy bounds

$$\begin{array}{ccccc}
\dfrac{\mathcal{U}^\infty(\mathcal{X}_i \wedge \mathcal{X}_j) < 0}{\mathcal{U}^\lambda(\mathcal{X}_i \wedge \mathcal{X}_j) < 0} & \dfrac{\mathcal{U}^\infty(\mathcal{X}_i \wedge \mathcal{X}_j) = 0}{\frac{2C}{p}} & \dfrac{\mathcal{U}^\infty(\mathcal{X}_i \wedge \mathcal{X}_j) > 0}{\mathcal{U}^\lambda(\mathcal{X}_i \wedge \mathcal{X}_j) < 0} & \dfrac{\mathcal{U}^\lambda(\mathcal{X}_i \wedge \mathcal{X}_j) = 0}{\frac{2C}{p} \cdot \frac{1-[1+C(m-1)](1-p)^{m-1}}{1-[1+p(m-1)](1-p)^{m-1}}} & \dfrac{\mathcal{U}^\infty(\mathcal{X}_i \wedge \mathcal{X}_j) > 0}{\mathcal{U}^\lambda(\mathcal{X}_i \wedge \mathcal{X}_j) > 0} \\
\end{array} \xrightarrow{\quad} p$$

Fig. 13. Comparison of the improved failure-free model to the limited budget model

Thus, Fig. 13 shows that if prize is less than $\frac{2C}{p}$ then the system is *secure* according to both models. If prize is greater than $\frac{2C}{p} \cdot \frac{1-[1+C(m-1)](1-p)^{m-1}}{1-[1+p(m-1)](1-p)^{m-1}}$ then the system is *insecure* according to both models. Only when the prize is between $\frac{2C}{p}$ and $\frac{2C}{p} \cdot \frac{1-[1+C(m-1)](1-p)^{m-1}}{1-[1+p(m-1)](1-p)^{m-1}}$ the limited budget model may produce result different from the result of the improved failure-free model.

We have experimented with various parameters and observed that the prize interval (13) becomes negligibly small – less than 1 €. In practice, as a rule, it is practically impossible to estimate the value of the protected assets with the precision of less than €1 and for this reason we think that the limited budget model may produce false-positive results in case analysts are unable to estimate prize with required precision and this makes us give preference to the failure-free models which provides reliable utility upper bounds.

Table 1 demonstrates an example of such calculations. It can be seen that already with rather small increase in budget (approximately 3 times greater than the costs of the attack steps) the prize must be estimated with precision less than €1 in order to ensure reliability of the results.

The first column in a table describes the monetary budget of the adversary. The second column describes the interval for possible prize values, the column named span shows the length of such an interval. Precision is the length of uncertainty interval divided by mean value.

5   Conclusions and Future Research

We have analyzed the 3 kinds of elementary games – the single attack game, the atomic OR and the atomic AND, assuming that the adversarial budget is limited. In the result of limiting adversarial budget the model and computations become reasonably complex that makes it doubtful that this approach is applicable for real-life case analysis. Additionally, in case of atomic AND we have to be able to estimate the *prize* parameter quite precisely – if we fail to do that, the analysis results will be unreliable. In practice it is very hard to estimate the cost of an asset or information with the desired precision and thus is it doubtful if it is reasonable to face the complexities of budget limitations and its false positive results which might happen in the case of AND type games.

The improved failure-free model is, on the contrary, less complex and provides reliable upper bounds. Due to the fact that when the move fails the player finds himself in the very same instance of the game results in the existence of non-adaptive strategies in the set of optimal strategies of the game and the ordering of the attack steps in non-adaptive optimal strategies is irrelevant. In the model

Table 1. *Initial setting: Prize: €30 Cost: €2 Probability: 0.3*

Lambda (#)	$\mathcal{P}$ Domain (€)	Span (€)	Deviation (€)	Precision (%)
2	(13.(3), 28.(8)]	15.(5)	±7.(7)	0.518519
3	(13.(3), 22.4074]	9.07407	±4.537035	0.302469
4	(13.(3), 19.242]	5.9087	±2.95435	0.196957
5	(13.(3), 17.4047]	4.07139	±2.035695	0.135713
6	(13.(3), 16.232]	2.89863	±1.449315	0.0966211
7	(13.(3), 15.4386]	2.10531	±1.052655	0.0701772
8	(13.(3), 14.8816]	1.54822	±0.77411	0.0516073
9	(13.(3), 14.4806]	1.14723	±0.573615	0.038241

with budget limitations the subset of non-adaptive strategies exists in the set of all strategies. Non-adaptive strategies are relatively easy to derive and compute. One of the open questions is to figure out how well the most optimal strategy from the subset of non-adaptive strategies $\mathcal{U}_{na}^{\lambda}(\mathcal{G})$ might approximate the optimal strategy from the set of all possible strategies $\mathcal{U}^{\lambda}(\mathcal{G})$. If $\mathcal{U}_{na}^{\lambda}(\mathcal{G})$ provides pretty good approximation to $\mathcal{U}^{\lambda}(\mathcal{G})$, then there exists infinitely small α such that:

$$\mathcal{U}_{na}^{\lambda}(\mathcal{G}) \leqslant \mathcal{U}^{\lambda}(\mathcal{G}) \leqslant \alpha \cdot \mathcal{U}_{na}^{\lambda}(\mathcal{G}) \leqslant \mathcal{U}^{\infty}(\mathcal{G}) \ .$$

If this holds, it might enable calculation of acceptably precise result without facing the complexity and the computational overhead introduced by the precise utility calculation routines.

Secondly, it would be interesting to see when the optimal move in certain states of the game changes by bouncing between the two possible moves thus following some pattern. Additionally, to verify the hypothesis that this might happen in the theoretical case when the ratio of the costs of the move is irrational.

The bigger the adversarial budget λ is, the more adversarial utility approaches the utility upper bound in the improved failure-free model. Optimal strategies in the improved failure-free model are non-adaptive and do not depend on the ordering of the attack steps. In the case of big λ optimal strategies are likely to behave non-adaptively as well in the limited budget model. This means that optimal move in certain states of the game is likely to bounce changing from one attack to another, but with increase in λ the optimal move remains the same. It also means that the utility of various strategies, beginning with different moves, become closer to each other with the increase in λ and there should exist infinitely small δ such that

$$|\mathcal{U}^{\lambda}(\mathcal{S}_i) - \mathcal{U}^{\lambda}(\mathcal{S}_j)| \leqslant \delta \ ,$$

where $\mathcal{S}_i$ and $\mathcal{S}_j$ are the two strategies from the set of all strategies of the game.

The improved failure-free model provides reliable utility upper bounds, however this results in systems that might be over-secured. It has not been studied how much extra cost the upper-bound oriented methods cause. The assumption that the adversarial budget is limited is natural, as this is what happens in reality. Models assuming limited budget model the adversarial strategic decision making in a better way, which is more close to the one likely to be observed in real life and the research on the adaptive strategies with limited budget is an important research area in quantitative security analysis based on attack trees.

References

1. Buldas, A., Laud, P., Priisalu, J., Saarepera, M., Willemson, J.: Rational choice of security measures via multi-parameter attack trees. In: López, J. (ed.) CRITIS 2006. LNCS, vol. 4347, pp. 235–248. Springer, Heidelberg (2006)
2. Buldas, A., Lenin, A.: New efficient utility upper bounds for the fully adaptive model of attack trees. In: Das, S.K., Nita-Rotaru, C., Kantarcioglu, M. (eds.) GameSec 2013. LNCS, vol. 8252, pp. 192–205. Springer, Heidelberg (2013)

3. Buldas, A., Stepanenko, R.: Upper bounds for adversaries' utility in attack trees. In: Grossklags, J., Walrand, J. (eds.) GameSec 2012. LNCS, vol. 7638, pp. 98–117. Springer, Heidelberg (2012)
4. Jürgenson, A., Willemson, J.: Computing exact outcomes of multi-parameter attack trees. In: Meersman, R., Tari, Z. (eds.) OTM 2008, Part II. LNCS, vol. 5332, pp. 1036–1051. Springer, Heidelberg (2008)
5. Jürgenson, A., Willemson, J.: On fast and approximate attack tree computations. In: Kwak, J., Deng, R.H., Won, Y., Wang, G. (eds.) ISPEC 2010. LNCS, vol. 6047, pp. 56–66. Springer, Heidelberg (2010)
6. Mauw, S., Oostdijk, M.: Foundations of attack trees. In: Won, D., Kim, S. (eds.) ICISC 2005. LNCS, vol. 3935, pp. 186–198. Springer, Heidelberg (2006)
7. Schneier, B.: Attack trees. Dr. Dobb's Journal of Software Tools 24(12), 21–22, 24, 26, 28–29 (1999)

FlipThem: Modeling Targeted Attacks with FlipIt for Multiple Resources

Aron Laszka[1], Gabor Horvath[2], Mark Felegyhazi[2], and Levente Buttyán[2]

[1] Institute for Software Integrated Systems (ISIS)
Vanderbilt University, Nashville, USA
[2] Department of Networked Systems and Services (HIT)
Budapest University of Technology and Economics (BME), Budapest, Hungary

Abstract. Recent high-profile targeted attacks showed that even the most secure and secluded networks can be compromised by motivated and resourceful attackers, and that such a system compromise may not be immediately detected by the system owner. Researchers at RSA proposed the FlipIt game to study the impact of such stealthy takeovers. In the basic FlipIt game, an attacker and a defender fight over a single resource; in practice, however, systems typically consist of multiple resources that can be targeted. In this paper, we present FlipThem, a generalization of FlipIt to multiple resources. To formulate the players' goals and study their best strategies, we introduce two control models: in the AND model, the attacker has to compromise all resources in order to take over the entire system, while in the OR model, she has to compromise only one. Our analytical and numerical results provide practical recommendations for defenders.

Keywords: FlipIt, game theory, advanced persistent threats, targeted attacks, attacker-defender games.

1 Introduction

In recent years, the world witnessed a series of high-profile targeted attacks against various targets [4,19,7,8,2,5,14,13]. These attacks showed that even the most secure and secluded networks can be compromised, and they induced an interesting discussion in the security industry and in the research community alike. An important lesson that the security community can learn from these incidents is that we must revisit some of the most fundamental assumptions which our systems rely on for security. In particular, one must make the assumption that motivated and resourceful attackers can fully compromise a system and gain access to its resources, and this may not be immediately detected by the system owner. The new challenge is to design security mechanisms that minimize the damage that such determined attackers can cause.

In order to help to address this challenge, researchers at RSA – which itself was a victim of a successful targeted attack in 2011 [18] – developed a game-theoretic modeling framework, called FlipIt [3,1]. FlipIt is an attacker-defender game

R. Poovendran and W. Saad (Eds.): GameSec 2014, LNCS 8840, pp. 175–194, 2014.

designed to study the problem of stealthy takeover of control over a critical resource. In FlipIt, control over the critical resource is obtained by "flipping" it for a certain cost, and the players receive benefits proportional to the total time that they control the resource. The payoff of each player is, therefore, determined by the difference between the benefit of controlling the resource and the cost of flipping it. Naturally, the goal of the players is to maximize their payoffs.

This is a simple, yet powerful model to study the strategic interaction of attackers and designers of security policies and mechanisms. Moreover, the basic model can be extended in different directions. For instance, in the basic FlipIt game, the players flip the resource without being able to observe who was in control before the flip. This model is ideal to study the security of a resource with off-line properties, such as passwords or cryptographic keys. In [16], Pham and Cid extend the basic model by giving the players the option to test if they control the resource before making a move, and use this extended model to study periodic security assessments and their positive effects. In [12,11], Laszka et al. propose and study another variation of the model, in which the defender's moves are non-stealthy, while the attacker's moves are non-instantaneous. Finally, researchers have also studied the FlipIt game in behavioral experiments, where human participants played against computerized opponents [15,17,6], which complement the theoretical work by showing the difficulty of finding optimal choices in games of timing.

In this paper, we propose a new generalization of the FlipIt game, which, to the best of our knowledge, has not been considered yet in the academic literature. Namely, we extend the basic FlipIt model, where the attacker and the defender fight over a single resource, to multiple resources. Accordingly, we call our generalized model the FlipThem game. In practice, compromising a system often requires more than attacking just a single component of it. Typically, successful takeovers consist of multiple steps, aiming at gradually escalating the privileges obtained by the attacker until he obtains full administrative access to the system. During this process, the attacker must gain control over a subset of available resources (e.g., he may be required to break a password *and* exploit a software vulnerability in an application). Hence, our model is closer to reality than the original FlipIt game, and, as we show in this paper, it is still amenable to mathematical analysis.

More specifically, we make the following contributions in this paper:
- We extend the FlipIt game to multiple resources. To formulate the players' goals, we introduce two control models: the AND and the OR control model. In the AND control model, the attacker needs to compromise all resources in order to take over the entire system, whereas in the OR control model, the attacker needs to control at least one resource (out of many available) to take over the entire system. More complex requirements on combinations of resources to be compromised for a successful take-over can be constructed by appropriate combination of these basic control models.
- As a first step to derive good multi-resource FlipThem strategies, we introduce two combinations of single-resource FlipIt strategies, namely the

independent and the synchronized combinations. In the independent case, the player flips each resource independently of the other resources, whereas in the synchronized case, the player always flips all resources together. We study and compare these two combinations, and derive analytical results for the players' gains.

- As a next step, to represent more complex multi-resource strategies, we introduce the Markov strategy class, where the decision to flip a resource (or set of resources) at a given time depends only on the times elapsed since the previous flips of the resources. We show how the best-response Markov strategy can be computed using a linear program. Using this linear program, we compare various defender strategies based on the resulting benefit for the defender.
- Finally, based on our analytical and numerical results, we provide practical recommendations for defenders. These recommendations can readily be used in practice where the assumptions of the FlipThem game apply.

It is important to note that, while the idea of generalizing FlipIt to multiple resources may seem straightforward, the exact mathematical treatment of FlipThem is not trivial at all. The reason for this is that FlipThem is more than just the collection of independent FlipIt instances. In general, the attacker and/or defender strategies in FlipThem do not handle the different resources independently from each other, and this dependence among the resources results in complex optimization problems when solving the game.

The organization of this paper is the following. In Section 2, we summarize the FlipIt game and the most important conclusions drawn in related work. In Section 3, we introduce FlipThem, the generalization of FlipIt for multiple resources. In Section 4, we show how single-resource FlipIt strategies can be combined into multi-resource strategies and compute the players' benefits for various combinations. In Section 5, we introduce the Markov strategy class and show how a best-response Markov strategy can be computed using a linear program. Finally, in Section 6, we discuss the implications of our results and provide practical recommendations for defenders.

2 The FlipIt Game

In this section, we summarize the FlipIt game and the most important conclusions drawn in related work. It is important to get familiar with the key concepts and notation of the original FlipIt game to understand our results for the multiple resources case. Table 1 contains the most important differences in notation between the original FlipIt game and our FlipThem game. Note that the assumptions of the FlipIt game are very different from those of the previous work in the field of game theory for security. For a detailed comparison between FlipIt and previous work, we refer the reader to [3].

FlipIt [3,1] is a two-player, non-zero-sum game modeling stealthy takeovers, in which both players are trying take control of a single resource. One of the players is called the *defender* (denoted by D), while the other player is called the

Table 1. List of Symbols

Symbol	Description
	FlipIt
c^i	player i's flipping cost
β^i	" asymptotic benefit rate
γ^i	" " gain rate
α^i	" " flip rate
Z^i	random variable representing the time since the last flip of player i
	FlipThem
N	number of resources
c_r^i	player i's flipping cost for resource r
α_r^i	" asymptotic flip rate for resource r
Z_r^i	rand. var. representing the time since the last flip of player i on resource r

attacker (denoted by A). The game starts at time $t = 0$ and continues indefinitely (that is, $t \to \infty$). In general, time can be both continuous and discrete, with most results being applicable to both cases. At any time instance, player i may choose to take control of the resource by "flipping" it, which costs her c^i. Then, the resource remains under the control of player i until the other player flips it. Consequently, at any given time instance, the resource is controlled by either one or the other player. The interesting aspect of the FlipIt game is that neither of the players knows who is in control. As a result, the players occasionally make unnecessary flips (i.e., flip the resource when it is already under their control) since they have to execute their flips "blindly". For an illustration of the game, see Figure 1.

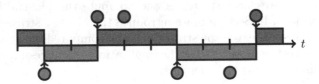

Fig. 1. An illustration of the FlipIt game with discrete flip timing. Blue and red disks represent the defender's and attacker's flips. Takeovers, that is, flips changing the player controlling the resource, are indicated by arrows. Blue and red shaded rectangles represent control of the resource by the defender and the attacker, respectively.

The state of the resource is represented by the time-dependent variables C^A and C^D: $C^A(t) = 1$ when the attacker controls the resource, and 0 otherwise; $C^D(t)$ is vice versa (i.e., $C^D(t) = 1 - C^A(t)$). Since the players can (and, as we will soon see, should) employ randomized strategies, both $C^D(t)$ and $C^A(t)$ are random variables. The variables $C^D(t)$ and $C^A(t)$ can be also expressed using the times elapsed since the last flips made by the players as

$$C^D(t) = I_{Z^D(t) \leq Z^A(t)} \qquad \text{and} \qquad C^A(t) = I_{Z^D(t) > Z^A(t)}, \qquad (1)$$

where Z^i is the time elapsed since the last flip of player i and I is the indicator function.

Player i's *asymptotic gain rate* γ^i is defined as the average fraction of time the resource is controlled by player i. Formally,

$$\gamma^i = \liminf_{t \to \infty} \frac{\int_0^t C^i(\tau)d\tau}{t} . \tag{2}$$

Note that player i's asymptotic gain is equal to the probability that the resource is controlled by player i at a random time instance. Formally,

$$\gamma^i = \Pr\left[C^i = 1\right] . \tag{3}$$

Player i's *asymptotic flip rate* α^i is defined as the average number of flips made by player i in a unit of time. Formally,

$$\alpha^i = \liminf_{t \to \infty} \frac{n^i(t)}{t} , \tag{4}$$

where $n^i(t)$ denotes the number of flips made by player i up to time t. Finally, player i's game-theoretic utility, called player i's *asymptotic benefit* β^i, is defined as the average fraction of time the resource is controlled by the player minus the average cost of flips. Formally,

$$\beta^i = \gamma^i - c^i \alpha^i . \tag{5}$$

Since takeovers are assumed to be stealthy in the FlipIt game, players do not automatically know when the other player has last moved. However, when a player makes a move (i.e., flips the resource), she might be able to receive some feedback. For example, when an attacker compromises a system, she may learn when the defender last updated the system (that could be attributed as a flip action), and use this information to plan her next move. In [3], three models are introduced for *feedback* received *during the game*:

- Non-adaptive (NA): The player does not receive any feedback when she moves.
- Last move (LM): The player learns the exact time of the other player's last flip.
- Full history (FH): The player learns the complete history of flips made by the other player.

Besides receiving feedback during the game, a player might also be able to receive information before the game starts. For example, an attacker might learn the defender's flip strategy and exploit this knowledge. In [3], two models are introduced for *information* received by a player *before the game starts*:

- Rate of Play (RP): The player knows the asymptotic flip rate α of the other player.

- Knowledge of Strategy (KS): Besides the asymptotic flip rate, the player knows additional information about the other player's strategy. For example, the player may know that the other player employs a renewal process to generate her flip sequence, and may also know the probability density function of the process. However, it is always assumed that the randomness of the other player's strategy remains secret; consequently, the player cannot know which realization of the renewal process will be used.

In our analysis of defender's strategies in Section 5, we assume a strong attacker model meaning that the attacker always has the Knowledge of Strategy. We assume that the attacker knows everything, except the randomness part of the defender's strategy. This complies with Kerckhoff's principle on security without obscurity.

2.1   Strategies

In this subsection, we summarize the most important strategies and the corresponding results from [3]. For a detailed analysis of these and some other strategies, we refer the interested reader to [3].

In this paper, we focus on non-adaptive strategies, which do not require feedback received by the player during the game. The rationale behind this is that

- defenders rarely know the exact strategies of the attackers (or even the identities of the attackers) in practice; thus, they have to use strategies that do not rely on feedback,
- defenders can choose randomized strategies that schedule their subsequent flips such that even an FH attacker has no more advantage than random guessing (see exponential strategy below), and
- in case of high-importance computer systems, attackers might have limited feedback options if they want to operate stealthily.

A renewal strategy is a non-adaptive strategy in which the time intervals between consecutive flips are generated by a renewal process. More formally, time intervals between consecutive moves are independent and identically distributed random variables, chosen according to a probability density function f. Renewal strategies include (but are not limited to) *periodic strategies* and *non-arithmetic renewal strategies*, which we discuss below.

A player can also choose to drop out of the game (i.e., never flip the resource), which is a rational decision if her expected benefit is less than zero for every strategy choice available to her. This can happen when her opponent's flipping cost is much lower and her opponent can afford to flip the resource extremely fast.

Periodic $\mathcal{P}$: A strategy is periodic if the time intervals between consecutive flips are constant, denoted by δ. It is assumed that a periodic strategy has a *random phase*, that is, the time of the first flip is chosen uniformly at random from $[0, \delta]$. A *periodic strategy with random phase* is characterized by the fixed time interval δ between consecutive flips. It is easy to see that the flip rate of a periodic strategy is $\alpha = \frac{1}{\delta}$. The periodic strategy of rate α is denoted by P_α, and the class of all periodic strategies is denoted by $\mathcal{P}$.

Periodic is probably the strategy most widely used in practice as most systems require passwords, cryptographic keys, etc. to be changed at regular intervals, for example, every thirty days or every three months. In [3], it was shown that the periodic strategy strongly dominates all other renewal strategies if the other player uses a periodic or non-arithmetic renewal strategy. Thus, the periodic strategy is a good choice for an attacker who plays against a non-adaptive (NA) defender.

However, due to its completely deterministic nature[1], the periodic strategy is a very poor choice for defenders who face an attacker observing the last move of the defender (LM attacker). An LM attacker can learn the exact time of the defender's next flip, and schedule her own flip to be immediately after that. Consequently, if flipping costs are of the same order of magnitude, an attacker can keep the resource permanently under her control (with negligible interrupts from the defender). Therefore, a defender facing an LM attacker has two options: if her flipping cost is much lower than that of the attacker, she can flip fast enough to force the attacker to drop out; otherwise, she has to use a randomized strategy, such as the following ones.

Non-Arithmetic Renewal $\mathcal{R}$: A renewal process is called *non-arithmetic* if there is no positive real number $d > 0$ such that interarrival times are all integer multiples of d. The renewal strategy generated by the non-arithmetic renewal process with probability density function f is denoted by R_f, and the class of all non-arithmetic renewal strategies is denoted by $\mathcal{R}$.

The class of non-arithmetic renewal strategies is very broad as there are an infinite number of possible probability density functions, even for a given flip rate. Of these probability density functions, the exponential is the most important one in the FlipIt game.

Exponential $\mathcal{E}$: An *exponential* (or *Poisson*) strategy is a non-arithmetic renewal strategy generated by a Poisson process. Formally, the interarrival times of the process follow an exponential distribution: $f(\tau) = \lambda e^{-\lambda \tau}$, where λ is the parameter characterizing the distribution. The flip rate of this strategy is simply $\alpha = \lambda$. The exponential strategy with rate λ is denoted by E_λ, and the class of all exponential strategies is denoted by $\mathcal{E}$.

The exponential strategy is of key importance, because the exponential distribution is the only *memoryless* continuous probability distribution. The memoryless property means that the conditional probability that we have to wait more than τ_1 time before the next flip, given that the time elapsed since the last flip is τ_2, is independent of τ_2. This implies that, if a defender uses an exponential strategy, an LM (or even an FH) attacker cannot learn *any* information regard-

---

[1] The random phase ensures that an NA opponent cannot determine the flip times of the player; however, if the opponent learns the exact time of at least one flip made by the player, she is able to determine the time of every flip.

ing the time of the defender's next flip. Consequently, the exponential strategy is a good choice for a defender facing an LM attacker.

3   The FlipThem Game: FlipIt on Multiple Resources

In this section, we generalize the FlipIt game for multiple resources as follows. There are N resources, identified by integer numbers $1, \ldots, N$. Each resource can be flipped individually and, as a result, becomes controlled by the flipping player. The cost of flipping resource r for player i is c_r^i. Each resource has to be flipped individually; i.e., if a player chooses to flip multiple resources at the same time, she still has to pay the flipping cost for each resource that she flips.

The goal of the attacker is to control the system of resources, while the goal of the defender is to prevent the attacker from doing so. The criterion for the attacker controlling the system can be defined in multiple ways, which makes the generalization non-straightforward: as we will later see, different formulations can lead to opposite results. In this paper, we study two elementary control models (see Figure 2 for an illustration):

- All resources [AND]: The attacker controls the system only if she controls *all* resources. Formally,

$$C^A(t) = Z_1^D(t) > Z_1^A(t) \wedge \ldots \wedge Z_N^D(t) > Z_N^A(t) . \tag{6}$$

  This models scenarios where the attacker has to compromise every resource in order to compromise her target.
- One resource [OR]: The attacker controls the system if she controls *at least one* resource. Formally,

$$C^A(t) = Z_1^D(t) > Z_1^A(t) \vee \ldots \vee Z_N^D(t) > Z_N^A(t) . \tag{7}$$

  This models scenarios where the attacker only has to compromise a single resource in order to compromise her target.

Similarly to the basic FlipIt game, the players receive benefits proportional to the time that they are controlling the system minus their costs of flipping the resources. More complex control models can be built by combining the AND and OR models in appropriate ways, but the study of that is left for future work.

Notice that, for non-adaptive strategies, the two control models are completely symmetric: the benefit of one player in one model is equivalent to the benefit of the other player in the other model. Consequently, for non-adaptive strategies, it suffices to compute the benefits only in one control model (the AND model in our paper) as the formulas for the other model can be derived readily.

In the following sections, we introduce and study various FlipThem (i.e., multi-resource) strategies, compute the resulting asymptotic benefits, and discuss which strategies should be chosen by the players. First, in Section 4, we study combinations of multiple single-resource strategies. Then, in Section 5, we propose a novel multi-resource strategy class, called the *Markov strategy* class.

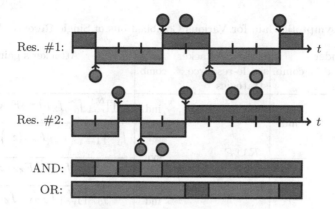

Fig. 2. An illustration of the FlipThem game with the AND and OR control models (see Figure 1 for graphical notations)

4   Combining Single-Resource Strategies

One of the challenges posed by FlipThem lies in the potentially complex structure of the strategies, which can use elaborate rules to exploit the dependence among the resources. A possible way of finding well-performing, yet analytically tractable multi-resource strategies is to combine multiple single-resource strategies that are known to perform well in the basic FlipIt game. In this section, we propose and study two combinations:

Independent: The player flips each resource independently of the other resources. More specifically, the player uses N independent single-resource strategies (i.e., processes), one for each resource, with each one having its own flip rate α_r^i. The asymptotic benefit of a player i using the independent combination is $\beta^i = \gamma^i - \sum_{r=1}^{N} c_r^i \alpha_r^i$.

- Synchronized: The player always flips all resources together. More specifically, the player uses only one single-resource strategy (i.e., process) for all of the resources, with a single flip rate α^i. The asymptotic benefit of a player i using the synchronized combination is $\beta^i = \gamma^i - \alpha^i \sum_{r=1}^{N} c_r^i$.

Since the AND and OR control models are symmetric, we only compute the asymptotic gains in the AND model in this paper. Formulas for the asymptotic gains in the OR model can be derived from our results readily. Furthermore, since the defender's asymptotic gain γ^D can be computed from the attacker's asymptotic gain γ^A using the simple formula $\gamma^D = 1 - \gamma^A$, we only compute the asymptotic gain of the attacker.

The proofs of the formulas can be found in the extended online version of this paper [10]. Here, we first show the more general results for the strategy class $\mathcal{R} \cup \mathcal{P}$ (Table 2); then, we analyze the game for the classes $\mathcal{E}$ and $\mathcal{P}$ (Table 3 and Figure 3).

Table 2 shows the attacker's asymptotic gain for various multi-resource strategies chosen by the defender and the attacker. The $\mathcal{R} \cup \mathcal{P}$ in the first and third

Table 2. Asymptotic Gain for Various Combinations of Single-Resource Strategies

Defender single-resource strategies	comb.	Attacker single-resource strategies	comb.	Attacker's gain γ^A
$\mathcal{R} \cup \mathcal{P}$	ind.	$\mathcal{R} \cup \mathcal{P}$	ind.	$\prod_{r=1}^{N} \int_0^{\infty} f_{Z_r^D}(z_r) F_{Z_r^A}(z_r) dz_r$
			syn.	$\int_0^{\infty} \prod_{r=1}^{N} \left(1 - F_{Z_r^D}(z)\right) f_{Z^A}(z) dz$ $\int_0^{\infty} f_{Z^D}(z) F_{Z^A}(z) dz$
	syn.		ind.	$\int_0^{\infty} \prod_{r=1}^{N} F_{Z_r^A}(z) f_{Z^D}(z) dz$

column indicates that we assume that the players use combinations of either non-arithmetic renewal ($\mathcal{R}$) or periodic ($\mathcal{P}$) single-resource strategies. The combinations used by the defender and the attacker are in the second and fourth columns, respectively. Finally, the attacker's gain γ^A for the given combinations is in the fifth column.

To express the attacker's gain, we use a notion similar to that of the basic FlipIt game. We let Z_r^i be the random variable representing the time elapsed since player i's last flip on resource r (we omit the index r and denote it by simply Z^i if the player uses a synchronized strategy). We denote the cumulative distribution and density functions of Z_r^i by $F_{Z_r^i}(z)$ and $f_{Z_r^i}(z)$. These functions can easily be computed from the generating distribution of any non-arithmetic renewal strategy (see the extended online version [10]).

It is noteworthy that, when both players use the synchronized combination, the game is equivalent to the basic FlipIt game (with $c^i = \sum_r c_r^i$): each player uses only one single-resource (i.e., basic FlipIt) strategy, and the state of all resources is the same as they are always flipped together. Consequently, the formula for the attacker's gain is identical to the corresponding formula in [3].

Table 3 shows the attacker's asymptotic gain for various combinations of exponential and periodic strategies. We selected these single-resource strategies because they are known to be optimal in some respect (see Section 2). The table is similar to Table 2, except that the synchronized defender against independent attacker case is omitted to keep the table simple (it can be found in the extended version of this paper [10]) and because it is not a good strategy for either of the players.

The table shows that the independent combination is generally better than the synchronized one for the defender, as her flip rates are added together in the former. This can be explained by the nature of the AND control model: since the defender only needs to control at least one resource, her best strategy is to flip one resource at a time. This forces the attacker to frequently flip all resources back as she cannot know which resources were flipped by the defender (since the exponential process is memoryless).

Table 3. Asymptotic Gain for Various Combinations of Exponential and Periodic Strategies

Defender		Attacker		Attacker's gain
single-resource strategy	comb.	single-resource strategy	comb.	γ^A
$\mathcal{E}$	ind.	$\mathcal{E}$	ind.	$\prod_{r=1}^{N} \dfrac{\alpha_r^A}{\alpha_r^A + \alpha_r^D}$
			syn.	$\dfrac{\alpha^A}{\alpha^A + \sum_{r=1}^{N} \alpha_r^D}$
	syn.			$\dfrac{\alpha^A}{\alpha^A + \alpha^D}$
	ind.	$\mathcal{P}$	ind.	$\prod_{r=1}^{N} \dfrac{\alpha_r^A}{\alpha_r^D}\left(1 - e^{-\frac{\alpha_r^D}{\alpha_r^A}}\right)$
			syn.	$\dfrac{\alpha^A}{\sum_{r=1}^{N} \alpha_r^D}\left(1 - e^{-\frac{\sum_{r=1}^{N} \alpha_r^D}{\alpha^A}}\right)$
	syn.			$\dfrac{\alpha^A}{\alpha^D}\left(1 - e^{-\frac{\alpha^D}{\alpha^A}}\right)$

The formulas also suggest that the attacker should choose the synchronized combination over the independent one. When both players use exponential single resource strategies, the attacker's gain decays exponentially as the number of resources increases ($\sim k^{-N}$) if she uses the independent combination, but only according to a power law ($\sim N^{-k}$) if she uses the synchronized one (given that flip rates stay the same). When the attacker uses the periodic single-resource strategy, the relationship between the number of resources and the attacker's gain is more complicated, but similar.

Figure 3 shows the attacker's asymptotic gain as a function of the number of resources for various combinations of exponential and periodic strategies. The plotted pairs of combinations are the following: both players use independent strategies (solid line —), the attacker uses synchronized while the defender uses independent strategy (dashed line - -), and both players use synchronized strategies (dotted line ····). The flip rates are assumed to be uniform, i.e., $\alpha^A = \alpha_r^A = \alpha^D = \alpha_r^D = 1$, $r = 1, \ldots, N$.

The figure shows that, for the given single-resource classes and parameters, the synchronized combination strongly dominates the independent one for the attacker. Again, this can be explained by the nature of the AND control model: since the attacker needs to control all resources, it makes sense to flip them all together. Otherwise, the probability that all resources become controlled by the attacker is very low. However, by using the synchronized combination, the attacker loses the freedom of choosing the flipping rate for each resource independently. Thus, when the heterogeneity of the attacker's flipping costs is very high, the independent combination may outperform the synchronized one.

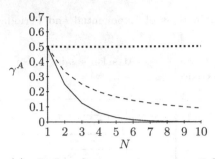

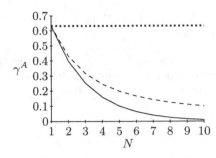

(a) Both players use exponential strategies.

(b) Defender uses exponential, attacker uses periodic strategy.

Fig. 3. The attacker's asymptotic gain as a function of the number of resources for various combinations of exponential and periodic strategies. Plotted pairs of combination are: both players use independent strategies (solid line), attacker uses synchronized strategies while defender uses independent strategies (dashed line), and both players use synchronized strategies (dotted line). In this figure, the flip rates are assumed to be uniform, i.e., $\alpha^A = \alpha_r^A = \alpha^D = \alpha_r^D = 1$, $r = 1, \ldots, N$.

The figure also supports our finding that the independent combination strongly dominates the synchronized one for the defender. Since a player has complete freedom in choosing her flip rates in the independent combination, this combination is better for the defender even for very heterogeneous flipping costs.

Finally, by comparing Subfigures 3a and 3b, we conclude that the periodic strategy dominates the exponential strategy as the attacker's gain is higher when she chooses the former.

5   The Markov Strategy Class

In the previous section, we studied how single-resource strategies can be combined into multi-resource strategies. However, such combinations represent only a tiny fraction of the actual multi-resource strategy space as there are an infinite number of multi-resource strategies that cannot be represented by such simple combinations. For example, a defender might choose to flip one resource periodically, then wait for a time interval chosen according to an exponential distribution, and then flip another resource. To model such complex multi-resource strategies, in this section, we introduce the *Markov* strategy class.

For the clarity of presentation, we derive results for two resources, yet the approach is applicable for any number of resources. Furthermore, as opposed to the basic model, we are going to use discrete time in this section. Note that the discrete time model can be very realistic as players typically do not flip their resources at arbitrary times. Examples are the change of passwords, cryptographic keys or the application of software updates. We denote the duration of a time step by Δ. Finally, we define the time-dependent age functions as follows. The random variables representing the number of time steps elapsed since the last flip of resource r by the attacker and the defender at time k are denoted by $Z_r^A(k)$ and $Z_r^D(k)$, respectively.

In the case of two resources, the attacker can perform one of the following actions in a given time slot:

- she does not flip any of the resources,
- she flips one of the resources,
- or she flips both resources.

If the decision which action to choose depends only on the times elapsed since the previous flips of the resources, then $\{(Z_1^A(k), Z_2^A(k)), k = 0, 1, \dots\}$ defines a Markov process. In this case, the behavior of the attacker can be characterized by the following joint distributions corresponding to the events that can happen in two consecutive time steps:

$$p_{i,j}^{(0)} = \Pr\left[Z_1^A(k) = i,\ Z_2^A(k) = j,\ Z_1^A(k+1) = i+1,\ Z_2^A(k+1) = j+1\right]$$

$$p_{i,j}^{(1)} = \Pr\left[Z_1^A(k) = i,\ Z_2^A(k) = j,\ Z_1^A(k+1) = 0,\ Z_2^A(k+1) = j+1\right]$$

$$p_{i,j}^{(2)} = \Pr\left[Z_1^A(k) = i,\ Z_2^A(k) = j,\ Z_1^A(k+1) = i+1,\ Z_2^A(k+1) = 0\right]$$

$$p_{i,j}^{(1,2)} = \Pr\left[Z_1^A(k) = i,\ Z_2^A(k) = j,\ Z_1^A(k+1) = 0,\ Z_2^A(k+1) = 0\right]\ ,$$

where $p_{i,j}^{(0)}$ is the probability that nothing is flipped in the next time step, $p_{i,j}^{(1)}$ ($p_{i,j}^{(2)}$) is the probability that only resource 1 (or 2) is flipped, while $p_{i,j}^{(1,2)}$ is the probability of both resources being flipped in the next time step.

We denote by M_p the Markov strategy generated by a Markov process with event probabilities $p = \{p_{i,j}^{(0)}, p_{i,j}^{(1)}, p_{i,j}^{(2)}, p_{i,j}^{(1,2)}$ for $i, j = 0, 1, \dots\}$, and by $\mathcal{M}$ the class of all Markov strategies. That is,

$$\mathcal{M} = \{M_p \mid p \text{ is a set of event probabilities}\}\ . \tag{8}$$

5.1   Linear Programming Solution

With these definitions and notations, we can define a linear program to determine the optimal probabilities $p_{i,j}^{(\bullet)}$. However, since linear programming problems can only be solved with a finite number of variables (in the general case), we have to restrict the game to a finite time horizon. The last time step we take into consideration is denoted by T.

The attacker wants to maximize her benefit β^A, which is composed of the asymptotic gain and the cost of the flips against both resources as

$$\beta^A = \max_p \left\{ \underbrace{\sum_{i=0}^{T}\sum_{j=0}^{T} q_{i,j} \Pr\left[Z_1^D > i, Z_2^D > j\right]}_{\gamma^A} \right. \tag{9}$$
$$\left. - c_1^A \underbrace{\left(\sum_{i=0}^{T}\sum_{j=0}^{T} p_{i,j}^{(1)} + p_{i,j}^{(1,2)}\right)\frac{1}{\Delta}}_{\alpha_1^A} - c_2^A \underbrace{\left(\sum_{i=0}^{T}\sum_{j=0}^{T} p_{i,j}^{(2)} + p_{i,j}^{(1,2)}\right)\frac{1}{\Delta}}_{\alpha_2^A} \right\},$$

where $q_{i,j}$ is the probability that the number of time steps since the attacker's last flips of resource 1 and 2 are i and j, respectively. This probability can be expressed easily as $q_{i,j} = p_{i,j}^{(0)} + p_{i,j}^{(1)} + p_{i,j}^{(2)} + p_{i,j}^{(1,2)}$; thus, the objective function given by (9) defines a linear relation with respect to $p_{i,j}^{(\bullet)}$.

As variables $p_{i,j}^{(\bullet)}$ must be valid probabilities, we need to apply the inequality constraints $p_{i,j}^{(0)} \geq 0$, $p_{i,j}^{(1)} \geq 0$, $p_{i,j}^{(2)} \geq 0$, $p_{i,j}^{(12)} \geq 0$; and we also need to ensure that the probabilities sum up to 1, that is, $\sum_{i=0}^{T} \sum_{j=0}^{T} p_{i,j}^{(0)} + p_{i,j}^{(1)} + p_{i,j}^{(2)} + p_{i,j}^{(1,2)} = 1$.

Further equality constraints are required to define the possible state transitions, yielding

$$q_{i,j} = p_{i-1,j-1}^{(0)} \text{ for } i > 0, j > 0, \qquad q_{0,0} = \sum_{i=0}^{T} \sum_{j=0}^{T} p_{i,j}^{(1,2)},$$

$$q_{0,j} = \sum_{i=0}^{T} p_{i,j-1}^{(1)} \text{ for } j > 0, \qquad q_{i,0} = \sum_{j=0}^{T} p_{i-1,j}^{(2)} \text{ for } i > 0, \qquad (10)$$

with $q_{i,j}$ given above.

Finally, we require that a resource is always flipped in the next time step if its age has reached the maximum age T:

$$p_{i,j}^{(0)} = 0 \text{ for } i = T \text{ or } j = T, \qquad p_{i,j}^{(1)} = 0 \text{ for } j = T, \qquad p_{i,j}^{(2)} = 0 \text{ for } i = T. \quad (11)$$

5.2   Results

The linear program defined above answers several questions regarding the **Flip-Them** game, including the following:

– What is the attacker's optimal strategy against a given defender strategy?
– What are the optimal flip rates maximizing the defender's benefit if the attacker always plays an optimal strategy?
– What is the Nash equilibrium of this game?

Solving the optimization problem using a linear programming based approach poses some challenges. In particular, the length of the time horizon T is limited by the capabilities of the linear program solver. For our examples, we used the built-in solver of MATLAB with $T = 30$ (resulting in 900 variables in case of two resources).[2] Note that the number of variables increases polynomially in the length of the time horizon and exponentially in the number of resources. Using custom software, the analysis can be extended to much larger values of T; however, the results we obtained with MATLAB are already revealing and useful.

In the rest of this section, we consider several numerical examples to demonstrate the usefulness of the model. In each of these examples, the attacker is assumed to be non-adaptive (NA), but she is assumed to know the strategy of the

---

[2] This can model, for example, the key update policy of a company over a duration of 2.5 years assuming that updates are defined by the granularity of a month.

defender (KS). The defender, however, has no information about the attacker. For the definitions and rationale behind these modeling choices, see Section 2.

Optimal Attack against a Given Defender Strategy. In this example, the defender flips the resources according to independent Poisson processes with parameters $\alpha_1^D = 1$ and $\alpha_2^D = 3$. The joint age function is then $\Pr\left[Z_1^D > i, Z_2^D > j\right] = e^{-\alpha_1^D i \Delta - \alpha_2^D j \Delta}$. The attacker's flip costs are $c_1^A = 0.1$ and $c_2^A = 0.05$. The discrete problem is solved with $T = 30$ and $\Delta = 0.03$.

At this point, we take the opportunity to introduce the conditional state transition probability matrices $\boldsymbol{P^{(0)}}, \boldsymbol{P^{(1)}}, \boldsymbol{P^{(2)}}$ and $\boldsymbol{P^{(1,2)}}$ that help to visualize and understand the strategy of the attacker. The entries of these matrices are

$$[\boldsymbol{P^{(\bullet)}}]_{i,j} = \frac{p_{i,j}^{(\bullet)}}{p_{i,j}^{(0)} + p_{i,j}^{(1)} + p_{i,j}^{(2)} + p_{i,j}^{(1,2)}} . \tag{12}$$

To simulate an attack, one has to follow the state of the attacker given by positions i, j in the matrices. In state (i, j), no flips occur with probability $[\boldsymbol{P^{(0)}}]_{i,j}$, and the next state of the attacker is $(i+1, j+1)$. With probability $[\boldsymbol{P^{(1)}}]_{i,j}$ (or $[\boldsymbol{P^{(2)}}]_{i,j}$), only resource 1 (or resource 2) is flipped in the next time step, and the next state of the system is $(0, j+1)$ (or $(i+1, 0)$). Finally, both resources are flipped in the next time step with probability $[\boldsymbol{P^{(1,2)}}]_{i,j}$, followed by a jump to state $(0, 0)$.

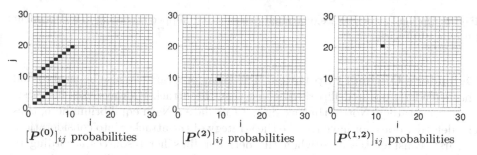

$[\boldsymbol{P^{(0)}}]_{ij}$ probabilities $[\boldsymbol{P^{(2)}}]_{ij}$ probabilities $[\boldsymbol{P^{(1,2)}}]_{ij}$ probabilities

Fig. 4. Optimal attack strategy against two resources flipped according to independent Poisson processes

By solving the linear program, we obtain the optimal strategy of the attacker, represented by the matrices depicted in Figure 4. In this particular example, the entries of all four matrices are all either 0 (represented by white squares) or 1 (black squares). Matrix $\boldsymbol{P^{(1)}}$ is not depicted as it has only 0 entries. By following the attacker's strategy in the above described manner, we have that she first waits 9 time steps (black squares on the diagonal of $[\boldsymbol{P^{(0)}}]$), then flips one resource (black square in $\boldsymbol{P^{(2)}}(9,9)$), waits another 10 time steps, and finally flips both resources (black square in $\boldsymbol{P^{(1,2)}}(20,11)$).

Thus, based on the matrices, a "periodic" attack can be identified with a period of $\delta = 20$. The resources are not flipped in a synchronized manner. Resource 2 is flipped at the 9th time step from the beginning of the period, while both resources are flipped at the end of the period.

If the defender flips both resources according to independent periodic strategies, the joint age process is given by $\Pr\left[Z_1^D > i, Z_2^D > j\right] = (1 - \alpha_1^D i\Delta)(1 - \alpha_2^D j\Delta)$, if $i\Delta < 1/\alpha_1^D$, $j\Delta < 1/\alpha_2^D$, and $\Pr\left[Z_1^D > i, Z_2^D > j\right] = 0$ otherwise. When keeping all parameters the same as before, the optimal strategy of the attacker is more complex in this case (see Figure 5). The period of her strategy is $\delta = 22$ now, and she flips solely resource 2 at time steps 6 and 13, while she flips both resources at time step 22, which also marks the end of her period.

It is noteworthy that the attacker's benefit is 0.265 in the Poisson case, but only 0.047 in the periodic case, which means that the periodic defense is less economical to attack (given, of course, that the attacker has no knowledge on the last move of the defender, thus it is of type NA).

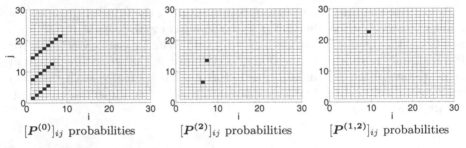

$[\boldsymbol{P}^{(0)}]_{ij}$ probabilities $[\boldsymbol{P}^{(2)}]_{ij}$ probabilities $[\boldsymbol{P}^{(1,2)}]_{ij}$ probabilities

Fig. 5. Optimal attack strategy against two resources flipped according to independent periodic strategies

Defender's Optimal Flip Rates. The linear program can also be used to find the defender's optimal flip rates given that the attacker always uses her best-response strategy. Notice that we do not calculate a Nash equilibrium here, thus the defender does not have to take the strategy of the attacker into consideration.

First, consider the case when the defender flips her resources according to independent Poisson processes. Assume that the attacker's flipping costs are $c_1^A = 0.1$ and $c_2^A = 0.2$. We solved the linear program with various combinations of α_1^D and α_2^D, and with two different settings for the parameters c_1^D and c_2^D. The results are shown in Figure 6. As the benefit of the attacker is the subject of optimization in the linear program, the corresponding plot is obviously smooth, and gives higher values for lower flip rates of the defender. The corresponding gain rates (which are not plotted due to the lack of space), however, are not smooth. As the defender's benefit is directly related to the attacker's gain rate, the plots of the defender's benefit are not smooth either. The maximum benefit for the defender is 0.222, obtained at $\alpha_1^D = 0.8, \alpha_2^D = 0.7$.

If the defender flips the resources according to independent periodic strategies, higher flip rates are required to maximize her benefit. The corresponding results

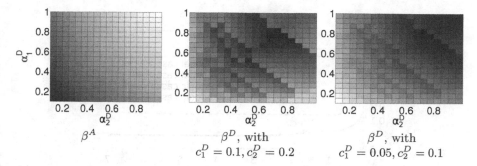

Fig. 6. Benefits of the attacker (β^A) and defender (β^D) for various flip rates of the defender (Poisson case). Darker shades of gray indicate higher benefit.

are depicted in Figure 7: the optimal flip rates are $\alpha_1^D = 0.9$, $\alpha_2^D = 1.2$, and her benefit $\beta^D = 0.61595$ is higher compared to the Poisson case. Observe that the attacker always drops out for higher flip rates, which is indicated by the white area on the plot of her benefit and also by the sharp line appearing on the plots of the defender's benefit.

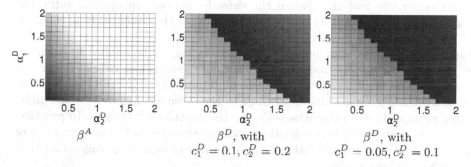

Fig. 7. Benefits of the attacker (β^A) and defender (β^D) for various flip rates of the defender (periodic case). Darker shades of indicate higher benefit.

Optimal Flip for a Fixed Budget. In this example, we assume that the defender has a fixed budget, and we are looking for the flip rates maximizing her benefit. By a fixed budget, we mean that the defender spends a fixed amount B on average on flipping her resources, thus $B = c_1^D \alpha_1^D + c_2^D \alpha_2^D$ is fixed, while the ratio of the flip rates $R = \alpha_1^D / \alpha_2^D$ is subject of optimization. Notice that B and R determine the flip rates uniquely as

$$\alpha_1^D = \frac{RB}{c_1^D R + c_2^D}, \qquad \alpha_2^D = \frac{B}{c_1^D R + c_2^D}. \tag{13}$$

The flip costs of the attacker and the defender are set to $c_1^A = 0.1, c_2^A = 0.05$ and $c_1^D = c_2^D = 0.001$, the total cost is $B = 0.004$, and we apply a finer discretization in this example with $T = 90$ and $\Delta = 0.01$.

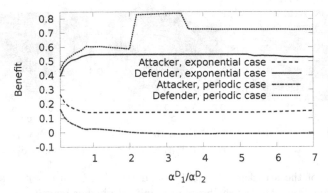

Fig. 8. Benefits of the attacker and the defender as functions of the ratio between the flip rates for the resources

Figure 8 depicts the players' benefits assuming that the attacker always flips according to her best-response Markov strategy. The optimal ratio R (from the defenders point of view) is 3.4 when she flips her resources periodically, and it is between 0.9 and 5.2 when she uses exponential strategies. By looking at the results closer, we find that, when the defender chooses an optimal ratio, the attacker uses a synchronized periodic attack against the resources in both cases.

Nash Equilibrium. The proposed linear program can be applied to calculate the optimal strategies of both the defender and the attacker. We can thus use a simple iterative algorithm to find a Nash equilibrium of the game. This algorithm starts with assigning a random strategy to the defender, followed by the alternating optimizations of the attacker's and the defender's strategies. In practice, however, we found that this algorithm does not converge in the vast majority of the cases, but it starts oscillating after a number of iterations, suggesting that no Nash equilibrium exists.

6   Concluding Remarks

Extending the `FlipIt` game to multiple resources requires modeling the players' goals as functions of the compromised resources. We selected the two most intuitive choices, namely the AND and OR control models, to represent the gains derived from controlling the resources. From the attacker's viewpoint, the AND control model represents the case when all resources need to be compromised to get access to the system. This is similar to the *total effort* model of security interdependence in the state-of-the-art [9,20]. The OR control model represents the case when the compromise of a single resource suffices to get access. This second choice relates to the *weakest link* model of interdependence [9,20].

We proposed two major classes of multi-resource strategies: combinations of single-resource strategies (independent processes and synchronized processes) and the Markov strategy class. Based on our result, we can formulate a set of recommendations for the defender. These recommendations can be readily used in practice where the assumptions of the FlipThem game apply, for example, when defining the key update strategy for a security infrastructure.

- In the AND control model, we found that the defender should use independent flipping strategies. In practice, this means that cryptographic keys should not be updated at the same time, but rather independently.
- On the other hand, in the OR control model, the defender should use synchronized flipping strategies. In practice, this means updating cryptographic keys synchronously. However, the defender needs to pay attention to the cost of updating keys in the OR control model. If these costs are very heterogeneous, the key update processes should remain synchronized, but with different update rates across the keys.
- If the attacker is non-adaptive, then the periodic defender strategy is a good choice according to our numerical results.[3] Periodic strategies have multiple advantageous properties such as higher benefits for the defender, robustness to optimization errors and ease of implementation in practice. However, periodic strategies perform poorly against an LM attacker [3]. Thus, the defender needs to carefully assess the potential information available to an attacker when choosing her strategy.
- Surprisingly, the defender's benefit is not a smooth or monotonic function of her flip rates, which makes optimization difficult in practice. Our numerical results imply that this observation holds for any combination of the periodic and the exponential strategy classes. The major reason behind this non-monotonous property is that, as the defender's flip rate reaches a threshold, the attacker drops out of the game. In realistic cases, the defender's flipping cost is much lower than the attacker's flipping cost, which causes the attacker to drop out.

Acknowledgment. This work is supported in part by the National Science Foundation (CNS-1238959).

References

1. Bowers, K.D., van Dijk, M., Griffin, R., Juels, A., Oprea, A., Rivest, R.L., Triandopoulos, N.: Defending against the unknown enemy: Applying FLIPIT to system security. In: Grossklags, J., Walrand, J. (eds.) GameSec 2012. LNCS, vol. 7638, pp. 248–263. Springer, Heidelberg (2012)
2. cnet.com: Comodo hack may reshape browser security (April 4, (2011), http://news.cnet.com/8301-31921_3-20050255-281.html
3. van Dijk, M., Juels, A., Oprea, A., Rivest, R.L.: FlipIt: The game of "stealthy takeover". Cryptology ePrint Archive, Report 2012/103 (2012)
4. Falliere, N., Murchu, L.O., Chien, E.: W32.Stuxnet Dossier (February 2011), http://www.symantec.com/connect/blogs/w32stuxnet-dossier

---

[3] This complies with the results of the basic FlipIt game for a single resource in [3].

5. Finkle, J., Shalal-Esa, A.: Hackers breached U.S. defense contractors (May 27, 2011), http://www.reuters.com/article/2011/05/27/us-usa-defense-hackers-idUSTRE74Q6VY20110527
6. Grossklags, J., Reitter, D.: How task familiarity and cognitive predispositions impact behavior in a security game of timing. In: Proceedings of the 27th IEEE Computer Security Foundations Symposium, CSF (2014)
7. Kaspersky Lab: Flame... the latest cyber-attack (May 2012), http://www.kaspersky.com/flame
8. Kaspersky Lab: The MiniDuke Mystery: PDF 0-day Government Spy Assembler 0x29A Micro Backdoor (February 2013), http://www.securelist.com/en/blog/208194129/The_MiniDuke_Mystery_PDF_0_day_Government_Spy_Assembler_0x29A_Micro_Backdoor
9. Laszka, A., Felegyhazi, M., Buttyán, L.: A survey of interdependent security games. Tech. Rep. CRYSYS-TR-2012-11-15, CrySyS Lab, Budapest University of Technology and Economics (November 2012)
10. Laszka, A., Horvath, G., Felegyhazi, M., Buttyán, L.: FlipThem: Modeling targeted attacks with FlipIt for multiple resources (extended version), http://www.crysys.hu/%7Elaszka/papers/laszka2014flipthem.pdf
11. Laszka, A., Johnson, B., Grossklags, J.: Mitigating covert compromises: A game-theoretic model of targeted and non-targeted covert attacks. In: Proceedings of the 9th Conference on Web and Internet Economics (WINE), pp. 319–332 (2013)
12. Laszka, A., Johnson, B., Grossklags, J.: Mitigation of targeted and non-targeted covert attacks as a timing game. In: Das, S.K., Nita-Rotaru, C., Kantarcioglu, M. (eds.) GameSec 2013. LNCS, vol. 8252, pp. 175–191. Springer, Heidelberg (2013)
13. Mandiant: APT1: Exposing one of China's cyber espionage units (February 18, 2013), http://www.mandiant.com/apt1
14. Menn, J.: Key Internet operator VeriSign hit by hackers (February 2, 2012), http://www.reuters.com/article/2012/02/02/us-hacking-verisign-idUSTRE8110Z820120202
15. Nochenson, A., Grossklags, J.: A behavioral investigation of the FlipIt game. In: Proceedings of the 12th Workshop on the Economics of Information Security, WEIS (2013)
16. Pham, V., Cid, C.: Are we compromised? Modelling security assessment games. In: Grossklags, J., Walrand, J. (eds.) GameSec 2012. LNCS, vol. 7638, pp. 234–247. Springer, Heidelberg (2012)
17. Reitter, D., Grossklags, J., Nochenson, A.: Risk-seeking in a continuous game of timing. In: Proceedings of the 13th International Conference on Cognitive Modeling (ICCM), pp. 397–403 (2013)
18. Rivner, U.: Anatomy of an attack (April 2011), http://blogs.rsa.com/anatomy-of-an-attack/
19. Symantec Security Response: W32.Duqu: The Precursor to the Next Stuxnet (October 18, 2011), http://www.symantec.com/connect/w32_duqu_precursor_next_stuxnet
20. Varian, H.: System reliability and free riding. In: Economics of Information Security, pp. 1–15. Springer (2004)

Secure Message Delivery Games for Device-to-Device Communications

Emmanouil Panaousis[1], Tansu Alpcan[2],
Hossein Fereidooni[3,*], and Mauro Conti[3,*]

[1] Queen Mary University of London, UK
e.panaousis@qmul.ac.uk
[2] The University of Melbourne, Australia
tansu.alpcan@unimelb.edu.au
[3] University of Padua, Italy
{hossein,conti}@math.unipd.it

Abstract. Device-to-Device (D2D) communication is expected to be
a key feature supported by next generation cellular networks. D2D can
extend the cellular coverage allowing users to communicate when
telecommunications infrastructure are highly congested or absent. In
D2D networks, any *message delivery* from a *source* to a *destination* relies
exclusively on intermediate devices. Each device can run different kinds
of *mobile security software*, which offer protection against viruses and
other harmful programs by using real-time scanning in every file entering the device. In this paper, we investigate the best D2D network path
to deliver a potentially malicious message from a source to a destination.
Although our primary objective is to increase security, we also investigate the contribution of energy costs and quality-of-service to the path
selection. To this end, we propose the *Secure Message Delivery* (SMD)
protocol, whose main functionality is determined by the solution of the
Secure Message Delivery Game (SMDG). This game is played between
the *defender* (i.e., the D2D network) which abstracts all legitimate network devices and the *attacker* which abstracts any adversary that can
inject different malicious messages into the D2D network in order, for
instance, to infect a device with malware. Simulation results demonstrate the degree of improvement that SMD introduces as opposed to a
shortest path routing protocol. This improvement has been measured in
terms of the defender's expected cost as defined in SMDGs. This cost
includes security expected damages, energy consumption incurred due
to messages inspection, and the quality-of-service of the D2D message
communications.

Keywords: game theory, security, device-to-device communications.

---

* Hossein Fereidooni and Mauro Conti are supported by the Project "Tackling Mobile
Malware with Innovative Machine Learning Techniques" funded by the University of
Padua. Mauro Conti is also supported by the EU Marie Curie Fellowship n. PCIG11-
GA-2012-321980 and by the MIUR-PRIN Project TENACE n. 20103P34XC.

R. Poovendran and W. Saad (Eds.): GameSec 2014, LNCS 8840, pp. 195–215, 2014.

1   Introduction

Nowadays, the vast demand for anytime-anywhere wireless broadband connectivity has posed new research challenges. As mobile devices are capable of communicating in both cellular (e.g., LTE) and unlicensed (e.g., IEEE 802.11) spectrum, the Device-to-Device (D2D) networking paradigm has the potential to bring several immediate gains. Networking based on D2D communication [1–4] not only facilitates wireless and mobile peer-to-peer services but also provides energy efficient communications, locally offloading computation, offloading connectivity and high throughput.

Another emerging feature of D2D is the establishment and use of multi-hop paths to enable communications among non-neighboring devices. In multi-hop D2D communications, messages are delivered from a source to a destination via intermediate devices, independently of operators' networks. Relay by device has been proposed by the Telecommunication Standardization Advisory Group (TSAG) in the International Telecommunication Union Telecommunication Sector (ITU-T).

A key question in *multi-hop D2D networks* is, which route should the originator of a message choose to send it to an intended destination? To motivate the application of our model, we emphasize in the need for *localized applications*. In particular, these applications run in a collaborative manner by groups of devices at a location where telecommunications infrastructures:

- are not presented at all, e.g., underground stations, airplanes, cruise ships, parts of a motorway, and mountains;
- have collapsed due to physical damage to the base stations or insufficient available power, e.g., areas affected by a disaster such as earthquake;
- are over congested due to an extremely crowded network, e.g., for events in stadiums, and public celebrations.

Furthermore, relay by device can be leveraged for commercial purposes such as advertisements and voucher distributions for instance in large shopping centers. This is considered a more efficient way of promoting businesses than other traditional methods such as email broadcasting and SMS messaging due to the immediate identification of the clients in a surrounding area. Home automation and building security are another two areas that multi-hop message delivery using D2D communications is likely to overtake our daily life in the near future. Lastly, multi-hop D2D could be leveraged towards the provision of anonymity against cellular operators as proposed in [12].

Due to the large number of areas D2D communications are applicable to, devices are likely to be an ideal target for attackers. Malware for mobile devices evolves in the same trend as malware for PCs. It can spread for instance through a Multimedia Messaging System (MMS) with infected attachments, or an infected message received via Bluetooth aiming at stealing users' personal data or credit stored in the device. An example of a well-known worm that propagates through Bluetooth was Cabir [7], which consists of a message containing an application file called `caribe.sis`. Mabir, a variant of Cabir, was

spread also via MMS by sending out copies of itself as a .sis file. Van Ruitenbeek et al. [8] investigated the effects of MMS viruses that spread by sending
infected messages to other devices. In addition, Bose and Shin [9] examined the
propagation of malware that spread via SMS or MMS messages and short-range
radio interfaces while Polla et al. [10] have made a thorough survey on mobile
malware.

1.1 Contributions

In this paper, we assume that each device has some host-based intrusion detection capabilities (e.g., antivirus). Therefore, a device would be able to detect
malicious application-level events as in [11]. We assume that each device has
its own detection rate which contributes towards the overall detection rate of
the routes that this device is on. To increase the level of security of a message delivery, the route with the highest detection capabilities must be selected
to relay the message to the destination. Apart from security, energy consumption is of crucial importance because devices (e.g., smartphones) usually impose
strict energy constraints. This becomes more important due to the limited CPU
and memory capabilities that devices have, which entail higher energy cost as
opposed to cases where no message inspection takes place.

In this paper, we propose the *Secure Message Delivery* (SMD) protocol. The
primary objective of this protocol is to choose the most secure path to deliver a
message from a sender to a destination in a multi-hop D2D network. SMD can
work on top of underlying physical and MAC layer protocols [5,6]. Apart from
security, SMD respects the energy costs and Quality-of-Service (QoS) of each
route. This happens by giving certain weights to each of the involved parameters
(security, energy, QoS) with more emphasis to be put on security.

We formulate *Secure Message Delivery Games* (SMDGs) in order to derive
an optimal behavior for the SMD. In these games, one or more adversaries,
abstracted by the *attacker*, aim at increasing the security damage, incurred to the
defender (i.e., network), by injecting malicious messages into the D2D network.
On the other hand, the defender chooses the "best route" for message delivery.
In SMDGs, the utility of the defender is influenced by: (i) the probability of
the delivered message to be correctly classified as malicious or benign before
it is delivered to the intended destination; (ii) the *energy cost* associated with
message forwarding, and *message inspection* on relay devices during message
delivery; and (iii) the QoS of the message communications on the chosen D2D
path.

The remainder of this paper is organized as follows. Section 2 summarizes
the most relevant related work within the intersection of game theory, security
and mobile distributed networking. In Section 3 we present the system model
whilst Section 4 formulates the SMDGs and it provides their solutions. In Section
5 the SMD routing protocol for D2D networks is described. We present some
preliminary simulation results in Section 6 for different number and types of
malicious messages distributions, and different D2D network profiles. Section
7 concludes this paper by summarizing its main contributions, limitations and
highlighting our plans for future work.

2   Related Work

The papers we discuss in this section have used game theory in favor of security in mobile distributed networks. These address different challenges including *secure routing* and *packet forwarding* [13,27,29–31], *trust establishment* [15,27], *intrusion detection* [15,21,23,24,26], and *optimization of energy costs* [17–19,22].

In [27], Sun et al. presented an information theoretic framework to evaluate trustworthiness in ad hoc networks and to assist malicious detection and route selection. According to their mechanism, a source node chooses a route to send a message to a destination by looking up the packet-forwarding nodes' trustworthiness, and selecting the most trustworthy route. Yu et al. examined, in [29], the dynamic interactions between "good" nodes and adversaries in mobile ad hoc networks (MANETs) as secure routing and packet forwarding games. They have derived optimal defense strategies and studied the maximum potential damage, which incurs when attackers find a route with maximum number of hops and they inject malicious traffic into it. Extension of the previous work is presented in [31], where Yu and Liu examined the issues of cooperation stimulation by modeling the interactions among nodes as multi-stage secure routing and packet forwarding games. In [30], the same authors focused on a two-player packet forwarding game stating that nodes must not help their opponents more than their opponents has helped them back. Felegyhazi et al. have studied in [13] the Nash equilibria of packet forwarding strategies with TFT (Tit-For-Tat) punishment strategy in a repeated game.

In [17], the authors presented a Bayesian hybrid detection approach to preserve the energy spent for intrusion detection. In the proposed static game, the defender fixes the prior probabilities about the types of his opponent. The dynamic game allows the defender to update his belief about his opponent's type based on new observed actions and the game history. The authors formulated the attacker/defender game model in both static and dynamic Bayesian game contexts, and investigated the equilibrium strategies of the two players. Lui et al. in [18] put forwarded a more comprehensive game framework and they used cross-feature analysis on feature vectors constructed from the training data to determine the actions of a potential attacker in each stage game. They proposed to use the equilibrium monitoring strategies to operate between a lightweight IDS and a heavyweight IDS. In [19], Marchang et al. proposed a game-theoretic model of IDS for MANETs. They have used game theory to model the interactions between the IDS and the attacker to determine whether it is essential to always keep the IDS running without impacting its effectiveness in a negative manner.

In [23], Patcha et al. provided a mathematical framework to analyze intrusion detection in MANETs. They model the interaction between an attacker and an individual node as a two player non-cooperative signaling game. The sender could be a *regular* or a *malicious* node. A receiving node equipped with an intrusion detection system (IDS) detects a "message/attack" with a probability depending on his belief, and the IDS updates the beliefs according to this message. However, it is not explained how the IDS updates the beliefs according to this message. The same authors have also reinforced the suitability of using game theory for modeling intrusion detection by giving a theoretically consistent model in [24]. They used the concept of multi-stage dynamic non-cooperative game with incomplete information to model intrusion detection in a network that uses host-based IDSs. A cooperative approach is proposed in [21] by Otrok et al. to detect and analyze intrusions in MANETs. The authors used the Shapley value to analyze the contribution of each node to the network security and proposed pre-defined security classes to decrease false positives. They also considered cache poisoning and malicious flooding attacks. Santosh et al. in [26], employed game theoretic approaches to detect intrusions and identify anomaly behaviors of nodes in MANETs. The authors aimed at building an IDS based on a cooperative scheme to detect intrusions in MANETs using game theoretic concepts.

In [15], Cho et al. developed a mathematical model to analyze and reveal the optimal rate to perform intrusion detection related tasks. They enhanced the *system reliability of group communication systems* in MANETs given information regarding operational conditions, system failures, and attacker behaviors. They have also discussed to prolong the system lifetime and cope with inside attacks. They proposed that intrusion detection should be executed at an optimal rate to maximize the mean time to failure of the system.

Finally in [22], Panaousis and Politis present a routing protocol that respects the energy spent by intrusion detectors on each route and therefore prolonging network lifetime. However, this protocol does not investigate the effect of different malicious messages. It rather takes a simplistic approach according to which the attacker either attacks or not a route.

As we have seen in this section, a substantial amount of game theoretic models for security in distributed mobile networks (e.g., mobile ad hoc networks) have been proposed in the literature. However, none of them addresses all aspects of security, QoS and energy efficiency at the same time. Motivated by this observation, our work contributes to the existing literature by bringing together these three aspects, under a generic but also customizable model provided by the SMDGs. Furthermore, our work defines the adversary's pure strategies to be a set of different malicious messages. And this is not an aspect of investigation of papers identified by our literature review. It is worth noting that we consider the work undertaken, in this paper here, as the first step towards a more complex and advanced game theoretic secure message delivery protocol for D2D networks.

3 System Model

This section presents our system model and its different components. We assume a multi-hop Device-to-Device (D2D) communication network that extends a cellular network (e.g., LTE Advanced) as illustrated in Fig. 1.

Data transmission takes place in the application layer in the form of data units called *messages*. Any device can be the source (s) of a message and each message has a final destination (d). When d is not within the transmission range

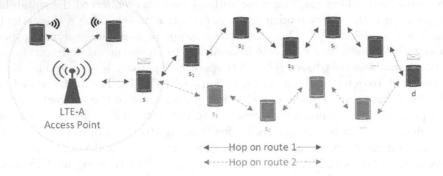

Fig. 1. Example of a D2D network

of **s**, a route must be established to allow message delivery. Therefore, there is an apparent need for the devices to collaborate to relay messages towards **d**.

We refer to the i-th mobile device by s_i, and define the set of all legitimate mobile devices in a mobile network as $S \triangleq \{s_i\}$. When the l-th type of message, denoted by m_l, has to be delivered to a destination device (**d**), a route must be chosen by **s** to serve that purpose. Formally, we denote route j by r_j. The devices on r_j must forward m_l towards **d**. We define the set of all routes from **s** to **d** as $R \triangleq \{r_j\}$, and the set of all devices that constitute r_j is expressed by S_j.

We denote the set of all different types of messages[1] by $\mathcal{M}$. This equals the union of the set of all malicious undetected messages ($\mathcal{M}_m$), and the set of all benign messages ($\mathcal{M}_b$). Therefore, $\mathcal{M} \triangleq \mathcal{M}_m \cup \mathcal{M}_b$. An *attack* is defined as the attempt of the attacker to harm **d** through the delivery of a malicious message. When m_l stays undetected prior to be delivered to **d**, we say that it causes harm $\mathcal{H}_l$, which is associated with the damage caused to an asset that the device holds (e.g., data loss). We also assume that any false alarm has loss equivalent to $\mathcal{F}$. The security effectiveness of a device against a malicious message is denoted by $\delta(s_i, m_l)$, and it is equivalent to the detection rate of an attack. The vector $\Delta(s_i) \triangleq \langle \delta(s_i, m_1), \dots, \delta(s_i, m_\psi) \rangle$ defines all the different values of security effectiveness of s_i with regard to the different messages. For more convenience, Table 1 summarizes the notation used in this paper.

3.1   Collaborative Detection

In our model, the aim of the devices is to detect malicious messages injected through an *entry point* into the D2D network. We assume that each device that receives a message is responsible for inspecting it by using its detection capabilities to the best level possible. Based on the results of the detection, the device updates the *confusion matrix* of the route. This is a right stochastic matrix, which holds the probability of the different messages being detected

---

[1] Very often, we use the terms *types of messages*, and *messages* interchangeably according to the context.

Table 1. Notation

S	Set of devices	s_i	device i
m_l	message l	$h^\star$	Maximum possible route length in hops
s	Message source	d	Message destination
P^A	Attacker	P^D	Defender
R	Set of routes from s to d	r_j	j-th route from s to d
S_j	Set of devices on r_j	$\mathcal{M}$	Set of messages
$\mathcal{M}_m$	Set of malicious messages	$\mathcal{M}_b$	Set of benign messages
$\delta(s_i, m_l)$	Security effectiveness of s_i against m_l	$\Delta(s_i)$	Security effectiveness vector of s_i
σ_i	Security energy cost of s_i	f_i	Forwarding energy cost of s_i
ϵ_i	Total message delivery energy cost of s_i	e_j	Total energy cost on r_j
T	Lifetime of a Nash message delivery plan	E	Vector of energy costs, $\forall\, r_j$ from s to d
$\mathbf{h_j}$	Number of hops on r_j	H	Vector of hops, $\forall\, r_j$ from s to d
$C^{(s_i)}$	Confusion matrix of s_i	$C^{(r_j)}$	Confusion matrix of r_j
$\mathcal{F}$	False alarm loss	$\mathcal{H}_l$	Security damage if m_l undetected
w_s	Security cost weight	w_{fa}	False alarm cost weight
w_e	Energy cost weight	w_q	QoS cost weight
D	Payoff matrix of P^D	A	Payoff matrix of P^A
d_{jl}	Utility of P^D for (r_j, m_l)	a_{jl}	Utility of P^A for (r_j, m_l)
$\mathbf{D}^*$	Nash message delivery plan	r^*	Nash route

correctly, being confused with other messages or being identified as benign. This matrix type was initially proposed in [28] (p. 100).

Each device that receives a message, follows exactly the same procedure until the message arrives at d. At this point, the confusion matrix should have taken the most accurate detection values (ideally is the identity matrix) due to all inspections undertaken by the devices on this route. Collaborative detection of a malicious message along a path requires forwarding state information, which includes results of the inspections previously conducted on the message. This prevents unnecessary duplication of inspections, thus saving energy.

3.2 Device Confusion Matrix

Given the set of messages $\mathcal{M}$, the linear mapping $C^{(s_i)}: \mathcal{M} \to \mathcal{M}$ describes the *detection capability* of s_i for a message received. This capability is modeled using a stochastic *device confusion matrix* as follows:

$$C^{(s_i)} \triangleq [C^{(s_i)}_{uv}]_{\psi \times \psi}, \text{ where } 0 \leq C^{(s_i)}_{uv} \leq 1, \ \forall u, v \in \{1, \dots, \psi\}. \tag{1}$$

A confusion matrix value $C^{(s_i)}_{uv}$ denotes the probability of a message u being reported as message v. If $m_u \neq m_v$, then the device confuses one message for another. Such misinterpretation is beneficial for the attacker because the attack associated with the message is not mitigated. If $m_u \in \mathcal{M}_m$, and $m_v \in \mathcal{M}_b$, $C^{(s_i)}_{uv}$ is the probability of the D2D network failing to report an attack. If $m_u \in$

$\mathcal{M}_b$, and $m_v \in \mathcal{M}_m$, then $C_{uv}^{(s_i)}$ is the probability of a *false alarm*. One of the objectives of the D2D network must be the confusion matrix to become the identify matrix (*no confusion*) by the time a message is delivered to d. In another sense, if the confusion matrix is the identity matrix, every single malicious message can be detected before it infects d. However this case is not likely to be achieved in practice due to, for instance, 0-day vulnerabilities, and other misclassification errors. To motivate the computation of confusion matrices we present the following example.

Example 1. Assume $S = \{s_1, s_2\}$, and $\mathcal{M} = \{m_1, m_2, m_3\}$. Also, $m_1, m_2 \in \mathcal{M}_m$, and $m_3 \in \mathcal{M}_b$. We also set the false alarm rate equal to 0.05 for both devices. The security effectiveness vectors are $\Delta(s_1) = \langle 0.5, 0.8 \rangle$ and $\Delta(s_2) = \langle 0.75, 0.6 \rangle$. We also assume that none device confuses a malicious message for another malicious message and therefore $C_{uv}^{(r_j)} = 0, \forall u \neq v, m_u, m_v \in \mathcal{M}_m$. Then the devices confusion matrices are the following:

$$C^{(s_1)} = \begin{pmatrix} 0.5 & 0 & 0.5 \\ 0 & 0.8 & 0.2 \\ 0.05 & 0.05 & 0.9 \end{pmatrix}, \ C^{(s_2)} = \begin{pmatrix} 0.75 & 0 & 0.25 \\ 0 & 0.6 & 0.4 \\ 0.05 & 0.05 & 0.9 \end{pmatrix}. \tag{2}$$

3.3   Route Confusion Matrix

Similarly, given the set of messages $\mathcal{M}$, the linear mapping $C^{(r_j)}: \mathcal{M} \to \mathcal{M}$ describes the final detection capability of the D2D network on r_j. This is the *route confusion matrix* for r_j derived from the confusion matrices of the devices that constitute this route. In the problem we examine, the order of detectors does not matter. Therefore, the confusion matrix for each combination can be computed prior to the message delivery.

An advanced way of deriving the route confusion matrix values is to use a boosting meta-algorithm such as *Adaboost* [16]. If we consider that each device detector is a weak classifier then boosting makes classifiers focusing on data that was previously misclassified. The underlying concept of Adaboost is that several weak classifiers can yield a strong classifier. The confusion matrix of a route is a representation of the weighted classifiers on the devices. It is worth mentioning here that boosting is effective only when all devices trust each other. For the boosting scheme to work there is a need for a broadcasting system which updates the classifiers and pre-sets confusion matrices for the combination of detectors. Nevertheless, such a system has to be implemented anyway for updating virus signatures and anomaly detector parameters. Thus, the update of the classifiers can be piggybacked on top of them.

A "naive" alternative to boosting can be a *linear combination algorithm* where each device contributes linearly to the final route detection capability by some weight determined by characteristics of the route (e.g., #hops).

3.4   Energy Costs and QoS

Each time a device receives a message it spends energy: (i) to detect any sign of malice (security energy cost, σ_i) and (ii) to forward a message towards **d** (forwarding energy cost, f_i). The former is determined by all required intrusion detection tasks undertaken during message inspection. The second is related to the energy spent for relaying the message towards the next-hop on the route from **s** to **d**. We denote by ϵ_i the *secure message delivery cost* incurred to a device during message delivery. Formally, we have that $\forall s_i \in S : \epsilon_i \triangleq \sigma_i + f_i$.

The total *route energy cost* on r_j, when a message is delivered over r_j, is denoted by $\mathbf{e_j}$, and it is derived by $\mathbf{e_j} = \sum_{s_i \in S_j} \epsilon_i$. The energy costs of all routes between **s** and **d** are given by the vector $\mathbf{E} \triangleq \langle \mathbf{e_1}, \ldots, \mathbf{e_\xi} \rangle$.

Apart from security and energy efficiency, QoS is an important consideration when deciding upon message delivery. We denote by $\mathbf{h_j}$ the number of hops on r_j. In this paper, we measure the QoS of a route as $\mathbf{h_j}/\mathbf{h}^\star$, where $h^\star \triangleq N_S - 1$, and N_S is the total number of devices in the D2D network. The number of hops of all routes $r_1, \ldots, r_\xi$ from **s** to **d** are given by $\mathbf{H} \triangleq \langle \mathbf{h_1}, \ldots, \mathbf{h_\xi} \rangle$.

In this paper, we assume a best effort message delivery service without acknowledgments. Along with having higher end-to-end delay due to this assumption, as the number of hops increases the probability of a message to be lost is higher. This is due to mobility, which is meant to be common in D2D networks. It is worth noting here that our model does not consider real-time multimedia communications because they require higher bandwidth than what a typical multi-hop D2D network provides.

3.5   Network Profiles

To allow the expression of different *network profiles*, we have defined an importance costs vector $[w_s, w_{fa}, w_e, w_q]$. By w_s, we denote the security importance weight which accounts for the level of importance the defender gives to some expected security damage (e.g., data theft); w_{fa} is the importance of the false alarm cost (i.e., cost for dropping an innocent message); w_e is the importance that the defender places into the energy cost which can influence the network lifetime and speed up network fragmentation; and w_q is the importance of the QoS for the defender which accounts for the message success delivery rate and end-to-end delay. This vector allows the network designer to define their *network profile* based on their requirements, measured in terms of security, energy preservation, and QoS.

4   Secure Message Delivery Games

In this section, we use game theory to model the interactions between a D2D network (the *defender*) and any adversarial entity (the *attacker*). The latter aims at launching an attack against a device by sending a malicious message to it through the network's entry point as depicted in Fig. 1. Formally, we define the set of players as $\mathcal{P} \triangleq \{P^D, P^A\}$.

The objective of P^D is to securely deliver a message to the intended destination d. By secure delivery we refer to the message being relayed through the network and collaboratively inspected by the devices on its way to d, in order to mitigate any security risk inflicted by P^A. Therefore the security objective of P^D is to correctly detect and filter out malicious messages before they reach their destination. Every request for message delivery to d defines a *Secure Message Delivery Game* (SMDG).

4.1 Game Characterization

The SMDG is a non-cooperative two-person zero-sum game. The explanation to the zero-sum nature of SMDG is that we have assumed that the attacker aims at inflicting the highest possible damage to the defender. We could model a game where the benefit of the attacker is smaller than the loss of the defender. However, we have left this for future work along with the investigation of different attacker profiles that are associated with different payoffs.

The defender primarily aims at delivering the message securely to d while the attacker aims at infecting d with some malware attached to a malicious message as we mentioned previously. The SMDG is a repeated game since players make their decisions once for a pair of $\langle d, T \rangle$, where T is a predefined timeout, and d is the destination device for which the game is played. Afterwards, they repeat the game for either every other destination or when T expires. The value of T may depend on the devices' mobility. For instance, high mobility dictates small T in order valid routes to be discovered.

In SMDG, the players make their decisions concurrently without any *order of play*. However, an order of play can be imposed as an alternative where the attacker becomes the *leader* and the defender the *follower* of a Stackelberg game. Nevertheless, this consideration is out of the scope of this paper.

4.2 Strategies and Payoffs

The pure strategies of P^D consists of all routes from s to d. Therefore, the action set of P^D is defined as $\mathcal{A}^D \triangleq R = \{r_1, r_2, \ldots, r_\xi\}$. On the other hand, the pure strategies of P^A are the different messages that P^A can choose to send to d. A message can be one of the following:

$$\{\texttt{malicious}_1, \ldots, \texttt{malicious}_n, \texttt{harmless}, \texttt{surveillance}\} \tag{3}$$

Then, the finite action set of the attacker is defined as:

$$\mathcal{A}^A \triangleq \mathcal{M} = \{m_1, \ldots, m_\psi\} = \{m_1, \ldots, m_n\} \cup \{\texttt{harmless}, \texttt{surveillance}\}.$$

We denote by $G_d \triangleq \langle D, A \rangle$ an $\xi \times \psi$ bi-matrix game where the P^D (i.e., row player) has a payoff matrix $D \in \mathbb{R}^{\xi \times \psi}$ and the payoff matrix of P^A (i.e. the column player) is denoted by $A \in \mathbb{R}^{\xi \times \psi}$.

P^D chooses as one of their pure strategies one of the rows of the payoff bi-matrix $(D, A) \triangleq (d_{j,l}, a_{j,l})_{(r_j, m_l) \in [\xi] \times [\psi]}$. For any pair of strategies, $(r_j, m_l) \in [\xi] \times [\psi]$, P^D, P^A have payoff values equivalent to $d_{j,l}$ and $a_{j,l}$, respectively. The payoff of the defender for a given pair of players' pure strategies (r_j, m_l) follows:

$$U_D(r_j, m_l) \triangleq d_{j,l} \triangleq -w_s(1 - C_{ll}^{(r_j)})\mathcal{H}_l - w_{f_a}(1 - C_{ll}^{(r_j)})\mathcal{F} - w_e e_j - w_q h_j. \quad (4)$$

Generally, the first term is the expected security damage (e.g., data theft) inflicted by the attacker due to malicious messages being undetected while the second term expresses the expected cost of the defender due to false alarms. This accounts for benign messages that are dropped due to being detected as malicious. The next to last term is the energy cost of the defender when message delivery takes place over r_j while the last term expresses the expected QoS experienced on this route. Since players act independently, we can enlarge the strategy spaces, so as to allow the players to base their decisions on the outcome of random events. Therefore we consider the mixed strategies of both P^D and P^A. The mixed strategy $\mathbf{D} \triangleq [q_1, \ldots, q_\xi]$ of the defender is a probability distribution over the different routes from s to d, where q_j is the probability of delivering a message via r_j. We refer to a mixed strategy of P^D as the *message delivery plan*. On the other hand, the attacker's mixed strategy $\mathbf{A} \triangleq [p_1, \ldots, p_\psi]$ is a probability distribution over the different messages, where p_l is the probability of choosing m_l.

When considering mixed strategies, the defender's objective is quantified by the utility function:

$$U_D(\mathbf{D}, \mathbf{A}) = \sum_{j=1}^{\xi} \sum_{l=1}^{\psi} q_j d_{j,l} p_l - -w_s [\sum_{m_l \in \mathcal{M}_m} \sum_{r_j \in R} q_j (1 - C_{ll}^{(r_j)}) p_l \mathcal{H}_l] -$$

$$w_{f_a} [\sum_{m_l \in \mathcal{M}_b} \sum_{r_j \in R} q_j (1 - C_{ll}^{(r_j)}) p_l \mathcal{F}] - w_e \mathbf{DE}^T - w_q \mathbf{DH}^T, \quad (5)$$

where $j \in \{1, \ldots, \xi\}$, $l \in \{1, \ldots, \psi\}$.

Because SMDG is a zero-sum game, the attacker's utility is given by $U_A(\mathbf{D}, \mathbf{A}) = -U_D(\mathbf{D}, \mathbf{A})$. This can be interpreted as, the attacker can cause the maximum damage to the defender.

4.3 Nash Equilibrium

SMDG is a two-person zero-sum game with finite number of actions for both players, and according to Nash [20] it admits at least a Nash Equilibrium (NE) in mixed strategies. Saddle-points correspond to Nash equilibria as discussed in [28] (p. 42).

The following result, from [14], establishes the existence of a saddle (equilibrium) solution in the games we examine and summarizes their properties.

Theorem 1 (Saddle point of the SMDG). *The Secure Message Delivery Game defined admits a saddle point in mixed strategies, $(\mathbf{D}^*, \mathbf{A}^*)$, with the property that*

$$\mathbf{D}^* = \arg\max_{\mathbf{D}}\min_{\mathbf{A}} U_D(\mathbf{D}, \mathbf{A}), \ \forall \mathbf{A} \ \ and \ \ \mathbf{A}^* = \arg\max_{\mathbf{A}}\min_{\mathbf{D}} U_A(\mathbf{D}, \mathbf{A}), \ \forall \mathbf{D}.$$

Then, due to the zero-sum nature of the game the following holds:

$$\max_{\mathbf{D}}\min_{\mathbf{A}} U_D(\mathbf{D}, \mathbf{A}) = \min_{\mathbf{A}}\max_{\mathbf{D}} U_D(\mathbf{D}, \mathbf{A}).$$

The pair of saddle point strategies $(\mathbf{D}^*, \mathbf{A}^*)$ are at the same time security strategies for the players, i.e., they ensure a minimum performance regardless of the actions of the other. Furthermore, if the game admits multiple saddle points (and strategies), they have the ordered interchangeability property, i.e., the player achieves the same performance level independent from the other player's choice of saddle point strategy.

Our results can be extended to non-zero sum, bi-matrix games. In the latter case, the existence of a NE is also guaranteed, but the additional properties hold only in the case where the attacker's utility is a negative affine transformation (NAT) of the defender's utility.

Definition 1. *The Nash message delivery plan, denoted by $\mathbf{D}^*$, is the probability distribution over the different routes, as determined by the NE of the SMDG.*

The minimax theorem states that for zero sum games NE and minimax solutions coincide. Therefore, $\mathbf{D}^* = \arg\min_{\mathbf{D}}\max_{\mathbf{A}} U_A(\mathbf{D}, \mathbf{A})$. This means that regardless of the strategy the attacker chooses, the *Nash message delivery plan* is the defender's security strategy that guarantees a minimum performance.

We can convert the original matrix game into a linear programming (LP) problem and make use of some of the powerful algorithms available for LP to derive the equilibrium. For a given mixed strategy $\mathbf{D}$ of P^D, P^A can cause a maximum damage to P^D by injecting a message $\widehat{m}$ into the D2D network. In that case, the utility of P^D is minimized and it is denoted by $U_D(\mathbf{D}, \widehat{m})$ (i.e., $U_D^{\min} = U_D(\mathbf{D}, \widehat{m})$). Formally, P^D seeks to solve the following LP:

$$\max_{\mathbf{D}} U_D(\mathbf{D}, \widehat{m})$$

$$\text{subject to} \begin{cases} U_D(\mathbf{D}, m_1) - U_D(\mathbf{D}, \widehat{m})e \geq 0 \\ \quad\vdots \\ U_D(\mathbf{D}, m_\psi) - U_D(\mathbf{D}, \widehat{m})e \geq 0 \\ \mathbf{D}e = 1 \\ \mathbf{D} \geq 0 \end{cases} \Rightarrow \begin{cases} \sum_{j=1}^{\xi} q_j d_{j,1} - U_D(\mathbf{D}, \widehat{m})e \geq 0 \\ \quad\vdots \\ \sum_{j=1}^{\xi} q_j d_{j,\psi} - U_D(\mathbf{D}, \widehat{m})e \geq 0 \\ \mathbf{D}e = 1 \\ \mathbf{D} \geq 0 \end{cases}$$

In this problem, e is a vector of ones of size ξ.

5   The Secure Message Delivery Protocol

In this section, we present the *Secure Message Delivery* (SMD) routing protocol whose routing decisions are taken according to the *Nash message delivery plan*. SMD increases security in a D2D network by mitigating the risk of adversaries harming legitimate devices via, for instance, malware attached to messages. SMD has been designed based on the mathematical findings of the SMDG and its main goal is to maximize $U_D(\mathbf{D}, \mathbf{A})$.

According to SMD, each time a request for message delivery to d is issued, s has to compute the *Nash message delivery plan* by solving an SMDG for this destination. To this end, the device uses its latest information about confusion matrices, QoS and energy costs. Then, the message is relayed and collaboratively inspected by the devices on its way to d. The objective of the network (i.e., P^D) is to correctly detect and filter out malicious messages before they infect d.

5.1 SMD Considerations

The SMD protocol takes routing decisions that increase the probability of detecting malicious messages. Apart from security, SMD utilizes standard approaches to take into account (i) the energy costs resulting from message forwarding and inspection, and (ii) the QoS of the chosen route. According to SMD, the devices maintain routing tables with at least three metrics per route:

- the route confusion matrix,
- the total expected energy cost on this route and,
- the shortest path in terms of number of hops (i.e., QoS).

If the only factor affecting the routing decision was security, then the route with the highest detection capability would be always chosen. This would result to a faster depletion of this route's energy as opposed to when a combination of different routes is chosen. Consequently, the D2D network would suffer fragmentation across the entire topology and consequently security would be reduced. This is the motivation behind considering energy costs upon path selection. Nevertheless, while the shortage of a device's battery can be solved by, for example, by using mobile solar cells as discussed in [4], and QoS might not be so much of a concern for message communications, secure message delivery remains a critical issue.

The formulation of the defender's utility function allows a device to decide how important the expected QoS and energy costs are compared to the expected security damage. For instance, the defender can decide to set the energy costs equal to 0 when a constant source of energy supply is available or to give a higher importance to security losses than QoS.

Due to the best effort nature of the communications (as a result of the multi-hop environment) the higher the number of hops (i.e., QoS) of a route the more likely a message is to be lost during its delivery via that route. QoS accounts for a successful message delivery rate and therefore the defender might never really want to ignore it. In general, SMD allows network designers to customize the protocol based on the *network profile* of the D2D network. In any case, all defender's preferences are reflected to the *Nash message delivery plan*.

5.2 Routing

Getting inspired by the functionalities of the well-known *Dynamic Source Routing* (DSR) [25] routing protocol, SMD consists of two main stages.

SMD - Stage I. In the first stage, s broadcasts a Route REQuest (RREQ_d) to discover routes towards d. Each device that receives a RREQ_d acts similarly by broadcasting it towards d and caches relevant information (i.e., originator of the request, ID of the RREQ_d). When d receives a RREQ_d, it prepares the RREP_d and sends it back towards s by using the reverse route which is built during the delivery of RREQ_d to d. Each RREP_d carries information about the route. This information includes the route confusion matrix (E_1), the total energy costs due to inspection and forwarding on this route (E_2), and the total number of hops (E_3). All three fields are updated while the RREP_d is traveling back to s.

Each device, involved in route discovery, that receives RREP_d, it updates E_1 by using boosting (e.g., Adaboost) or simply a linear combination algorithm without learning features. The same device (e.g., s_i) updates E_2 by adding its total energy cost ϵ_i to the route energy cost. Lastly, E_3 is increased by 1 in every hop from s to d.

Data: s, d, m_l
Result: m_l delivered
STAGE 1:
s seeks for a route to d by broadcasting RREQ_d
if *device s_i receives* RREQ_d **then**
  if $s_i \neq d$ **then**
    $s \leftarrow s_i$
    Execute Algorithm 1
  else
    Send an RREP_d back towards s using the reverse route r_j
  end
end
STAGE 2:
if *device s_i receives* RREP_d **then**
  if $s_i \neq s$ **then**
    Update $C^{(r_j)}$, e_j, h_j
    Attach $\langle C^{(r_j)}, e_j, h_j \rangle$ to the RREP_d
    Relay RREP_d back towards s
  else
    Cache $\langle C^{(r_j)}, e_j, h_j \rangle$ to the routing table
    break;
  end
end
s: Derive the *Nash message delivery plan* $\mathbf{D}^*$
s: Choose r^* probabilistically as dictated by $\mathbf{D}^*$
s: Deliver m_l to d over r^*

Algorithm 1. SMD Stages

According to SMD, after s sends a RREQ_d it has to await for some timeout T_{req}. Within this period s aggregates RREP_d messages and updates its routing table with information from those messages.

SMD - Stage II. In the second stage, s uses its routing table to solve the SMDG by computing the *Nash message delivery plan* **D***. The latter has a lifetime equivalent to T, as defined earlier. Then, s probabilistically selects a route according to **D*** to deliver the message to d. The chosen route is called the *Nash route* and it is denoted by r^*. Note that for the same d and before T expires, s uses the same **D*** to derive r^*, upon a message delivery request. Algorithm 1 summarizes the main SMD functionalities.

It is worth noting here that the complexity of the SMD protocol measured in terms of the number of messages exchanged in performing route discovery is $\mathcal{O}(2N_S)$, where N_S is the total number of devices in the D2D network.

6 Performance Evaluation

6.1 Simulation Parameters

In this section, we evaluate the performance of SMD by simulating 30 devices and 6 routes between s and d. The number of devices per route is selected randomly and the maximum number of devices per route has been set to 10. The number of malicious messages vary from 2 to 20 with an incremental step of 2.

We consider different network profiles to assess the performance of the SMD protocol. Note here that the network profile refers to the preference of the D2D network in terms of security (i.e., risk appetite), QoS (i.e., delay in message delivery), energy cost (i.e., spent for message inspection and message forwarding), and false alarm (probability of dropping benign messages) as determined by the *cost importance vector*.

We have used a uniform random generator to create the security effectiveness values for all devices. From these values the simulator creates all devices' confusion matrices. Then, we derive the route confusion matrices by using the Algorithm 2. Note that Algorithm 2 is executed by each device at the step of Algorithm 1 where $C^{(r_j)}$ is updated. This is a linear algorithm (less efficient than boosting due to lack of learning features) which allows us to get some preliminary results about the performance of SMD. This algorithm implements a weighted method according to which each device contributes to the route security effectiveness by

Table 2. The importance cost vectors used in our simulations

Network Profile	w_s	w_{fa}	w_e	w_q	Network Profile	w_s	w_{fa}	w_e	w_q
Security	10	0.5	0	0	Security & Energy Efficiency	5	0.5	5	0
Security & QoS	5	0.5	0	5	Security & QoS & Energy Efficiency	4	0.5	3	2.5

(Device security effectiveness) × *(1/Maximum number of hops in the network).*

The final route detection capability not only depends on the detection capability of each device on the route but also on the number of devices. As a result of this, the longer a route is the better its final security effectiveness.

After the route confusion matrices have been derived, the simulator computes the *Nash message delivery plan* for each of the network profiles presented in Table 2. We evaluate the performance of SMD by measuring the defender's expected cost when s uses SMD instead of a shortest path routing protocol. According to the latter, s chooses the path with the minimum number of hops to d. For each message delivery and protocol used we compute the defender's total expected cost which includes security, false alarm, energy and QoS costs.

Data: $C^{(s_i)}, C^{(r_j)}$
Result: Updated $C_{uv}^{(r_j)}$
for $u \in \mathcal{M}$ **do**
    for $v \in \mathcal{M}$ **do**
        if $u \in \mathcal{M}_m$ **then**
            if $v == u$ **then**
                $C_{uv}^{(r_j)} \leftarrow C_{uv}^{(s_i)}/h^\star + C_{uv}^{(r_j)}$
            end
            if $v \in \mathcal{M}_b$ **then**
                $C_{uv}^{(r_j)} \leftarrow 1 - C_{uu}^{(r_j)}$
            else
                // probability a malicious message u to be confused
                   with another malicious message
                $C_{uv}^{(r_j)} \leftarrow 0$
            end
        end
        if $u \in \mathcal{M}_b$ **then**
            if $v \notin \mathcal{M}_b$ **then**
                // f_a: device false alarm rate
                $C_{uv}^{(r_j)} \leftarrow f_a/h^\star + C_{uv}^{(r_j)}$
                $f_a^{route} \leftarrow C_{uv}^{(r_j)}$
            else
                // f_a^{route}: route false alarm rate
                $C_{uv}^{(r_j)} \leftarrow 1 - f_a^{route}$
            end
        end
    end
end

Algorithm 2. How a device s_i updates the route confusion matrix

---

We have considered 10 Cases each representing a different attacker's action set akin to different number of available malicious messages namely; $2, 4, \ldots, 20$.

For each Case we have simulated 1,000 message deliveries for a fixed network topology and we refer to the run of the code for the pair ⟨Case,#message deliveries⟩ by the term Experiment. We have repeated each Experiment for 25 independent network topologies to compute the standard deviation. We do that for all 10 Cases and each type of *attacker profile*.

In this paper we consider 2 different attacker profiles; *Uniform* and *Nash*. A *Uniform* attacker chooses any of the available messages with the same probability whilst a *Nash* attacker plays the attack mixed strategy given by the NE of the SMDG. Therefore, we have totally simulated

$$10 \ (Cases) \times 1,000 \ (Message \ deliveries) \times 25 \ (Runs \ of \ each \ experiment) \times 2$$
$$(Attacker \ profiles) = 500,000 \ Message \ deliveries.$$

Per message delivery, the simulator chooses an attack sample from the attack probability distribution which is determined by the attacker profile. The simulator aggregates the cost values of each Experiment for both SMD and the shortest path routing protocol.

6.2   Simulation Results

We have plotted the improvement on the total expected defender's cost when SMD is chosen as opposed to the shortest path routing protocol. The plots illustrate different number of available malicious messages, attacker profiles and importance cost vectors, in Figures 2 and 3.

From both figures we notice that SMD outperforms the shortest path routing protocol with the highest improvement to be achieved under the "Security" network profile. From Fig. 2 we notice that the average values of this improvement fluctuate approximately within the range [30%, 43%]. The second best performance is achieved under the "Security & QoS" network profile and it is only slightly better than the improvement we get under the "Security and Energy Efficiency" profile. The lowest improvement is noticed under the "Security & QoS & Energy Efficiency" network profile with the mean values to be within the range [10%, 18%]. We notice the same trends for a Nash attacker as illustrated in Fig. 3. One difference in the results is that under the network profile Security & QoS the difference in improvement compared to the Security & Energy Efficiency is more pronounced as opposed to the scenarios with a Nash attacker. We also notice that for all network profiles SMD improves the defender's expected cost in a greater degree in the presence of a Uniform Attacker rather than a Nash attacker although the defender chooses the Nash routing plan in either cases (since it minimizes the maximum potential cost inflicted by the attacker). This is due to the attacker maximizing the minimum defender's expected cost at the NE as stated in Theorem 1. On the other hand, the uniform attacker follows a naive distribution to inject different messages into the D2D network and therefore achieving a worse performance than the Nash attacker.

As a generic comment, the more focused objectives SMD has the higher the improvement of the defender's expected cost is, compared to a shortest path protocol. We also notice that the standard deviation is large in all Experiments.

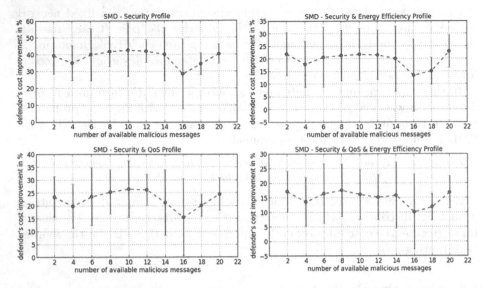

Fig. 2. Simulation results in presence of a uniform attacker

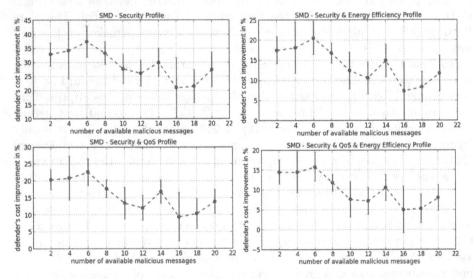

Fig. 3. Simulation results in presence of a Nash attacker

This can be explained by looking at the results from the different Experiments in more detail. By doing so, we noticed that occasionally the same routes are chosen by both SMD and the shortest path routing protocol. This can be explained by the number of available routes being only 6 in our simulations here. The generic trends demonstrate the improvement that SMD introduces even without the use of a boosting algorithm. These preliminary results are promising and we have plans for further investigations when a boosting algorithm (e.g., Adaboost) is used and a larger number of devices and routes are given. In addition, we are planning to examine different mobility levels and see how these affect the expected defender's cost under different network profiles with SMD.

7   Conclusions

In this paper we have investigated secure message delivery for device-to-device networks in a hostile environment with possible malicious behavior. We have formulated *Secure Message Delivery Games* (SMDGs) to study the interactions between the *defender* (i.e., device-to-device network), and different adversaries, which are abstracted by the player called *attacker*. The defender seeks the "best route" to deliver a message from a source device to a destination device whilst the latter aims to harm the destination with mobile malware attached to a message. The defender solves an SMDG to derive the *Nash message delivery plan* (i.e., Nash mixed strategy). Then, the defender probabilistically chooses a route according to this plan and delivers the message to the destination. Due to the multi-hop nature of the network, intermediate devices relay the message towards the destination. Apart from forwarding, the relaying devices are responsible for the inspection of the message to identify malicious signs and therefore providing security for the D2D message communications.

We have proposed the *Secure Message Delivery* (SMD) routing protocol which takes routing decisions according to the *Nash message delivery plan*. Apart from security, the protocol respects energy costs and end-to-end delay with the ability to be customized to consider each objective at a different degree. We have undertaken simulations to show how much SMD improves the defender's expected utility compared to a shortest path routing protocol. We believe this improvement will be more pronounced when we implement boosting techniques for the computation of the final intrusion detection capabilities (i.e., confusion matrices) of the routes. We have also plans to take into account the remaining energy of each route in the utility function of the defender, and investigate the impact of mobility to the results. Lastly, future work will consider a network-wide extension of the per-message game where the attacker aims to spread a mobile malware while the defender is attempting to stop it.

References

1. Doppler, K., Rinne, M., Wijting, C., Ribeiro, C.B., Hugl, K.: Device-to-device communication as an underlay to LTE-advanced networks. IEEE Communications Magazine 47(12), 42–49 (2009)
2. Feng, D., Lu, L., Yuan-Wu, Y., Ye Li, G., Li, S., Feng, G.: Device-to-device communications in cellular networks. IEEE Communications Magazine 52(4), 49–55 (2014)
3. Fodor, G., Dahlman, E., Mildh, G., Parkvall, S., Reider, N., Miklos, G., Turanyi, Z.: Design aspects of network assisted device-to-device communications. IEEE Communications Magazine 50(3), 170–177 (2012)

4. Nishiyama, H., Ito, M., Kato, N.: Relay-by-Smartphone: Realizing Multihop Device-to-Device Communications. IEEE Communications Magazine 52(4), 56–65 (2014)
5. Jianting, Y., Chuan, M., Hui, Y., Wei, Z.: Secrecy-Based Access Control for Device-to-Device Communication Underlaying Cellular Networks. IEEE Communications Magazine 17(11), 2068–2071 (2013)
6. Daohua, Z., Swindlehurst, A.L., Fakoorian, S.A.A., Wei, X., Chunming, Z.: Device-to-device communications: The physical layer security advantage. IEEE Communications Magazine, 1606–1610 (2014)
7. F-Secure: Bluetooth-Worm:SymbOS/Cabir, http://www.f-secure.com/v-descs/cabir.shtml (accessed June 2004)
8. Van Ruitenbeek, E., Courtney, T., Sanders, W.H., Stevens, F.: Quantifying the effectiveness of mobile phone virus response mechanisms. In: Proc. of the 37th Annual IEEE/IFIP International Conference on Dependable Systems and Networks (DSN), pp. 790–800 (2007)
9. Bose, A., Shin, K.G.: On mobile viruses exploiting messaging and bluetooth services. In: Proc. of the Securecomm and Workshops, pp. 1–10 (2006)
10. La Polla, M., Martinelli, F., Sgandurra, D.: A survey on security for mobile devices. IEEE Communications Surveys and Tutorials 15(1), 446–471 (2013)
11. Miettinen, M., Halonen, P., Hatonen, K.: Host-based intrusion detection for advanced mobile devices. In: Proc. of the 20th International Conference on Advanced Information Networking and Applications (AINA), vol. 2, pp. 72–76 (2006)
12. Ardagna, C.A., Conti, M., Leone, M., Stefa, J.: An anonymous end-to-end communication protocol for mobile cloud environments. IEEE Transactions on Services Computing (2014)
13. Felegyhazi, M., Buttyan, L., Hubaux, J.-P.: Nash equilibria of packet forwarding strategies in wireless ad hoc networks. IEEE Transactions on Mobile Comput. 5(5), 463–476 (2006)
14. Basar, T., Olsder, G.J.: Dynamic noncooperative game theory, 2nd edn. London Academic Press (1995)
15. Cho, J.H., Chen, I.R., Feng, P.G.: Effect of intrusion detection on reliability of mission-oriented mobile group systems in mobile ad hoc networks. IEEE Transactions on Reliability 59(1), 231–241 (2010)
16. Freund, Y., Schapire, R.E.: A decision-theoretic generalization of on-line learning and an application to boosting. Journal of Computer and System Sciences, 119–139 (1997)
17. Liu, Y., Comaniciou, C., Man, H.: A Bayesian game approach for intrusion detection in wireless ad hoc networks. In: Proc. of the Workshop on Game theory for Communications and Networks, GameNets (2006)
18. Liu, Y., Comaniciou, C., Man, H.: Modelling misbehaviour in ad hoc networks: A game theoretic approach for intrusion detection. International Journal of Security and Networks 1(7), 243–254 (2006)
19. Marchang, N., Tripathi, R.: A game theoretical approach for efficient deployment of intrusion detection system in mobile ad hoc networks. In: Proc. of the International Conference on Advanced Computing and Communications (ADCOM), pp. 460–464 (2007)
20. Nash, J.F.: Equilibrium points in n-person games. Proc. of the National Academy of Sciences 36(1), 48–49 (1950)
21. Otrok, H., Debbabi, M., Assi, C., Bhattacharya, P.: A cooperative approach for analyzing intrusions in mobile ad hoc networks. In: Proc. of the International Conference on Distributed Computing Systems Workshops, ICDCSW (2007)

22. Panaousis, E.A., Politis, C.: A game theoretic approach for securing AODV in emergency mobile ad hoc networks. In: Proc. of the 34th IEEE Conference on Local Computer Networks (LCN), Zurich, Switzerland, pp. 985–992 (2009)
23. Patcha, A., Park, J.M.: A game theoretic approach to modeling intrusion detection in mobile ad hoc networks. In: Proc. of the 5th Annual IEEE SMC Information Assurance Workshop, pp. 280–284 (2004)
24. Patcha, A., Park, J.M.: A game theoretic formulation for intrusion detection in mobile ad hoc networks. International Journal of Network Security 2(2), 131–137 (2006)
25. Johnson, D.B., Maltz, D.A.: Dynamic source routing in ad hoc wireless networks. In: Mobile Computing, pp. 153–181. Springer US (1996)
26. Santosh, N., Saranyan, R., Senthil, K.P., Vetriselvi, V.: Cluster based co-operative game theory approach for intrusion detection in mobile ad-hoc grid. In: Proc. of the International Conference on Advanced Computing and Communications (ADCOM), pp. 273–278 (2008)
27. Sun, Y.L., Yu, W., Han, Z., Liu, K.J.R.: Information theoretic framework of trust modeling and evaluation for ad hoc networks. IEEE Journal on Selected Areas of Communication 24(2), 305–317 (2006)
28. Alpcan, T., Basar, T.: Network Security: A Decision and Game-Theoretic Approach. Cambridge University Press (2010)
29. Yu, W., Ji, Z., Liu, K.J.R.: Securing cooperative ad-hoc networks under noise and imperfect monitoring: Strategies and game theoretic analysis. IEEE Transactions on Information Forensics and Security 2(2), 240–253 (2007)
30. Yu, W., Liu, K.J.R.: Game theoretic analysis of cooperation stimulation and security in autonomous mobile ad hoc networks. IEEE Transactions on Mobile Computing 6(5), 507–521 (2007)
31. Yu, W., Liu, K.J.R.: Secure cooperation in autonomous mobile ad-hoc networks under noise and imperfect monitoring: A game-theoretic approach. IEEE Transactions on Information Forensics and Security 3(2), 317–330 (2008)

A Three-Stage Colonel Blotto Game: When to Provide More Information to an Adversary

Abhishek Gupta[1], Tamer Başar[1], and Galina A. Schwartz[2,*]

[1] Coordinated Science Laboratory,
University of Illinois at Urbana-Champaign,
Urbana, Illinois, USA
{gupta54,basar1}@illinois.edu
[2] Department of Electrical Engineering and Computer Science,
University of California at Berkeley,
Berkeley, California, USA
schwartz@eecs.berkeley.edu

Abstract. In this paper, we formulate a three-player three-stage Colonel Blotto game, in which two players fight against a common adversary. We assume that the game is one of complete information, that is, the players have complete and consistent information on the underlying model of the game; further, each player observes the actions taken by all players up to the previous stage. The setting under consideration is similar to the one considered in our recent work [1], but with a different information structure during the second stage of the game; this leads to a significantly different solution.

In the first stage, players can add additional battlefields. In the second stage, the players (except the adversary) are allowed to transfer resources among each other if it improves their expected payoffs, and simultaneously, the adversary decides on the amount of resource it allocates to the battle with each player subject to its resource constraint. At the third stage, the players and the adversary fight against each other with updated resource levels and battlefields. We compute the subgame-perfect Nash equilibrium for this game. Further, we show that when playing according to the equilibrium, there are parameter regions in which (i) there is a net positive transfer, (ii) there is absolutely no transfer, (iii) the adversary fights with only one player, and (iv) adding battlefields is beneficial to a player. In doing so, we also exhibit a counter-intuitive

* The corresponding author for this article is Abhishek Gupta. Research was supported in part by AFOSR MURI Grant FA9550-10-1-0573, and in part by NSA through the Information Trust Institute of the University of Illinois. The research by Galina Schwartz is supported by NSF grant CNS-1239166, which provides funding for a frontier project FORCES (Foundations of Resilient CybEr-Physical Systems), NSF grant CNS-0910711, and by TRUST (Team for Research in Ubiquitous Secure Technology), which receives support from the NSF (#CCF-0424422) and the following organizations: AFOSR (#FA9550-06-1-0244), BT, Cisco, DoCoMo USA Labs, EADS, ESCHER, HP, IBM, iCAST, Intel, Microsoft, ORNL, Pirelli, Qualcomm, Sun, Symantec, TCS, Telecom Italia, and United Technologies. The authors would like to thank the anonymous referees for helpful suggestions.

R. Poovendran and W. Saad (Eds.): GameSec 2014, LNCS 8840, pp. 216–233, 2014.

property of Nash equilibrium in games: extra information to a player in the game does not necessarily lead to a better performance for that player. The result finds application in resource allocation problems for securing cyber-physical systems.

1   Introduction

The Colonel Blotto game is a complete-information static non-cooperative two-player game, in which two resource-constrained players fight against each other on a fixed number of battlefields. The players decide on the allocation of resources on each battlefield subject to their resource constraints. On each battlefield, the player deploying the maximum resource is declared the winner of that battlefield and accrues certain payoff. The goal of each player is to maximize the expected total number of battlefields that he/she wins.

The setup of the Colonel Blotto game is found naturally in several engineering and economic systems. Consider, for example, a data center with multiple servers under attack from a hacker. Each server can be viewed as a battlefield with the data center and the hacker viewed as the two players. Each player has limited computational resource to deploy – the data center deploys resource for securing the servers, and the hacker deploys resource for hacking the servers. The resulting game is captured by the Colonel Blotto game. Similarly, the competition between two research companies that are deploying their resources in different projects can also be analyzed within the framework of the Colonel Blotto game.

The Colonel Blotto game in which both players have equal resources and there are three battlefields was first solved in [2]. This result was later extended to the case of symmetric resources and arbitrary number of battlefields in [3]. In the same paper, the authors computed the Nash equilibrium for the case of asymmetric resources and two battlefields. However, Colonel Blotto game with asymmetric resources and three or more battlefields remained open until 2006, when Roberson established the existence of a Nash equilibrium in mixed strategies, and computed the (mixed) equilibrium strategies of the players in [4]. A similar setup was also considered in [5], in which the resource levels of both players were considered to be equal.

The work of Roberson sparked great interest in the field; numerous theoretical extensions of the game followed after 2006. In particular, [6] and [7] considered two-stage Colonel Blotto games. In [6], the authors identified situations in which adding battlefields during the first stage of the game is beneficial to the players. In [7], the authors considered a three-player Colonel Blotto game, in which the first two players fight against a common adversary. They have identified conditions under which forming a coalition could be beneficial to both players. However, they do not obtain a Nash equilibrium of the game.

Applications of the Colonel Blotto game have also received attention. References [8] and [9] studied phishing attacks and defense strategies over the internet. References [10] and [11] conducted experimental studies of the Colonel Blotto game with human subjects, and proposed a novel decision procedure, which the authors called *multi-dimensional reasoning*. Another interesting experimental paper is [12], where the authors study social interactions using a Facebook application called "Project Waterloo", which allows users to invite both friends and strangers to play Colonel Blotto against themselves.

Recently, we have formulated in [1] a three-stage Colonel Blotto game with hierarchical information structure, in which two players fight against a common adversary. In that paper, the problem formulation was as follows: At the first stage, the players may add battlefields. At the second stage, the game has a hierarchical information structure; the players may transfer some resources to each other, and the adversary has access to the amount of resource transferred. Based on this information, the adversary decides on its allocation of resources for the battles against the two players. At the third stage, the adversary fights two battles against the two players with the updated resource levels and battlefields. We further assumed that this is a game of complete information, that is, at any stage, all players including the adversary have access to all the information that has been generated in the past stage(s), and this is common knowledge.

This paper also considers a similar setup as in [1], but with a different information structure. In [1], we had assumed that the adversary has access to the information about the amount of resources that are transferred between the players during the second stage. In this paper, on the other hand, we assume that the adversary does not have access to that information. In other words, the transfer between the two players and the resource allocation of the adversary towards the two battles happen *simultaneously*[1]. This leads to a very different Nash equilibrium. One of the primary goals of this paper is to underscore the importance of information structure in the allocation of resources in a class of Colonel Blotto games. Furthermore, this study also provides insight on "what information about the formation of a strategic alliance should be made public" in such games. The information that is made public in a strategic alliance between two cyber-physical systems may have severe repercussions on the security and vulnerabilities of those systems if they are attacked by a strategic adversary. For the setting considered in this paper, if the information about the transfer between the first two players is provided to the adversary, then the adversary ends up with a lower total expected payoff. In other words, the first two players are better off by making the information about their strategic alliance public.

1.1   Outline of the Paper

We formulate the three-stage three-player Colonel Blotto game problem and identify several outstanding issues in Section 2. Thereafter, we recall the Nash equilibrium and the equilibrium expected payoffs to the players in the classical static two-player Colonel Blotto game in Section 3. The discussion in this section

---

[1] In decision problems, when decision makers act *simultaneously*, then it does not necessarily mean that they act at the same time instant; it simply means that a decision maker may not have access to the action of the other decision maker who may have acted in the past. The two cases require the same analysis.

is based on [4]. In Section 4, we compute the subgame-perfect Nash equilibrium of the game formulated in Section 2 for three specific cases. We also discuss and comment on the Nash equilibrium obtained in that section. We provide some concluding discussions and state the future directions that the research can take in Section 5.

Before we discuss the general setup of the game, we introduce a few notations in the next subsection.

1.2 Notations

For a natural number N, we use $[N]$ to denote the set $\{1, \ldots, N\}$. $\mathbb{R}_+$ and $\mathbb{Z}_+$ denote, respectively, the sets of all non-negative real numbers and non-negative integers. Let $\mathcal{X}_i, i \in [N]$ be non-empty sets. If $x_1 \in \mathcal{X}_1, \ldots, x_N \in \mathcal{X}_N$ are elements, then $x_{1:N}$ denotes the sequence $\{x_1, \ldots, x_N\}$. Similarly, $\mathcal{X}_{1:N}$ denotes the product space $\mathcal{X}_1 \times \cdots \times \mathcal{X}_N$.

2 Problem Formulation

In this section, we formulate a three-player three-stage Colonel Blotto game. The first two players are fighting against an adversary, call it A, who is the third player in the game. Henceforth, we use Player 3 and A interchangeably to refer to the adversary. Each player is endowed with some resources at the beginning of the game. We use β_i and α, respectively, to denote the initial endowment of the resources of Player $i \in \{1, 2\}$ and the adversary. At the beginning of the game, for every Player $i \in \{1, 2\}$, there are $n_i \geq 3$ battlefields, each with payoff v_i, at which the battle between Player i and the adversary will take place.

During the first stage of the game, each of the first two players may add additional battlefields at some cost. During the second stage of the game, the first two players may exchange resources among themselves if it improves their payoffs, while the adversary decides on the allocation of resources to fight against the first two players. At the final stage, the players fight against the adversary with updated battlefields and resources. We consider here a game of complete and perfect information.

2.1 Information Structures and Strategies of the Players

At the first stage of the game, the players know all the parameters and the model of the game, and we assume that this is common knowledge. Player $i \in \{1, 2\}$ decides on $m_i \in \mathbb{Z}_+$, the number of battlefields he/she wants to add to the existing set of battlefields and pays a total cost of cm_i^2. The adversary does not take any action at the first stage.

At the second stage of the game, all players, including the adversary, observe the number of battlefields (m_1, m_2) that were added. At this stage, the first two players decide on the transfers: Player i chooses a function $t_{i,j} : \mathbb{Z}_+^2 \to [0, \beta_i]$ which takes the number of battlefields added by the players, (m_1, m_2), as input, and outputs the amount of resource he/she transfers to Player $j \neq i$, where $i, j \in \{1, 2\}$. These functions have to satisfy the following constraint:

$$t_{i,j}(m_1, m_2) \leq \beta_i, \quad \text{for all } m_1, m_2 \in \mathbb{Z}_+, j \neq i.$$

The adversary does not observe the transfers among the players, and decides on functions $\alpha_i : \mathbb{Z}_+^2 \to [0, \alpha]$ with the constraint

$$\alpha_1(m_1, m_2) + \alpha_2(m_1, m_2) \leq \alpha, \quad \text{for all } m_1, m_2 \in \mathbb{Z}_+^2.$$

We use r_i to denote the amount of resource available to Player $i \in \{1, 2\}$ after the redistribution of resources. This is given by

$$r_i := r_i(t_{1,2}, t_{2,1}) = \beta_i + (t_{j,i} - t_{i,j}) \quad i \neq j, i, j \in \{1, 2\}.$$

For a given triple $\alpha_i, r_i \in \mathbb{R}_+$ and $m_i \in \mathbb{Z}_+$, let us define the sets

$$\mathcal{A}_i(\alpha_i, m_{1:2}) := \left\{ \{\alpha_{i,k}\}_{k=1}^{n_i+m_i} \subset \mathbb{R}_+ : \sum_{k=1}^{n_i+m_i} \alpha_{i,k} = \alpha_i(m_1, m_2) \right\},$$

$$\mathcal{B}_i(r_i, m_i) := \left\{ \{\beta_{i,k}\}_{k=1}^{n_i+m_i} \subset \mathbb{R}_+ : \sum_{k=1}^{n_i+m_i} \beta_{i,k} = r_i(t_{1,2}, t_{2,1}) \right\}.$$

At the final stage of the game, Player i and the adversary play the usual static two-player Colonel Blotto game on $n_i + m_i$ battlefields, with Player i having r_i and adversary having α_i amounts of resource. Thus, given the resource levels r_i of Player i, $i = 1, 2$, and α_i of the adversary, the action spaces of Player i and the adversary are, respectively, $\mathcal{B}_i(r_i, m_i)$ and $\mathcal{A}_i(\alpha_i, m_i)$. If $n_i \geq 3$, then there is no pure strategy Nash equilibrium of Player i and the adversary at the final stage. Thus, given the resource levels of the players, Player i and the adversary, respectively, decide on probability measures $\mu_i \in \wp(\mathcal{B}_i(r_i, m_i))$ and $\nu_i \in \wp(\mathcal{A}_i(\alpha_i, m_i))$ over their respective action spaces.

Henceforth, we use $\gamma^i := \{\gamma_1^i, \gamma_2^i, \gamma_3^i\}$ to denote the strategy of Player $i \in \{1, 2, A\}$, which is defined as follows:

$$\gamma_1^i := m_i, \quad \text{for } i \in \{1, 2\},$$
$$\gamma_2^1(m_1, m_2) := \{t_{1,2}(m_1, m_2)\}, \quad \gamma_2^2(m_1, m_2) := \{t_{2,1}(m_1, m_2)\},$$
$$\gamma_2^A(m_1, m_2) := \{\alpha_1(m_1, m_2), \alpha_2(m_1, m_2)\}$$
$$\gamma_3^i(m_1, m_2, t_{1,2}, t_{2,1}) := \{\mu_i\}, \quad i \in \{1, 2\},$$
$$\gamma_3^A(m_1, m_2, t_{1,2}, t_{2,1}) := \{\nu_1, \nu_2\}.$$

Thus, each γ^i is a collection of functions; we denote the set of all such γ^is by Γ^i.

2.2 Payoff Functions of the Players

Consider the game between Player i and the adversary at the third stage of the game. Let us use $\beta_{i,k}$ and $\alpha_{i,k}$ to denote, respectively, the amounts of resource

Player i and the adversary deploy on battlefield $k \in [n_i + m_i]$. On every battlefield $k \in [n_i + m_i]$, the player who deploys maximum amount of resource wins and receives a payoff v_i. In case of a tie, the players share the payoff equally[2]. We let $p_{i,k}(\beta_{i,k}, \alpha_{i,k})$ denote the payoff that Player i receives on the battlefield k, which we take to be given by

$$p_{i,k}(\beta_{i,k}, \alpha_{i,k}) = \begin{cases} v_i & \beta_{i,k} > \alpha_{i,k}, \\ \frac{v_i}{2} & \beta_{i,k} = \alpha_{i,k}, \\ 0 & \text{otherwise}, \end{cases}$$

for $i \in \{1, 2\}$ and $k \in [n_i + m_i]$. The payoff to the adversary on a battlefield k in the battle with Player i is given by

$$p_{i,k}^A(\beta_{i,k}, \alpha_{i,k}) = v_i - p_{i,k}(\beta_{i,k}, \alpha_{i,k}).$$

We take the expected payoff functionals of Player i and the adversary as

$$\pi_i(\gamma^{1:3}) = \mathbb{E}\left[\sum_{k=1}^{n_i+m_i} p_{i,k}(\beta_{i,k}, \alpha_{i,k})\right] - cm_i^2, \quad i \in \{1, 2\},$$

$$\pi_A(\gamma^{1:3}) = \mathbb{E}\left[\sum_{i=1}^{2}\sum_{k=1}^{n_i+m_i} p_{i,k}^A(\beta_{i,k}, \alpha_{i,k})\right],$$

where the expectation is taken with respect to the probability induced on the random variables $\{\beta_{i,k}, \alpha_{i,k}\}_{i,k}$ by the choice of strategies of the players in the game. The model of the game and the payoff functions are common knowledge among the players. The Colonel Blotto game formulated above is referred to as $\mathbf{CB}(\underline{n}, \underline{\beta}, \underline{\alpha}, \underline{v}, c)$.

We now define the Nash equilibrium of the game formulated above. The set of strategy profiles $\{\gamma^{1*}, \gamma^{2*}, \gamma^{A*}\}$ is said to form a Nash equilibrium of the game if it satisfies

$$\pi_i(\gamma^{1:2*}, \gamma^{A*}) \geq \pi_i(\gamma^i, \gamma^{-i*}, \gamma^{A*}), \qquad i \in \{1, 2\}$$
$$\pi_A(\gamma^{1:2*}, \gamma^{A*}) \geq \pi_A(\gamma^{1:2*}, \gamma^A)$$

for all possible $\gamma^i \in \Gamma^i$, $i \in \{1, 2, A\}$, where $\gamma^{1:2} := \{\gamma^1, \gamma^2\}$.

The set of all subgame-perfect Nash equilibria (SPNE) of a complete information game is a subset of all Nash equilibria of the game, and they can be obtained using a dynamic programming type argument (for precise definition, see [13]). In Section 4, we compute the SPNE of the game formulated above.

---

[2] It should be noted that if players play according to the Nash equilibrium strategies on the battlefields, then the case of both players having equal resource on a battlefield has a measure zero. Therefore, in equilibrium, the tie breaking rule does not affect the equilibrium expected payoffs.

2.3 Research Questions and Solution Approach

At the outset, it is not clear what kind of solution we would expect in such a game. We are particularly interested in investigating the conditions on the parameters of the game, under which the following scenarios are possible:

1. There is a positive transfer from one player to another. Since this is a non-cooperative game, the transfer should *increase or maintain* the payoffs to both players - the player who transfers resources and the player who accepts the transfer.
2. There is no transfer among the players at the second stage.
3. The adversary allocates all its resource to fight only one player.
4. The players have an incentive to add new battlefields.

We first recall some relevant results on the two-players static Colonel Blotto game from [4]. Solving the general problem formulated above is somewhat difficult due to the discontinuity of the expected payoff functions in the endowments of the players in the static game. Therefore, we restrict our attention to a subset of all possible parameter regions in order to keep the analysis tractable. We compute the parameter regions which feature the scenarios listed above.

3   Relevant Results on the Static Two-Player Colonel Blotto Game

In this section, we recall the two-player Colonel Blotto game considered in [4]. The setting is that of two agents, and for clarity, we call them agents in this section. Agent $i \in \{1, 2\}$ is endowed with certain amount of resources, denoted by $r_i \in \mathbb{R}_+$. There is a total of n battlefields over which the agents fight. Define $\mathcal{R}_i := \{a \in \mathbb{R}_+^n : \sum_{k=1}^n a_k \leq r_i\}$ and let $\partial \mathcal{R}_i$ be the boundary of the region $\mathcal{R}_i$. The action space of Agent i is $\mathcal{R}_i$. Each agent decides on a mixed strategy over its action space, that is, a probability distribution over its action space, denoted by $\mu_i \in \wp(\mathcal{R}_i)$.

On each battlefield, the agent who deploys maximum resources wins, and accrues a payoff denoted by $v \in \mathbb{R}_+$ [3]. In case both agents deploy equal amount of resources, then each accrue a payoff of $\frac{v}{2}$.

For a strategy of Agent i, μ_i, let $\mathrm{Pr}_{\#}^k \mu_i$ denote the marginal of μ_i on the kth battlefield. Since any agent winning a battlefield is dependent only on the amount of resources deployed by both agents, for a given strategy tuple of the agents (μ_1, μ_2), the expected payoff to Agent i on battlefield $k \in [n]$ is dependent solely on the marginal distributions $(\mathrm{Pr}_{\#}^k \mu_1, \mathrm{Pr}_{\#}^k \mu_2)$.

For this game, we assume that all the parameters defined above is common knowledge among the agents. We denote this game by $\mathbf{SCB}(\{1, r_1\}, \{2, r_2\}, n, v)$. Let us now recall the following result from [4].

---

[3] Typically, v is taken to be $\frac{1}{n}$ in the static Colonel Blotto game.

Theorem 1. *For the static Colonel Blotto game* $SCB(\{1, r_1\}, \{2, r_2\}, n, v)$ *with* $n \geq 3$, *there exists a Nash equilibrium* $(\mu_1^\star, \mu_2^\star)$ *with unique payoffs to each agent.*

The set of all Nash equilibria of the game $\mathbf{SCB}(\{1, r_1\}, \{2, r_2\}, n, v)$ is denoted by $\mathrm{NE}(\mathbf{SCB}(\{1, r_1\}, \{2, r_2\}, n, v))$. Note that we do not claim uniqueness of Nash equilibrium of the game $\mathbf{SCB}(\{1, r_1\}, \{2, r_2\}, n, v)$. However, for any $i \in \{1, 2\}$, there exists a unique measure $\nu \in \wp([0, r_i])$ such that if $\mu_i^\star$ and $\tilde{\mu}_i^\star$ are two Nash equilibrium strategies of Agent i, then $\mathrm{Pr}_{\#}^k \mu_i^\star = \mathrm{Pr}_{\#}^l \tilde{\mu}_i^\star = \nu$ for all $l, k \in [n]$. In other words, the marginals on any two battlefields under any two equilibrium strategies for a agent are the same, and this marginal is unique.

We have the following result on the expected payoffs of the agents when playing under Nash equilibrium strategies in the game $\mathbf{SCB}(\{1, r_1\}, \{2, r_2\}, n, v)$.

Lemma 1. *Consider the static Colonel Blotto game* $SCB(\{1, r_1\}, \{2, r_2\}, n, v)$ *with* $n \geq 3$. *Let* $P^i(SCB(\{1, r_1\}, \{2, r_2\}, n, v))$ *denote the expected payoff to Agent i when both agents act according to Nash equilibrium strategies. If r_1 and r_2 are such that $\frac{1}{n-1} \leq \frac{r_1}{r_2} \leq n-1$, then the expected payoffs to the agents under Nash equilibrium strategies* $(\mu_1^\star, \mu_2^\star)$ *are*

$$
P^1(SCB(\{1, r_1\}, \{2, r_2\}, n, v)) = \begin{cases} nv\left(\frac{2}{n} - \frac{2r_2}{n^2 r_1}\right) & \text{if } \frac{1}{n-1} \leq \frac{r_1}{r_2} < \frac{2}{n} \\ nv\left(\frac{r_1}{2r_2}\right) & \text{if } \frac{2}{n} \leq \frac{r_1}{r_2} \leq 1 \\ nv\left(1 - \frac{r_2}{2r_1}\right) & \text{if } 1 < \frac{r_1}{r_2} \leq \frac{n}{2} \\ nv\left(1 - \frac{2}{n} + \frac{2r_1}{n^2 r_2}\right) & \text{if } \frac{n}{2} < \frac{r_1}{r_2} < n-1 \end{cases},
$$

$$
P^2(SCB(\{1, r_1\}, \{2, r_2\}, n, v)) = nv - P^1(SCB(\{1, r_1\}, \{2, r_2\}, n, v)).
$$

If $r_1 = 0$, then $P^1(SCB(\{1, 0\}, \{2, r_2\}, n, v)) = 0$.

Remark 1. Note that for fixed r_2, n and v, $r_1 \mapsto P^1(\mathbf{SCB}(\{1, r_1\}, \{2, r_2\}, n, v))$ is a concave monotonically increasing function in the parameter region $\frac{1}{n-1} \leq \frac{r_1}{r_2} \leq n - 1$. This is also illustrated in Figure 1 for a specific set of parameters. Furthermore, $r_1 \mapsto P^1(\mathbf{SCB}(\{1, r_1\}, \{2, r_2\}, n, v))$ is a non-decreasing function on $\mathbb{R}_+$ (note that here we do not restrict the range of r_1). This is a consequence of the result in [4]. □

This completes our revisit of the results for two-player static Colonel Blotto game from [4].

4   SPNE of the Game

We now consider the three-stage Colonel Blotto game formulated in Section 2. To ease exposition, let us write $t := t_{1,2} - t_{2,1}$, which denotes the net transfer from Player 1 to Player 2. This can be negative if Player 2 transfers more resources than Player 1. We further define $r_1 := r_1(t) = \beta_1 - t$ and $r_2 := r_2(t) = \beta_2 + t$

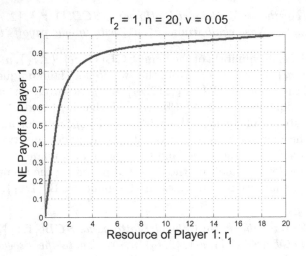

Fig. 1. For a fixed resource $r_2 = 1$ of Agent 2, the payoff to Agent 1 is a concave function of its endowment of resources r_1. Here, $n = 20$ and $v = \frac{1}{n}$.

to denote, respectively, the resource levels of Player 1 and Player 2 after the transfer is complete.

As stated previously, we are interested in computing the subgame-perfect Nash equilibrium for the three-stage game. At the final stage of the game, all players know the resource levels of all players and the resource allocation of the adversary for the battles against the other two players. All players also know the updated number of battlefields over which the battle is to be fought. Thus, the game at the final stage comprises two instances of the static Colonel Blotto game recalled in the previous section. This insight results in the following lemma.

Lemma 2. *At the final stage, Player $i \in \{1, 2\}$ and the adversary will play a static Colonel Blotto game $\boldsymbol{SCB}(\{1, r_i\}, \{A, \alpha_i\}, n_i + m_i, v_i)$. Thus, the SPNE strategy pair of Player i and the adversary at the third (last) stage is $(\mu_i^\star, \nu_i^\star) \in NE(\boldsymbol{SCB}(\{1, r_i\}, \{A, \alpha_i\}, n_i + m_i, v_i))$.*

In the light of the lemma above, to compute the SPNE of the game, we need to compute (i) at the second stage, the allocation functions of the adversary $\{\alpha_1^\star, \ldots, \alpha_N^\star\}$, the transfer functions $\{t_{1,2}^\star\}$ and $\{t_{2,1}^\star\}$ of the first two players, and (ii) at the first stage, the battlefields added by the first two players $m_1^\star$ and $m_2^\star$.

As noted in the previous section, the expected payoff functions of the players in the static Colonel Blotto game are computed in four different parameter regions. Thus, for the game at hand, we have a total of 64 different cases to consider. To ease the exposition, we consider here only four of these cases. These cases comprise games in which, when players act according to Nash equilibrium at the first stage (so that m_1, m_2 are fixed and are common knowledge), the ratio of the adversary's allocation of resource for the battle with Player i and Player i's resources after the transfer is complete lie in the interval $(\frac{2}{n_i+m_i}, \frac{n_i+m_i}{2})$. This simplification leads us to the following four cases:

1. $2/n_1 < \alpha_1/r_1 < 1$ and $2/n_2 < \alpha_2/r_2 < 1$
2. $2/n_1 < \alpha_1/r_1 < 1$ and $2/n_2 < r_2/\alpha_2 < 1$
3. $2/n_1 < r_1/\alpha_1 < 1$ and $2/n_2 < r_2/\alpha_2 < 1$
4. $2/n_1 < r_1/\alpha_1 < 1$ and $2/n_2 < \alpha_2/r_2 < 1$

It should be noted that the second and third cases are essentially the same with only the indices of the first two players interchanged. Thus, we only focus on three cases, Cases 1, 2 and 4, with the understanding that the result for Case 3 can directly be obtained from the result of Case 2.

In the next subsection, we compute the reaction curves of the players (also called best response strategies) for the game at the second stage. Thereafter, we compute the SPNE of the game in the sequel for all the three cases.

4.1 Reaction Functions of the Players

In this subsection, we compute the best response strategies of the players in the game.

Preliminary Notations: We use the following notations to describe the allocation strategy of the adversary for various cases:

$$a_1(m_1, m_2, t) := \frac{\alpha}{1 + \sqrt{\frac{(n_2+m_2)v_2(\beta_2+t)}{(n_1+m_1)v_1(\beta_1-t)}}},$$

$$\lambda_1(m_1, m_2, t) := \sqrt{\frac{(n_2+m_2)v_2(\beta_1-t)(\beta_2+t)}{(n_1+m_1)v_1}},$$

$$d(m_1, m_2, t) := \begin{cases} \alpha & \text{if } \frac{(n_1+m_1)v_1}{\beta_1-t} > \frac{(n_2+m_2)v_2}{\beta_2+t} \\ 0 & \text{if } \frac{(n_1+m_1)v_1}{\beta_1-t} < \frac{(n_2+m_2)v_2}{\beta_2+t} \\ \alpha \ w.p. \ p \in (0,1) & \text{if } \frac{(n_1+m_1)v_1}{\beta_1-t} = \frac{(n_2+m_2)v_2}{\beta_2+t} \end{cases}.$$

Note that in the definition of $d(m_1, m_2, t)$, the probability p can take any value in the interval $(0,1)$. The next lemma computes the reaction curves of the players at the second stage of the game.

Lemma 3. *Consider a game* $\mathbf{CB}(\underline{n}, \beta, \alpha, \underline{v}, c)$. *For a* $t \in [-\beta_2, \beta_1]$, *let* $r_1 = \beta_1 - t$ *and* $r_2 = \beta_2 + t$. *Fix* $m_1, m_2 \in \mathbb{Z}_+$. *The reaction curves of the players at the second stage are given by the following expressions in various cases:*

1. *If* $\frac{2}{n_1+m_1} < \frac{\alpha}{\beta_1-t} < 1$ *and* $\frac{2}{n_2} < \frac{\alpha}{\beta_2+t} < 1$, *then*

$$\alpha_1^*(m_1, m_2, t) = d(m_1, m_2, t).$$

2. *If* $\frac{2}{n_i+m_i} < \frac{r_i}{a_i(m_1, m_2, t)} < 1$, $i = 1, 2$, *then*

$$\alpha_1^*(m_1, m_2, t) = a_1(m_1, m_2, t).$$

3. *If* $\frac{2}{n_1+m_1} < \frac{\alpha-\lambda_1(m_1,m_2,t)}{\beta_1-t} < 1$ *and* $\frac{2}{n_2+m_2} < \frac{\beta_2+t}{\lambda_1(m_1,m_2,t)} < 1$, *then*

$$\alpha_1^*(m_1, m_2, t) = \alpha - \lambda_1(m_1, m_2, t).$$

In all cases, if α_1 is a constant (or dependent only on m_1, m_2), then

$$t_{1,2}^*(m_1, m_2, \alpha_1) = \begin{cases} 0 & \text{if } \alpha_1 > 0, \\ t_{1,2} \in [0, \beta_1] & \text{if } \alpha_1 = 0. \end{cases}$$

$$t_{2,1}^*(m_1, m_2, \alpha_1) = \begin{cases} 0 & \text{if } \alpha_1 < \alpha, \\ t_{2,1} \in [0, \beta_2] & \text{if } \alpha_1 = \alpha. \end{cases}$$

Proof: The proof is available in [14, Lemma 4, p. 13], but we recall it here for the convenience of the reader.

Since Player i and the adversary are going to play a static Colonel Blotto game $\mathbf{SCB}(\{i, r_i\}, \{A, \alpha_i\}, n_i + m_i, v_i)$ at the final stage of the game, the expected payoff functions to the players are given by the result of Lemma 1 (that are dependent on the ratio r_i/α_i).

The reaction function for the adversary is the best response strategy of the adversary given the strategy of the other two players. Towards this end, fix $m_{1:2}$ and t and define $e_i := (n_i + m_i)v_i$ for $i = 1, 2$. The expected payoff function to the adversary as a function of the adversary's allocation α_1 to the battle with Player 1 for the three cases are

$$\text{Case 1: } \pi_A(\alpha_1) = \frac{e_1\alpha_1}{2(\beta_1 - t)} + \frac{e_2(\alpha - \alpha_1)}{2(\beta_2 + t)},$$

$$\text{Case 2: } \pi_A(\alpha_1) = e_1\left(1 - \frac{(\beta_1 - t)}{2\alpha_1}\right)$$
$$+ e_2\left(1 - \frac{(\beta_2 + t)}{2(\alpha - \alpha_1)}\right),$$

$$\text{Case 3: } \pi_A(\alpha_1) = \frac{e_1\alpha_1}{2(\beta_1 - t)} + e_2\left(1 - \frac{(\beta_2 + t)}{2(\alpha - \alpha_1)}\right).$$

In Cases 2 and 3, the payoff to the adversary π_A is a concave function of α_1, since the second derivative of π_A with respect to α_1 is strictly negative. One can set the first derivative of π_A to zero to get the optimal value of α_1 as a function of m_1, m_2, and t. The fact that $d(m_1, m_2, t)$ maximizes the payoff π_A in Case 1 can be verified easily. This completes the proof of the lemma.   ∎

Having now computed the reaction functions of the players at the second stage of the game, we now compute the SPNE strategies of the players below.

4.2   The Case of Weakest Adversary

We now turn our attention to computing SPNE of the game for Case 1, in which the adversary has the least amount of resources among all players.

Preliminary Notation for Theorem 2. Let $\bar{m}_1 = \arg\max_{m_1 \in \mathbb{Z}_+} m_1 v_1 - c m_1^2$ and $\bar{m}_2 = \arg\max_{m_2 \in \mathbb{Z}_+} m_2 v_2 - c m_2^2$. Define

$$\bar{t}_{1,2}(m_1, m_2) = \frac{(n_2 + m_2)v_2\beta_1 - (n_1 + m_1)v_1\beta_2}{(n_1 + m_1)v_1 + (n_2 + m_2)v_2},$$

$$\bar{t}_{2,1}(m_1, m_2) = \frac{(n_1 + m_1)v_1\beta_2 - (n_2 + m_2)v_2\beta_1}{(n_1 + m_1)v_1 + (n_2 + m_2)v_2},$$

$$\zeta_1 = \bar{t}_{2,1}(0, \bar{m}_2) \qquad \zeta_2 = \bar{t}_{1,2}(\bar{m}_1, 0).$$

Theorem 2. *Consider a game* $\mathbf{CB}(\underline{n}, \beta, \alpha, \underline{v}, c)$ *with* $\alpha < \min\{\beta_1, \beta_2\}$ *and* $\frac{2}{n_i} < \frac{\alpha}{\beta_i}$ *for both* $i \in \{1, 2\}$. *If the parameters of the game satisfy either*

$$\frac{(n_1 + \bar{m}_1)v_1}{\beta_1} < \frac{n_2 v_2}{\beta_2}, \qquad \left(1 - \frac{\alpha}{2(\beta_2 + \zeta_2)}\right)v_2 < c,$$

$$\frac{2}{n_1 + \bar{m}_1} < \frac{\alpha}{\beta_1 - \zeta_2} < 1, \qquad \frac{2}{n_2} < \frac{\alpha}{\beta_2 + \zeta_2} < 1,$$

or $\quad \frac{n_1 v_1}{\beta_1} > \frac{(n_2 + \bar{m}_2)v_2}{\beta_2}, \qquad \left(1 - \frac{\alpha}{2(\beta_1 + \zeta_1)}\right)v_1 < c,$

$$\frac{2}{n_2 + \bar{m}_2} < \frac{\alpha}{\beta_2 - \zeta_1} < 1, \qquad \frac{2}{n_1} < \frac{\alpha}{\beta_1 + \zeta_1} < 1,$$

then there is a family of SPNEs for this game, given by

$$\alpha_1^\star(m_1, m_2) = d(m_1, m_2, 0),$$

$$t_{1,2}^\star(m_1, m_2) = \begin{cases} t \in [0, \bar{t}_{1,2}(m_1, m_2)) & \text{if } \frac{(n_1 + m_1)v_1}{\beta_1} < \frac{(n_2 + m_2)v_2}{\beta_2} \\ 0 & \text{otherwise} \end{cases}$$

$$t_{2,1}^\star(m_1, m_2) = \begin{cases} t \in [0, \bar{t}_{2,1}(m_1, m_2)) & \text{if } \frac{(n_1 + m_1)v_1}{\beta_1} > \frac{(n_2 + m_2)v_2}{\beta_2} \\ 0 & \text{otherwise} \end{cases}$$

$$m_1^\star = \begin{cases} \bar{m}_1 & \text{if } \frac{(n_1 + \bar{m}_1)v_1}{\beta_1} < \frac{n_2 v_2}{\beta_2} \\ 0 & \text{otherwise} \end{cases},$$

$$m_2^\star = \begin{cases} \bar{m}_2 & \text{if } \frac{n_1 v_1}{\beta_1} > \frac{(n_2 + \bar{m}_2)v_2}{\beta_2} \\ 0 & \text{otherwise} \end{cases}.$$

Proof: The reaction curves of the players are given as in Lemma 3. It is easy to see that for given m_1 and m_2, the (family of) Nash equilibria stated above are the best response strategies of each other. Now, maximizing the payoff functionals of Players 1 and 2 over m_1 and m_2 given $\alpha_1^\star, t_{1,2}^\star$ and $t_{2,1}^\star$, we get the result. The sufficient conditions on the parameters ensure that Players 1 and 2 and the adversary's allocation have appropriate ratios if all players act according to the SPNE. ∎

Remark 2. Along the equilibrium path, one player has an incentive to add battlefields and transfer some (or none) of its resource to the other player.   □

Remark 3. In the theorem above, if $v_1 < c$, then $\bar{m}_1 = 0$. Similarly, if $v_2 < c$, then $\bar{m}_2 = 0$.   □

4.3   Other Cases

We now consider other scenarios, where the adversary may have comparable or large endowment of resources as compared to any other player in the game.

Preliminary Notation for Theorem 3

$$s_i := \sqrt{v_i \beta_i} \left(\sqrt{n_j v_j \beta_j} \right), \quad i, j \in \{1, 2\}, i \neq j,$$

$$c_1 := v_1 \left(1 - \frac{\alpha}{2\beta_1} \right) + \left(\sqrt{n_1 + 1} - \sqrt{n_1} \right) \frac{\sqrt{n_2 v_2 \beta_2 v_1}}{2\sqrt{\beta_1}}$$

$$c_2 := \left(\sqrt{n_2 + 1} - \sqrt{n_2} \right) \frac{\sqrt{v_2 \beta_2 n_1 v_1}}{2\sqrt{\beta_1}}.$$

Theorem 3. *Consider a game* $\mathbf{CB}(\underline{n}, \underline{\beta}, \alpha, \underline{v}, c) \in$. *The SPNE of the game is given as:*

1. *If* $\frac{2}{n_i + m_i} < \frac{\beta_i}{a_i(m_1, m_2, 0)} < 1$, $i = 1, 2$ *and*

$$c > \frac{1}{2\alpha} \max_{i \in \{1, 2\}} \left(v_i \beta_i + \left(\sqrt{n_i + 1} - \sqrt{n_i} \right) s_i \right),$$

   then $\alpha_1^\star(m_1, m_2) = a_1(m_1, m_2, 0)$.
2. *If* $\alpha > \lambda_1(m_1, m_2, 0)$, $\frac{2}{n_1 + m_1} < \frac{\alpha - \lambda_1(m_1, m_2, 0)}{\beta_1} < 1$, $\frac{2}{n_2 + m_2} < \frac{\beta_2}{\lambda_1(m_1, m_2, 0)} < 1$, *and* $c > \max\{c_1, c_2\}$, *then*

$$\alpha_1^\star(m_1, m_2) = \alpha - \lambda_1(m_1, m_2, 0).$$

In both cases, $t_{1,2}^\star(m_1, m_2) = t_{2,1}^\star(m_1, m_2) = 0$ *and* $m_1^\star = m_2^\star = 0$.

Proof: Given the best response strategies of the players as in Lemma 3, one can just check that the given strategies indeed form a SPNE of the game. Furthermore, the sufficient conditions on c merely ensure that adding any battlefield leads to lower expected payoffs to the first two players.   ∎

Remark 4. In the statement of both cases in Theorem 3 above, the sufficient conditions on c are not hard constraints. If the value of c is small, then adding battlefields may be beneficial to one or both players. The Nash equilibrium at the second stage of the game remains unchanged (as long as the restrictions on the parameters are met).   □

4.4 Discussions on Equilibrium Strategies

In Theorem 2, we see that the amount of resource one player transfers to another could take any value in a set. This is due to the fact that the adversary does not attack the player who makes the transfer (hence, his payoff is not affected by making the transfer) if everyone plays according to the equilibrium.

Figure 2 shows that for a specific set of parameters, a transfer takes place from one player to another in certain regions of β_1 and β_2. It is interesting to note that there is a transfer from Player 1 to Player 2 even when the resource level of Player 1 is significantly small as compared to the resource level of Player 2.

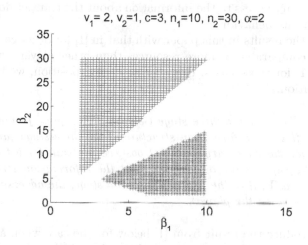

Fig. 2. For fixed parameters $v_1 = 2, v_2 = 1, c = 3, n_1 = 10, n_2 = 30$, and $\alpha = 2$, Player 1 transfers to Player 2 in the red region, whereas Player 2 transfers to Player 1 in the blue region. There is no addition of battlefield by any player (see also Remark 3) in the colored region. In the white region, transfer may or may not occur. See Theorem 2 for a complete characterization.

On the other hand, in Theorem 3, where the adversary has comparable or more resources than other players, the SPNE is unique, and there is no transfer among the first two players. There are two reasons why we see no transfer among the players as SPNE strategies in both cases. The first reason is that the adversary divides its resources into two positive parts, and allocates each of the two parts to the battle with one of the other two players. Since both players are fighting against the adversary, the best response strategies of the first two players are not to transfer their resources to the other player (see Lemma 3). The second reason, which is more subtle, is that the adversary does not observe the value of the transfer among the other two players (or in other words, all players act simultaneously in the second stage). If we allow the adversary to access information on the amount of resource transferred between the players, then the SPNE may feature a transfer *even if the adversary allocates positive resources to fight both players.* A few such cases are investigated in [7] and our earlier work [1]. However, [7] does not compute the Nash equilibrium strategies (or SPNE) of the players under such a setting.

In all cases, if the cost for adding battlefields is sufficiently high, then the first two players do not add any battlefield.

We now outline the differences in the behaviors of the players in this game as compared to the one studied in [1]. For the case when the adversary is weakest (that is, has the least amount of resources among all players), the players act according to the same behavior as proved in [1]. This is because the adversary deploys all its resource to fight against only one player in this case. So, whether or not a transfer occurs, the behavior of the adversary remains unchanged. Thus, giving the adversary access to the information about the transfer does not result in any change in its behavior.

To compare the results in this paper with that in [1] for other cases, when the adversary has comparable or more resources than other players, we will recall the result for [1] for those cases. However, to ease exposition, we introduce the following definition.

Definition 1. *Consider the three-stage Colonel Blotto game formulated in Section 2. We say that the information structure of the three-stage game is* **N** *if at the second stage, the adversary does not observe the transfer between the first two players. We say, on the other hand, that the information structure of the three-stage game is* **T**, *if at the end of the second stage, the adversary has access to the transfer between the players.* □

We now reproduce the result from [1] below for the case when adversary has comparable or more resources as compared to other players.

Preliminary Notation for Theorem 4

$$\bar{t}_1(m_1, m_2) := \frac{(\beta_1 - \beta_2)}{2} - \frac{(\beta_1 + \beta_2)}{2}\sqrt{\frac{(n_1 + m_1)v_1}{(n_1 + m_1)v_1 + (n_2 + m_2)v_2}},$$

$$w_1(m_1, m_2) := (n_1 + m_1)v_1 + \sqrt{(n_1 + m_1)v_1((n_1 + m_1)v_1 + (n_2 + m_2)v_2)},$$

$$\bar{m}_1 := \arg \max_{m_1 \in \mathbb{Z}_+} m_1 v_1 \left(1 - \frac{\alpha}{2(\beta_1 + \beta_2)}\right) - cm_1^2,$$

$$\zeta_1(m_1, m_2) := \frac{4(n_1 + m_1)v_1\alpha^2}{(n_2 + m_2)v_2(\beta_1 + \beta_2)^2}.$$

Theorem 4 ([1]). *Consider a game* **CB**$(\underline{n}, \underline{\beta}, \alpha, \underline{v}, c)$ *with information structure* **T** *in which the adversary has access to the information about the transfer of resources among the first two players at the second stage of the game. The SPNE of the game is given as follows:*

1. *Assume $c > \frac{\beta_1+\beta_2}{4\alpha} \max\{w_1(1,0) - w_1(0,0), v_2\}$ and let $\bar{t}_1 := \bar{t}_1(m_1, m_2)$. If $\frac{2}{n_i+m_i} < \frac{r_i(t)}{a_i(m_1,m_2,t)} < 1$, $i = 1, 2$, then*

$$\alpha_1^\star(m_1, m_2, t) = a_1(m_1, m_2, t),$$

$$t_{1,2}^\star(m_1, m_2) = \begin{cases} \bar{t}_1 \ \ if \ \frac{\beta_1-\beta_2}{2\beta_1\beta_2} > \sqrt{\frac{(n_1+m_1)v_1}{(n_2+m_2)v_2}} \\ 0 \ \ otherwise \end{cases}$$

$$t_{2,1}^\star(m_1, m_2) = 0, \qquad m_1^\star = m_2^\star = 0.$$

2. *If $c > \frac{(\beta_1+\beta_2)v_2}{4\alpha}$, $\frac{2}{n_1+m_1} < \frac{\alpha-\lambda_1(m_1,m_2,t)}{(\beta_1-t)} < 1$, and $\frac{2}{n_2+m_2} < \frac{(\beta_2+t)}{\lambda_1(m_1,m_2,t)} < 1$, then*

$$\alpha_1^\star(m_1, m_2, t) = \alpha - \lambda_1(m_1, m_2, t),$$

$$t_{1,2}^\star(m_1, m_2) = \begin{cases} \frac{\beta_1-\zeta_1(m_1,m_2)\beta_2}{\zeta_1(m_1,m_2)+1} \ \ if \ \frac{\beta_1+\beta_2}{2\alpha} > \sqrt{\frac{(n_1+m_1)v_1\beta_2}{(n_2+m_2)v_2\beta_1}} \\ 0 \qquad\qquad\quad\ otherwise. \end{cases}$$

$$t_{2,1}^\star(m_1, m_2) = 0, \qquad m_1^\star = \bar{m}_1, \qquad m_2^\star = 0.$$

∎

An interesting distinction in the behaviors of the players in the games with two different information structures is as follows: In the game with information structure **N**, the players do not transfer resources among themselves. In contrast, the game with information structure **T** features a transfer. The reason for this behavior is the following. With information structure **T**, the adversary, after observing the transfer, allocates more resource to fight against Player 2 as compared to what it allocates in the game with information structure **N**. Thus, in the game with information structure **T**, the transfer makes *both* Players 1 and 2 better off[4], while the adversary loses in terms of the expected payoff[5],[6].

Remark 5. The analysis above exposes a counterintuitive feature of games. One may be led into thinking that the extra information about the transfer to the adversary should make him better off, but this, clearly, is not the case in the game with information structure **T**. *Consistent with the results of [15], in games, extra information to a player does not necessarily result in a better performance for that player!* □

---

[4] Note here that since this is a non-cooperative game, if the transfer does not improve the expected payoffs to both Players 1 and 2, then either the receiving player will not accept the transfer, or the donating player will not initiate a transfer. The fact that a positive transfer is a Nash equilibrium implies that the transfer it increases or maintains the expected payoffs to both players.

[5] We assume that the parameters of the game are such that the sufficient conditions on parameters are satisfied, enabling us to make this comparison.

[6] Since the Colonel Blotto game is a constant-sum game, the sum of total expected payoffs for all the players (including the adversary) is a constant. Thus, if Players 1 and 2 increase their expected payoffs, then this leads to a decrease in the expected payoff to the adversary.

5   Conclusion

We formulated a three-stage three-player Colonel Blotto (non-cooperative) game in which the first two players fight against a common adversary. The first two players could add battlefields at some cost and they can form a coalition and transfer resources among each other if it improves their expected payoffs. We computed subgame-perfect Nash equilibria of the game. We found that if the adversary is weakest, that is, has the least endowment of resources, then it attacks only one of the two players (when playing under Nash equilibrium). The player who does not suffer an attack can transfer some of its resources to the other player. If the adversary has comparable or more resources than the other two players, then there is no transfer of resources among those two players when playing under Nash equilibrium. In all cases, additional battlefields are created by the first two players if the cost for adding them is sufficiently low.

The result gives a qualitative picture of how players should behave in order to secure cyber-physical systems. In case the cyber-physical systems under attack have significantly more resources (computational or physical) as compared to the attacker, then it is in their best interest to share their resources to secure themselves. On the other hand, if the adversary is as mighty as the systems, then it is in the best interests for the systems to use all their resources to secure themselves.

Furthermore, we see that adding battlefields could result in a better payoff. Consider, for example, a data center which acts to reduce the threat of data compromise. If adding additional servers for storing data is cheap, then it is in its best interest to keep small amount of data in different servers. In doing so, even if a certain number of data servers are compromised, the amount of compromised data will be less.

For the future, our plan is to extend the analysis to general N-player games in which an adversary fights against $N-1$ players. We expect that a similar type of result (as obtained in this paper) will also hold for the general N-player setup – (i) if the adversary is weakest, then it will attack only one player, whereas other players may or may not transfer the resource to the player under attack, and (ii) if the adversary has comparable or more resources, then the adversary will fight against a set (or all) of the players, and the players will not transfer any resource among each other. Another possibility is to analyze the N-player M-adversary Colonel Blotto game, where M adversaries can collaborate (non-cooperatively) for battles against N players, who themselves can collaborate among each other (non-cooperatively).

Incomplete information static Colonel Blotto game is also an important problem that requires further investigation, in which the existence of a Nash equilibrium has not been established yet.

References

1. Gupta, A., Schwartz, G., Langbort, C., Sastry, S.S., Başar, T.: A three-stage Colonel Blotto game with applications to cyberphysical security. In: Proc. 2014 American Control Conference (ACC), pp. 3832–3837 (June 2014)
2. Borel, E., Ville, J.: Applications de la théorie des probabilités aux jeux de hasard. J. Gabay (1938)
3. Gross, O., Wagner, R.: A continuous Colonel Blotto game. RAND Project No. RM-408, Santa Monica, CA (June 1950)
4. Roberson, B.: The Colonel Blotto game. Economic Theory 29(1), 1–24 (2006)
5. Kvasov, D.: Contests with limited resources. Journal of Economic Theory 136(1), 738–748 (2007)
6. Kovenock, D., Mauboussin, M.J., Roberson, B.: Asymmetric conflicts with endogenous dimensionality. Korean Economic Review 26, 287–305 (2010)
7. Kovenock, D., Roberson, B.: Coalitional Colonel Blotto games with application to the economics of alliances. Journal of Public Economic Theory 14(4), 653–676 (2012)
8. Chia, P.H., Chuang, J.: Colonel Blotto in the phishing war. In: Baras, J.S., Katz, J., Altman, E. (eds.) GameSec 2011. LNCS, vol. 7037, pp. 201–218. Springer, Heidelberg (2011)
9. Chia, P.H.: Colonel Blotto in web security. In: The Eleventh Workshop on Economics and Information Security, WEIS Rump Session (2012)
10. Arad, A., Rubinstein, A.: Colonel Blotto's top secret files. Levine's Working Paper Archive 926159280 (2009)
11. Arad, A., Rubinstein, A.: Colonel Blotto's top secret files: Multi-dimensional iterative reasoning in action. Tech. rep., Working paper (2010)
12. Kohli, P., Kearns, M., Bachrach, Y., Herbrich, R., Stillwell, D., Graepel, T.: Colonel Blotto on Facebook: The effect of social relations on strategic interaction. In: Proceedings of the 3rd Annual ACM Web Science Conference, WebSci 2012, pp. 141–150. ACM, New York (2012)
13. Fudenberg, D., Tirole, J.: Game Theory. MIT Press (1991)
14. Gupta, A., Schwartz, G., Langbort, C., Sastry, S.S., Başar, T.: A three-stage Colonel Blotto game with applications to cyberphysical security. Tech. Rep. UCB/EECS-2014-19. EECS Department, University of California, Berkeley (March 2014), http://www.eecs.berkeley.edu/Pubs/TechRpts/2014/EECS-2014-19.html
15. Başar, T., Ho, Y.-C.: Informational properties of the Nash solutions of two stochastic nonzero-sum games. Journal of Economic Theory 7(4), 370–387 (1974)

Toward Optimal Network Topology Design for Fast and Secure Distributed Computation*

Ji Liu and Tamer Başar

Coordinated Science Laboratory
University of Illinois at Urbana-Champaign,
1308 W Main Street, Urbana, Illinois, USA
{jiliu,basar1}@illinois.edu

Abstract. A typical distributed computation problem deals with a network of multiple agents and the constraint that each agent is able to communicate only with its neighboring agents. Two important issues of such a network are the convergence rate of the corresponding distributed algorithm and the security level of the network against external attacks. In this paper, we take algebraic connectivity as an index of convergence rate, which works for consensus and gossip algorithms, and consider certain type of external attacks by using the expected portion of the infected agents to measure the security level. Extremal examples and analysis show that fast convergence rate and high security level require opposite connectivity of the network. Thus, there has to be a trade-off between the two issues in the design of network topology. This paper aims to provide an approach to design a network topology which balances between convergence rate and security. A class of tree graphs, called extended star graphs, are considered. The optimal extended star graph is provided under appropriate assumptions.

Keywords: security, topology design, external attack, algebraic connectivity, distributed computation.

1 Introduction

Over the past few decades, there has been considerable interest in developing algorithms for information distribution and computation among members of interactive agents via local interactions [1–3]. Recently, distributed computation and decision making problems of all types have arisen naturally. Notable among these are consensus problems [4], distributed averaging [5], multi-agent coverage problems [6], rendezvous problems [7], multi-sensor localization [8], and multi-robot formation control [9]. These problems have found applications in a variety of fields including sensor networks, robotic teams, social networks [10], and electric power grids [11]. Thus, distributed control has become an active area of research. Compared with traditional centralized control, distributed control is

---

* This research was supported in part by the U.S. Air Force Office of Scientific Research (AFOSR) MURI grant FA9550-10-1-0573.

R. Poovendran and W. Saad (Eds.): GameSec 2014, LNCS 8840, pp. 234–245, 2014.

believed to be more promising for large-scale complex networks because of its fault tolerance and cost saving features, and its ability to accommodate various physical constraints such as limitations on sensing, computation, and communication.

One of the most important problems in distributed control and computation is the consensus problem [4, 12–16]. In a typical consensus seeking process, the agents in a given group all try to agree on some quantity by communicating what they know only to their neighboring agents. One particular type of consensus process, whose goal is to compute the average of the initial values of the quantity of interest at the agents, is called distributed averaging [5]. There are three different approaches to the distributed averaging problem: linear iterations [5, 17], gossiping [18, 19], and double linear iterations [20] which are also known as push-sum [21], weighted gossip [22], and ratio consensus [23]. Recently, based on the ideas of consensus and distributed averaging, various algorithms have been proposed for more general distributed computation scenarios, such as convex optimization [24], constrained consensus [25], voting [26], liner programming [27], linear algebraic equations [28], and Nash equilibrium seeking [29].

Of particular interest is the rate at which these algorithms converge. It is well known that the convergence rate depends on the network topology of the neighbor relationships among the agents [30]. The neighbor relationships are often described by a graph $\mathbb{G}$ in which vertices correspond to agents and edges indicate neighbor relationships. We assume that $\mathbb{G}$ is an undirected graph without self loops. Thus, the neighbors of an agent i have the same labels as the vertices in $\mathbb{G}$ which are adjacent to vertex i. For convergence to be possible, it is clearly necessary that $\mathbb{G}$ be a connected graph, and we make this assumption in this paper. We focus on tree graphs as they satisfy the least restrictive connectivity condition required for distributed computation.

Another issue which has received much attention lately is the security of a network. The effects of external attacks such as Byzantine attacks were studied in [31] for the consensus process. Modified consensus protocols were proposed in [32] and [33] for persistent disturbances and malicious agents, respectively. Privacy-preserving distributed averaging was considered in [34]. Potential-theoretic strategies were investigated in [35] for robust distributed averaging in the presence of adversarial intervention. Node capture and cloning attacks in a wireless sensor network were considered in [36]. Randomized and strategic attacks for general multiagent networks were studied in [37]. A complementary line of research is called competitive contagion in social networks [38–40].

A natural question that arises is that of the "best" topology of neighbor graph $\mathbb{G}$ for a fixed number of agents. So far the topology design works in the literature have only considered the convergence rate issue [41–43]. Although some new concepts have been proposed to measure the "robustness" of a network [44–46], very few papers take convergence rate and security into account together. The ultimate goal of this work is to provide an approach to design a network topology which is both fast and secure for distributed computation.

1.1 Preliminaries

An undirected graph is *connected* if there is a path between every pair of distinct vertices in the graph. A connected graph is called a *tree* if any two vertices are connected by exactly one path. In other words, a tree is a connected graph without cycles. A path graph and a star graph are two examples of a tree. The *degree* of a vertex in a graph is the number of edges incident to the vertex. A *leaf vertex* is a vertex of degree 1. The *distance* between two vertices in a graph is the number of edges in a shortest path connecting them. The *diameter* of a graph is the largest distance between any pair of two vertices in the graph.

Given an n-vertex undirected graph $\mathbb{G} = (\mathcal{V}, \mathcal{E})$, where $\mathcal{V} = \{1, 2, \ldots, n\}$ denotes the vertex set and $\mathcal{E} \subset \{(i, j) \mid i, j \in \mathcal{V}, \ i \neq j\}$ denotes the edge set, its *Laplacian matrix* is an $n \times n$ matrix defined by $L = D - A$, where D is the $n \times n$ diagonal matrix whose ith diagonal entry equals the degree of vertex i and A is the adjacency matrix of $\mathbb{G}$ whose ijth entry equals 1 if $(i, j) \in \mathcal{E}$ and 0 otherwise. It is well known that L is always positive semi-definite with an eigenvalue at 0. Its second smallest eigenvalue, denoted by $a(\mathbb{G})$, is called the *algebraic connectivity* of $\mathbb{G}$. The algebraic connectivity of $\mathbb{G}$ is positive if and only if $\mathbb{G}$ is connected.

1.2 Organization

The remainder of this paper is organized as follows. In Section 2, we consider a specific randomized gossip algorithm to illustrate, for a fixed neighbor graph, that algebraic connectivity can be used to measure the convergence rate of the algorithm. In Section 3, we consider a certain type of external attacks and take the expected portion of infected agents in a network as an index of the security level of the network. The trade-off between convergence rate and security is discussed in Section 4. In Section 5, we focus on extended star graphs, a class of tree graphs, and derive the optimal extended star graph with the diameter constraint. The paper ends with a couple of illustrative examples in Section 6, and some concluding remarks in Section 7.

2 Convergence Rate

It is well known that in both discrete- and continuous-time linear consensus processses, algebraic connectivity determines convergence rate when the neighbor graph does not change over time [12]. The following randomized gossip algorithm illustrates that algebraic connectivity also works for gossiping.

Consider a network of $n > 1$ agents labeled 1 to n. Each agent i has control over a real-valued scalar quantity x_i, called a gossip variable, which the agent is able to update from time to time. We say that a gossip occurs at time $t \in \{1, 2, \ldots\}$ between agents i and j if the values of both agents' gossip variables at time $t + 1$ equal the average of their values at time t; in other words, in this case $x_i(t + 1) = x_j(t + 1) = \frac{1}{2}(x_i(t) + x_j(t))$. If agent i does not gossip at time t, its gossip variable does not change; thus, in this case $x_i(t+1) = x_i(t)$. Each agent can gossip only with its neighbors and is allowed to gossip with at most one of its neighbors at one time. Assume that at each time t, only one pair of neighboring agents are activated to gossip. Each pair of neighboring agents has an equal probability $\frac{1}{m}$ of being activated, where m is the number of edges in $\mathbb{G}$. In the case when $\mathbb{G}$ is a tree, $m = n-1$.

For the above randomized gossiping algorithm, the agents' gossip variable update rules can be written in a state form. Toward this end, for each pair of agents $(i,j) \in \mathcal{E}$, where $\mathcal{E}$ denotes the set of edges, let $A_{(i,j)}$ be the matrix that describes the updating rule when agents i and j are the only pair to gossip. Then, $x(t+1) = A_{\sigma(t)}x(t)$, where x is the state vector $x = [x_1 \ x_2 \ \cdots \ x_n]'$ and $\sigma : \{1, 2, ...\} \to \mathcal{E}$ is a switching signal whose value at time t is the index representing the randomly chosen pair of agents at time t. Let $\bar{A}(\mathbb{G})$ denote the mean of the independent and identically distributed matrices $A_{\sigma(t)}$. It has been shown that the rate of convergence in mean square is governed by $\lambda_2(\bar{A}(\mathbb{G}))$, the second largest eigenvalue of $\bar{A}(\mathbb{G})$ [18]. It is straightforward to show that $\bar{A}(\mathbb{G}) = I - \frac{1}{m}L(\mathbb{G})$, where $L(\mathbb{G})$ denotes the Laplacian matrix of the neighbor graph $\mathbb{G}$. Then, we can express the eigenvalues of $\bar{A}(\mathbb{G})$ in terms of those of $L(\mathbb{G})$. In particular, $\lambda_2(\bar{A}(\mathbb{G})) = 1 - \frac{a(\mathbb{G})}{m}$. Thus, algebraic connectivity also determines the convergence rate for this gossiping process in that the larger algebraic connectivity is, the faster is the convergence.

It is well known that among all graphs with n vertices, the complete graph has the maximum algebraic connectivity. Thus, the complete graph achieves the fastest convergence. If we focus on tree graphs, more can be said. It has been shown in [47] that among all trees with n vertices, the path has the minimum algebraic connectivity and the star attains the maximum algebraic connectivity, which leads to the following result.

Lemma 1. *Suppose that the neighbor graph $\mathbb{G}$ is a tree with n vertices. Then, the above randomized gossip algorithm achieves fastest convergence when $\mathbb{G}$ is a star and slowest convergence when $\mathbb{G}$ is a path.*

3   External Attack

Next we consider the effects of topology under security constraints. Consider a network of $n > 1$ agents whose neighbor graph is $\mathbb{G}$ with n vertices. Suppose that there is an external attacker who is able to infect one of the agents in the network. We assume that each agent has probability p with which it will be infected when it is under attack. In other words, each agent is immune to an attack with probability $1 - p$. It is also assumed that if an agent i is immune to an attack, it will not be infected any longer. Once an agent is infected at time t, then each of its neighbors will be under attack at the next time $t + 1$.

In distributed averaging, for example, the initial values of the infected agents may deviate from their true values, thus negatively affecting the final agreement among the agents. It follows that the fewer are the infected agents in the network, the more accurate would the final agreement be. Thus, a rational attacker will aim to attack that agent which leads to the largest expected number of infected agents, and a topology designer should minimize this number.

Suppose that an attacker chooses to attack agent i at the initial time. In this case, it is possible to explicitly express the expected number of infected agents. Toward this end, let $N(d)$ denote the number of those vertices in the graph whose distance to vertex i equals d and let $E(i)$ denote the expected number of infected agents in the network when agent i is initially attacked. Then, it is straightforward to verify that

$$E(i) = p + \sum_{d=1}^{D} N(d)p^d$$

where D is the diameter of the underlying graph $\mathbb{G}$. The attacker will seek the agent, say i^*, achieving the following maximum:

$$E_{\max} = \max_{i \in \mathcal{V}} E(i) = E(i^*)$$

where $\mathcal{V}$ is the vertex set of $\mathbb{G}$. It is worth noting that i^* may not be unique.

The value of $E_{\max}$ can be viewed as a security level of a network. In particular, the larger $E_{\max}$ is, the lower is the security level of the network.

Lemma 2. *Suppose that the neighbor graph $\mathbb{G}$ is a tree with n vertices. Then, the security level against the above external attack achieves lowest level when $\mathbb{G}$ is a star and highest level when $\mathbb{G}$ is a path.*

The proof of this lemma is fairly simple and thus is omitted.

4   Trade-Off

From the preceding discussion and results, the topology of the underlying neighbor graph has opposite effects on convergence rate and security. In particular, when we focus on tree graphs, star graphs are the best for convergence rate but the worst for security, and path graphs are the best for security but the worst for convergence rate. Thus, if we take convergence rate and security into account together, there must be some optimal tree structure between the extremes of the star graphs and path graphs. Notwithstanding this, the following questions remain. What are the criteria to evaluate convergence rate and security together? What is the best tree topology based on those criteria? These are the questions which will be considered next.

Roughly speaking, the better is the connectivity of the topology, the faster is the convergence rate, but the lower is the security level against the attack. To be more precise, the following result shows that adding an additional edge to a given graph will increase the algebraic connectivity and thus accelerate the convergence.

Lemma 3. [47] *Let $\mathbb{G}$ be a non-complete graph and $\mathbb{G}'$ be the graph obtained from $\mathbb{G}$ by joining two non-adjacent vertices of $\mathbb{G}$ with an edge. Then, $a(\mathbb{G}') \geq a(\mathbb{G})$.*

But the next lemma says that adding more edges to a given graph will decrease the security level.

Lemma 4. *Let $\mathbb{G}$ be a non-complete graph and $\mathbb{G}'$ be the graph obtained from $\mathbb{G}$ by joining two non-adjacent vertices of $\mathbb{G}$ with an edge. Then, $E_{max}(\mathbb{G}') \geq E_{max}(\mathbb{G})$.*

Proof of Lemma 4: Suppose that $\mathbb{G}'$ is obtained from $\mathbb{G}$ by joining two vertices i and j which are non-adjacent in $\mathbb{G}$. Then, for any two distinct vertices i and j, the distance between i and j in $\mathbb{G}'$ is shorter than that in $\mathbb{G}$. Let $E'(v)$ denote the expected number of infected agents when agent v in $\mathbb{G}'$ is initially attacked. Since $p \in [0,1]$, $E'(v) \geq E(v)$ for any vertex v. Therefore, $E_{max}(\mathbb{G}') \geq E_{max}(\mathbb{G})$. □

In the sequel, we will focus on tree graphs because they satisfy the least restrictive connectivity condition for distributed computation. It has been shown in [47] that for any tree graph $\mathbb{T}$ with n vertices, $a(\mathbb{T}) \leq 1$ with equality if and only if $\mathbb{T}$ is the star graph. We use $b(\mathbb{T})$ to denote the maximum expected portion of infected agents in the network. Then,

$$b(\mathbb{T}) = \frac{E_{max}(\mathbb{T})}{n} \leq 1$$

Our final goal is to find the optimal tree which maximizes the following function:

$$\rho(\mathbb{T}) = \frac{a(\mathbb{T})}{b(\mathbb{T})}$$

5   Extended Stars

In this section, we consider "extended stars", a class of tree graphs. We say that a tree graph is an *extended star* if it is composed of $k > 1$ paths connected at one end to a common vertex. We call the common vertex the *root* of the extended tree and each path a *branch* of the extended tree at the root. An extended star is an ordinary path when $k = 2$. Fig. 1 shows an example of an extended star which has 4 path branches at the root v. An extended star graph is called *uniform* if its k paths have the same length l. Thus, a uniform extended star is an ordinary star when $l = 1$.

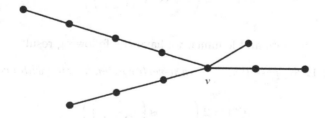

Fig. 1. An extended star

The main result of this paper is as follows.

Theorem 1. *Suppose that* $\mathbb{T}$ *is an n-vertex extended star with even diameter* D. *Let* q *and* r *respectively denote the unique integer quotient and remainder of* n *divided by* $\frac{D}{2}$. *If* $r \neq 0$, *then* $\rho(\mathbb{T})$ *attains the maximum if* $\mathbb{T}$ *has* q *branches of length* $\frac{D}{2}$ *and one branch of length* r. *If* $r = 0$, *then* $\rho(\mathbb{T})$ *attains the maximum if* $\mathbb{T}$ *is a uniform extended star having* q *branches of length* $\frac{D}{2}$.

To prove this theorem, we need the following concepts and lemmas.

In [48], tree graphs are partitioned into two classes, Type I and Type II, according to their eigenvectors corresponding to algebraic connectivity, the second smallest eigenvalue of the Laplacian matrix. To be more precise, a tree is of Type I if the eigenvector has at least one zero entry; in this case, the vertex corresponding to the zero entry is called a *characteristic vertex*. A tree is of Type II if the eigenvector does not have any zero entry.

Lemma 5. [49] *Let* $\mathbb{T}$ *be a Type I tree with characteristic vertex* v. *Suppose that* $\mathbb{T}'$ *is the tree obtained from* $\mathbb{T}$ *by adjoining a new leaf vertex to* v. *Then,* $a(\mathbb{T}') = a(\mathbb{T})$. *In particular,* $\mathbb{T}'$ *is a Type I tree with characteristic vertex* v.

Lemma 6. [50] *Let* $\mathbb{T}$ *be a Type I tree with characteristic vertex* v *and* $\mathbb{T}'$ *be the graph obtained from* $\mathbb{T}$ *by taking any subtree of a branch at* v *and joining it on at vertex* v. *Then,* $\mathbb{T}'$ *is a Type I tree and* $a(\mathbb{T}') = a(\mathbb{T})$.

In [51], a tree $\mathbb{T}$ is called *multi-symmetric* if there is a vertex v such that all the branches of $\mathbb{T}$ at v can be partitioned into finite classes which satisfy the following conditions: (1) each class has two or more branches; (2) any two branches from the same class are isomorphic; (3) any two branches from different classes are not isomorphic. In particular, v is called the *center* of $\mathbb{T}$.

Lemma 7. [51] *Every multi-symmetric tree is of Type I and its center is the characteristic vertex of the tree.*

We also need the following lemmas.

Lemma 8. [52] *Let* $\mathbb{G}$ *be a graph and* $\mathbb{G}'$ *be the graph obtained from* $\mathbb{G}$ *by adding a leaf vertex to a vertex of* $\mathbb{G}$. *Then,* $a(\mathbb{G}') \leq a(\mathbb{G})$.

Lemma 9. [53] *Let* $\mathbb{T}$ *be an n-vertex tree with diameter* D. *Then,*

$$a(\mathbb{T}) \leq 2 \left(1 - \cos \left(\frac{\pi}{D+1} \right) \right)$$

With these concepts and lemmas, we have the following result.

Proposition 1. *Suppose that* $\mathbb{T}$ *is an n-vertex extended star with even diameter* D. *Then,*

$$a(\mathbb{T}) = 2 \left(1 - \cos \left(\frac{\pi}{D+1} \right) \right)$$

Proof of Proposition 1: Suppose that $\mathbb{T}$ is an n-vertex extended star with $k > 1$ branches at root v. It must have two branches of length $\frac{D}{2}$ since its diameter is even and equals D. Let $\mathbb{T}'$ be an extended star obtained from $\mathbb{T}$ by adding each branch of $\mathbb{T}$ to the root v one more time. Then, $\mathbb{T}'$ is a multi-symmetric tree. By Lemma 7, $\mathbb{T}'$ is a Type I tree with characteristic vertex v. Moreover, by Lemma 8, $a(\mathbb{T}') \le a(\mathbb{T})$.

Note that $\mathbb{T}'$ has $2k$ branches at v. Keep two of them which are of length $\frac{D}{2}$. We construct another extended star $\mathbb{T}''$ by taking each leaf vertex of the remaining $2k - 2$ branches and adding it to v until all the $2k - 2$ branches are of length 1. Thus, $\mathbb{T}''$ is an extended star which has 2 branches of length $\frac{D}{2}$ and all the remaining branches are of length 1. By Lemma 6, $\mathbb{T}''$ is still of Type I with characteristic vertex v and $a(\mathbb{T}'') = a(\mathbb{T}')$.

Note that $\mathbb{T}''$ can also be obtained from a path of length D by adding leaf vertices to v. Thus,

$$a(\mathbb{T}'') = 2\left(1 - \cos\left(\frac{\pi}{D+1}\right)\right)$$

Since $a(\mathbb{T}') \le a(\mathbb{T})$ and $a(\mathbb{T})$ cannot exceed $a(\mathbb{T}'')$ by Lemma 9, it follows that $a(\mathbb{T}) = a(\mathbb{T}'')$, which completes the proof. $\square$

Now we are in a position to prove the main result.

Proof of Theorem 1: We provide a constructive proof for the theorem. Since $\mathbb{T}$ is an extended star on n vertices with even diameter D, it must have two branches of length $\frac{D}{2}$. For each of the remaining vertices, it can be added to either the root v or the end of an existing branch whose length is less than $\frac{D}{2}$. From Proposition 1, no matter where each vertex is added, the construction process will not change the algebraic connectivity. Thus, each vertex should be added to the place which leads to the smallest $b(\mathbb{T})$, the expected portion of the final infected agents. Note that the best position of $\mathbb{T}$ for the initial attack is the root v. By the definition of $b(\mathbb{T})$, the vertex should be added as far as possible to the root v so as to minimize $b(\mathbb{T})$. Thus, if there is an existing branch whose length is less than $\frac{D}{2}$, the vertex should be added to the longest of such branches. Otherwise, the vertex can only be added to the root v. From this construction process, the optimal extended star has q branches of length $\frac{D}{2}$ and one branch of length r. In particular, if $r = 0$, the optimal extended star is uniform. $\square$

6   Discussion

In this section, we first provide an example to show that the result of Theorem 1 does not hold when the diameter is odd. Suppose that $\mathbb{T}$ is an 8-vertex extended star with diameter $D = 5$. With this constraint, there are only two possible extended stars which are shown in Fig. 2. We write $\mathbb{T}_1$ and $\mathbb{T}_2$ respectively for the left and right extended stars in Fig. 2. It is straightforward to compute that $a(\mathbb{T}_1) = 0.2434$, $a(\mathbb{T}_2) = 0.2538$, $b(\mathbb{T}_1) = (p + 3p^2 + 3p^3 + p^4)/8$, and $b(\mathbb{T}_2) = (p + 4p^2 + 2p^3 + p^4)/8$. Thus, in this case, the relation between $\rho(\mathbb{T}_1)$ and $\rho(\mathbb{T}_2)$ depends on the value of p.

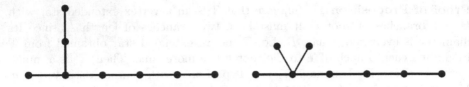

Fig. 2. Two 8-vertex extended stars with diameter 5

Next we consider general tree graphs. In general, the algebraic connectivity $a(\mathbb{G})$ and the security level $b(\mathbb{G})$ cannot reach the optimal values for the same graph $\mathbb{G}$. Thus, it is natural to introduce a trade-off index in topology design. To be more precise, given the set of all trees with n vertices, we aim to find the optimal tree which minimizes the following function:

$$f(\mathbb{T}) = (1 - \beta)a(\mathbb{T}) - \beta b(\mathbb{T})$$

where β is a constant in the interval $[0, 1]$ which represents the trade-off between convergence rate and security. In the special cases when β equals 0 and 1, only convergence rate and security are considered, respectively. With the above target function, it follows that the optimal tree will depend on the value of β, as well as the value of p. For example, assuming that $n = 8$, the computations show that the optimal trees are as given in Fig. 3 and Fig. 4, with different values of β and p.

Fig. 3. The optimal 8-vertex tree with $\beta = 0.5$ and $p = 0.5$

Fig. 4. The optimal 8-vertex tree with $\beta = 0.85$ and $p = 0.7$

7   Concluding Remarks

An approach to design a network topology for distributed computation which balances between convergence rate and security has been introduced. As a first step, a class of tree graphs, extended star graphs, have been considered and the optimal extended star graph has been derived under a diameter constraint. The optimal topology of more general tree graphs is a subject of future study.

References

1. Tsitsiklis, J.N.: Problems in Decentralized Decision Making and Computation. PhD thesis, Department of Electrical Engineering and Computer Science, MIT (1984)
2. Bertsekas, D.P., Tsitsiklis, J.N.: Parallel and Distributed Computation: Numerical Methods. Prentice Hall (1989)
3. Lynch, N.A.: Distributed Algorithms. Morgan Kaufmann Publishers (1997)
4. Jadbabaie, A., Lin, J., Morse, A.S.: Coordination of groups of mobile autonomous agents using nearest neighbor rules. IEEE Transactions on Automatic Control 48(6), 988–1001 (2003)
5. Xiao, L., Boyd, S.: Fast linear iterations for distributed averaging. Systems and Control Letters 53(1), 65–78 (2004)
6. Cortés, J., Martínez, S., Karataş, T., Bullo, F.: Coverage control for mobile sensing networks. IEEE Transactions on Robotics and Automation 20(2), 243–255 (2004)
7. Lin, J., Morse, A.S., Anderson, B.D.O.: The multi-agent rendezvous problem. Part 1: The Synchronous case. SIAM Journal on Control and Optimization 46(6), 2096–2119 (2007)
8. Hu, L., Evans, D.: Localization for mobile sensor networks. In: Proceedings of the 10th Annual International Conference on Mobile Computing and Networking, pp. 45–57 (2004)
9. Krick, L., Broucke, M.E., Francis, B.A.: Stabilisation of infinitesimally rigid formations of multi-robot networks. International Journal of Control 82(3), 423–439 (2009)
10. Liu, J., Hassanpour, N., Tatikonda, S., Morse, A.S.: Dynamic threshold models of collective action in social networks. In: Proceedings of the 51st IEEE Conference on Decision and Control, pp. 3991–3996 (2012)
11. Dörfler, F., Chertkov, M., Bullo, F.: Synchronization in complex oscillator networks and smart grids. Proceedings of the National Academy of Sciences 110(6), 2005–2010 (2013)
12. Olfati-Saber, R., Murray, R.M.: Consensus problems in networks of agents with switching topology and time-delays. IEEE Transactions on Automatic Control 49(9), 1520–1533 (2004)
13. Moreau, L.: Stability of multi-agent systems with time-dependent communication links. IEEE Transactions on Automatic Control 50(2), 169–182 (2005)
14. Ren, W., Beard, R.W.: Consensus seeking in multiagent systems under dynamically changing interaction topologies. IEEE Transactions on Automatic Control 50(5), 655–661 (2005)
15. Kashyap, A., Başar, T., Srikant, R.: Quantized consensus. Automatica 43(7), 1192–1203 (2007)
16. Touri, B., Nedić, A.: Product of random stochastic matrices. IEEE Transactions on Automatic Control 59(2), 437–448 (2014)
17. Xiao, L., Boyd, S., Lall, S.: A scheme for robust distributed sensor fusion based on average consensus. In: Proceedings of the 4th International Conference on Information Processing in Sensor Networks, pp. 63–70 (2005)
18. Boyd, S., Ghosh, A., Prabhakar, B., Shah, D.: Randomized gossip algorithms. IEEE Transactions on Information Theory 52(6), 2508–2530 (2006)
19. Liu, J., Mou, S., Morse, A.S., Anderson, B.D.O., Yu, C.: Deterministic gossiping. Proceedings of the IEEE 99(9), 1505–1524 (2011)

20. Liu, J., Morse, A.S.: Asynchronous distributed averaging using double linear iterations. In: Proceedings of the 2012 American Control Conference, pp. 6620–6625 (2012)
21. Kempe, D., Dobra, A., Gehrke, J.: Gossip-based computation of aggregate information. In: Proceedings of the 44th Annual IEEE Symposium on Foundations of Computer Science, pp. 482–491 (2003)
22. Bénézit, F., Blondel, V., Thiran, P., Tsitsiklis, J.N., Vetterli, M.: Weighted gossip: distributed averaging using non-doubly stochastic matrices. In: Proceedings of the 2010 IEEE International Symposium on Information Theory, pp. 1753–1757 (2010)
23. Domínguez-García, A.D., Cady, S.T., Hadjicostis, C.N.: Decentralized optimal dispatch of distributed energy resources. In: Proceedings of the 51st IEEE Conference on Decision and Control, pp. 3688–3693 (2012)
24. Nedić, A., Ozdaglar, A.: Distributed sub-gradient methods for multi-agent optimization. IEEE Transactions on Automatic Control 58(6), 48–61 (2009)
25. Nedić, A., Ozdaglar, A., Parrilo, P.A.: Constrained consensus and optimization in multi-agent networks. IEEE Transactions on Automatic Control 55(4), 922–938 (2010)
26. Bénézit, F., Thiran, P., Vetterli, M.: The distributed multiple voting problem. IEEE Journal of Selected Topics in Signal Processing 5(4), 791–804 (2011)
27. Bürger, M., Notarstefano, G., Bullo, F., Allgöwer, F.: A distributed simplex algorithm for degenerate linear programs and multi-agent assignments. Automatica 48(9), 2298–2304 (2012)
28. Mou, S., Liu, J., Morse, A.S.: A distributed algorithm for solving a linear algebraic equation. In: Proceedings of the 51st Annual Allerton Conference on Communication, Control, and Computing, pp. 267–274 (2013)
29. Gharesifard, B., Cortés, J.: Distributed convergence to Nash equilibria in two-network zero-sum games. Automatica 49(6), 1683–1692 (2013)
30. Olfati-Saber, R., Fax, J.A., Murray, R.M.: Consensus and cooperation in networked multi-agent systems. Proceedings of the IEEE 95(1), 215–233 (2007)
31. Pasqualetti, F., Bicchi, A., Bullo, F.: Consensus computation in unreliable networks: a system theoretic approach. IEEE Transactions on Automatic Control 57(1), 90–104 (2012)
32. Yucelen, T., Egerstedt, M.: Control of multiagent systems under persistent disturbances. In: Proceedings of the 2012 American Control Conference, pp. 5264–5269 (2012)
33. LeBlanc, H., Zhang, H., Koutsoukos, X., Sundaram, S.: Resilient asymptotic consensus in robust networks. IEEE Journal on Selected Areas in Communications 31(4), 766–781 (2013)
34. Manitara, N.E., Hadjicostis, C.N.: Privacy-preserving asymptotic average consensus. In: Proceedings of the 2013 European Control Conference, pp. 760–765 (2013)
35. Khanafer, A., Touri, B., Başar, T.: Robust distributed averaging on networks with adversarial intervention. In: Proceedings of the 52nd IEEE Conference on Decision and Control, pp. 7131–7136 (2013)
36. Zhu, Q., Bushnell, L., Başar, T.: Game-theoretic analysis of node capture and cloning attack with multiple attackers in wireless sensor networks. In: Proceedings of the 51st IEEE Conference on Decision and Control, pp. 3404–3411 (2012)
37. Acemoglu, D., Malekian, A., Ozdaglar, A.: Network security and contagion (2013) (submitted)
38. Chasparis, G.C., Shamma, J.S.: Control of preferences in social networks. In: Proceedings of the 49th IEEE Conference on Decision and Control, pp. 6651–6656 (2010)

39. Goyal, S., Kearns, M.: Competitive contagion in networks. In: Proceedings of the 44th Annual ACM Symposium on Theory of Computing, pp. 759–774 (2012)
40. Etesami, S.R., Başar, T.: Complexity of equilibrium in diffusion games on social networks. In: Proceedings of the 2014 American Control Conference, pp. 2065–2070 (2014)
41. Cao, M., Wu, C.W.: Topology design for fast convergence of network consensus algorithms. In: Proceedings of the 2007 IEEE International Symposium on Circuits and Systems, pp. 1029–1032 (2007)
42. Kar, S., Moura, J.M.F.: Sensor networks with random links: topology design for distributed consensus. IEEE Transactions on Signal Processing 56(7), 3315–3326 (2008)
43. Rafiee, M., Bayen, A.M.: Optimal network topology design in multi-agent systems for efficient average consensus. In: Proceedings of the 49th IEEE Conference on Decision and Control, pp. 3877–3883 (2010)
44. Zhang, H., Sundaram, S.: Robustness of complex networks with implications for consensus and contagion. In: Proceedings of the 51st IEEE Conference on Decision and Control, pp. 3426–3432 (2012)
45. Bamieh, B., Jovanović, M.R., Mitra, P., Patterson, S.: Coherence in large-scale networks: dimension-dependent limitations of local feedback. IEEE Transactions on Automatic Control 57(9), 2235–2249 (2012)
46. Chapman, A., Mesbahi, M.: Semi-autonomous consensus: network measures and adaptive trees. IEEE Transactions on Automatic Control 58(1), 19–31 (2013)
47. Fiedler, M.: Algebraic connectivity of graphs. Czechoslovak Mathematical Journal 23(2), 298–305 (1973)
48. Fiedler, M.: A property of eigenvectors of nonnegative symmetric matrices and its application to graph theory. Czechoslovak Mathematical Journal 25(4), 619–633 (1975)
49. Grone, R., Merris, R.: Algebraic connectivity of trees. Czechoslovak Mathematical Journal 37(112), 660–670 (1987)
50. Kirkland, S., Neumann, M., Shader, B.L.: Characteristic vertices of weighted trees via perron values. Linear and Multilinear Algebra 40(4), 311–325 (1996)
51. Mitchell, L.H.: A characterization of tree type. Rose-Hulman Undergraduate Mathematics Journal 4(2) (2003)
52. Fallat, S.M., Kirkland, S., Pati, S.: Maximizing algebraic connectivity over unicyclic graphs. Linear and Multilinear Algebra 51(3), 221–241 (2003)
53. Grone, R., Merris, R., Sunder, S.: The Laplacian spectrum of a graph. SIAM Journal on Matrix Analysis and Applications 11(2), 218–238 (1990)

An Economic Model and Simulation Results of App Adoption Decisions on Networks with Interdependent Privacy Consequences

Yu Pu and Jens Grossklags

College of Information Sciences and Technology
The Pennsylvania State University, University Park, PA, USA
{yxp134,jensg}@ist.psu.edu

Abstract. The popularity of third-party apps on social network sites and mobile networks emphasizes the problem of the interdependency of privacy. It is caused by users installing apps that often collect and potentially misuse the personal information of users' friends who are typically not involved in the decision-making process. In this paper, we provide an economic model and simulation results addressing this problem space. We study the adoption of social apps in a network where privacy consequences are interdependent. Motivated by research in behavioral economics, we extend the model to account for users' other-regarding preferences; that is, users care about privacy harms they inflict on their peers.

We present results from two simulations utilizing an underlying scale-free network topology to investigate users' app adoption behaviors in both the initial adoption period and the late adoption phase. The first simulation predictably shows that in the early adoption period, app adoption rates will increase when (1) the interdependent privacy harm caused by an app is lower, (2) installation cost decreases, or (3) network size increases. Surprisingly, we find from the second simulation that app rankings frequently will not accurately reflect the level of interdependent privacy harm when simultaneously considering the adoption results of multiple apps. Given that in the late adoption phase, users make their installation decisions mainly based on app rankings, the simulation results demonstrate that even rational actors who consider their peers' well-being might adopt apps with significant interdependent privacy harms. Our findings complement the usable privacy and security studies which show that users install privacy-invasive apps because they are unable to identify and understand apps' privacy consequences; however, we show that fully-informed and rational users will likely fall for privacy-invasive apps as well.

Keywords: Economic Model, Simulation, Interdependent Privacy, Other-Regarding Preferences, Scale-Free Networks, Social Network Sites, Mobile Networks, Third-Party Apps, App Adoption.

R. Poovendran and W. Saad (Eds.): GameSec 2014, LNCS 8840, pp. 246–265, 2014.

1   Introduction

Over the last ten years, we have witnessed the rapidly increasing popularity of social network sites, with Facebook being the most successful entity. In order to expand its service and functionality, Facebook opened its platform to allow outside developers to interact with users through so-called third-party Facebook applications (or social *apps*). Those applications gained worldwide popularity ever since their emergence. Similarly, the most important mobile platforms such as Android and iOS have enabled outside developers to create app content which met significant success in the marketplace.

Despite their high adoption rates, third-party apps pose privacy risks to users when they collect and potentially use user information. Some well-acknowledged issues are apps collecting more information than needed for their stated purposes [1,2]; and users demonstrating very little understanding of or ability with the management of app permissions [3,4].

A newly addressed problem associated with app permissions is the *interdependency of privacy*, which refers to the phenomenon that in an interconnected setting, the privacy of individual users not only depends on their own behaviors, but is also affected by the decisions of others [5].[1] The interdependent privacy issue is caused by users installing apps that often collect and potentially misuse the personal information of users' friends who are typically not involved in the decision-making process.

Research has not yet adequately investigated the problem of interdependent privacy, in particular, from an economic perspective. Most closely related to our work, Biczók and Chia aim to define interdependent privacy and to provide initial evidence from the Facebook permission system for social apps. They further develop a game-theoretic model to analyze users' app adoption decisions under the scenario of interdependent privacy. However, their study is limited to cases where two users are engaged in the decision-making over the adoption of one app, and therefore does not consider the complex dynamics of today's app adoption behaviors. To address this literature gap, we follow an economic approach to study how large groups of users, who are connected in a complex social network, act in an interdependent privacy scenario.

We develop an app adoption model of a rational consumer who considers cost of app adoption, benefits of an app, and the privacy consequences associated with an app adoption decision. Individuals in our model do not only consider personal costs and benefits of their decision. Instead, we consider that consumers have different levels of concern about the consequences of their adoption decisions for their peers. To accomplish this objective, we utilize the theory of

---

[1] In the security context, several studies have considered the interdependency of decision-making, but those models are less applicable to the app adoption scenario [6,7]. For a survey of the results in the area of interdependent security see [8].

other-regarding preferences which is well-established in psychology and economics, and has been demonstrated in various experimental studies [9,10]. In a nutshell, the theory of other-regarding preferences allows us to model users so that they consider their peers' utility when making adoption decisions.

In our research, we take a graph-theoretical approach and simulate app adoption decisions in scale-free networks to represent an approximate version of real social networks. More specifically, we conduct two simulations to investigate individuals' app adoption behaviors in two phases. One phase is the start-up period of new apps, the other phase is the later app adoption stage. More precisely, the first simulation, which considers the iterative/sequential adoption process of social apps, is used to study users' app adoption behaviors when an app is initially introduced. The second simulation, which is about comparing early adoption results of multiple apps, allows us to establish popularity rankings of the early adoption of those apps. We use those rankings to draw conclusions about the likely adoption processes of the considered apps in later adoption phases which are then heavily influenced by rankings [11].

As expected, we find that in the initial adoption phase, app adoption rates will increase when (1) the interdependent privacy harm caused by an app is lower, (2) installation cost decreases, or (3) network size increases. In the second simulation, interestingly, we find that app rankings frequently will not accurately reflect the level of interdependent privacy harm when considering the adoption results of multiple apps. Our analysis implies that in the later adoption period, even rational actors who consider their peers' well-being might adopt apps with invasive privacy practices. This helps us to explain why some apps that cause significant interdependent privacy issues are nevertheless highly popular on actual social network sites and mobile networks.

The paper is structured as follows. In Section 2, we discuss research on privacy consequences of installing third-party applications on social networking sites and mobile platforms. In Section 3, we develop our economic model of app adoption behavior. In Section 4, we describe our simulation setup. In Sections 5 and 6, we present our simulation results. Finally, we conclude in Section 7.

2   Related Work

2.1   Third-Party Applications on Social Network Sites

Primary motivators for our study are incidents that highlight the potential negative privacy and security consequences of third-party app adoption on social network sites. Several studies have documented how third-party apps are utilized to extract and to transfer user information not only to third-party app developers but also to advertising and data firms [12,13,14]. These studies are highly valuable because in most cases it is difficult to observe data practices once users have authorized third-parties to access their profiles (and their friends' profiles).

To understand the problem space from a more user-centered perspective, several research papers focus on the disclosure and authorization procedures associated with third-party apps. User studies document the concerns users have about app adoption, and their misunderstandings about the access of third-party developers to their profiles [3,15,16]. Similarly, the impact of interface improvements of the authorization dialogues for third-party apps on user behavior has been investigated in several user studies (see, for example, [17,18]).[2]

Table 1. Most frequently requested Facebook permissions explicitly involving information of users' friends (abbreviated table from Wang et al. [17])

Permission	Number of apps requesting permission	Percentage of apps requesting permission	Total times a permission is requested by apps
friends_birthday	206	2.19%	19,237,740
friends_photos	214	2.27%	13,051,340
friends_online_presence	121	1.29%	10,745,500
friends_location	104	1.11%	8,121,000
friends_hometown	21	0.22%	5,862,500
friends_work_history	86	0.91%	5,260,660
friends_education_history	14	0.15%	3,564,500
friends_activities	22	0.23%	3,448,300
friends_about_me	17	0.18%	3,328,000
friends_interests	13	0.14%	3,163,500
user_work_history	73	0.78%	2,961,900
friends_relationships	3	0.03%	2,912,000
friends_photo_video_tags	32	0.34%	2,423,340
friends_likes	36	0.38%	2,385,960
friends_checkins	6	0.06%	1,350,000
friends_relationship_details	4	0.04%	741,000
friends_videos	2	0.02%	230,400

A selected number of studies have focused on measuring aspects of the permissions system for third-party apps on social network sites [1,21,22]. These studies identify the most requested permissions, and the average number of permissions for all apps and specific categories. In Table 1, we summarize data that relates to the sharing of other users' information from a study by Wang et al. [17]. They find that specific permissions (except for basic information and email) are only used by a subset of all apps. However, due to the popularity of the over 9000 surveyed apps, the impact of these data collection and usage practices is significant. As a result, even though less than 1% of the apps request the friends' employment history, this nevertheless means that the data is accessible to third-party developers (and potentially other parties) in over 5 Million cases.

In aggregate, these studies document many obstacles that users have to overcome to identify privacy consequences of social apps, and to implement their privacy preferences in practice during the app adoption process. We complement these studies by showing that even from a rational consumer perspective

---

[2] Already in the context of desktop computing, user studies have investigated how to inform users more effectively about third-party apps which collect personal information and potentially allow for privacy-invasive practices [19,20].

the severity of privacy intrusions does not always translate into a low ranking of an app in comparison to more privacy-friendly offerings.

2.2   Third-Party Applications on Mobile Networks

Security and privacy issues associated with third-party applications on mobile networks are increasingly gaining importance. Particularly troublesome, app developers have been trying to use unwitting users' devices for spam and unwanted costly premium services [23]. More broadly, measurements studies found that most apps include permission requests that enable potentially dangerous practices [24].

User studies of the utilized permission systems document comprehension and usability problems that are largely similar to the results in the Facebook context (see, for example, [4,25]). As a response, technological measures to help users to manage permissions on mobile systems have been proposed. For example, Beresford et al. introduced a system to disable information requests made by a mobile application and to disable unwanted permissions [26].

Similar to the context of apps on social network sites, mobile applications gain access in various ways to information of friends (or contacts more generally). Apps with multi-platform functionality that have access, for example, to a user's Facebook account will be able to share the same information also in the mobile context. However, apps will frequently enrich this data with additional information gathered in the mobile context. For example, the new Facebook mobile app has caused a stir due to the requirement to access a user's SMS and MMS (i.e., personal and professional communications with other users) [27].

Security firm BitDefender audited over 800000 apps in the Android Play Store and found that apps frequently require access to information that impacts friends and other contacts. For example, almost 10% of the surveyed apps can read your contact list, and a sizable minority leak a user's phone book and call history [28].

For iOS devices, security firm Zscaler discovered when it scanned the 25 most popular apps across five categories, that 92% require access to a user's address book, and 32% go through a user's calendar [29].

These examples highlight that the problem of sharing the information of friends or other contacts without their explicit consent goes well beyond the context of applications on social network sites. We aim to better understand the reasons for such sharing behaviors by developing an economic model that focuses on the adoption of apps with different interdependent privacy consequences.

3   Model Overview

The framework of our model builds on the local network effects research by Sundararajan [30]. His model studies the Bayes-Nash equilibria of a network game in which heterogeneous agents connected in a social network consider the purchase of a product with network effects. Individuals are rational and make their decisions based on a well-specified payoff function. A person who does not purchase the product receives zero payoff, while the payoff of a purchaser is influenced by the actions of her peers, her own valuation type of the product, and the product cost.

Although, we use the basic structure of the payoff function of Sundararajan's model, the focus of our analysis is quite different. Sundararajan studied individuals' purchasing behavior in a scenario where all decisions are made simultaneously, while our goal is to discover users' behavior when they can make adoption decisions sequentially. In addition, we do not only consider positive network effects, but also integrate one specific type of negative network effects: *interdependent privacy harm*. We further consider individuals to have other-regarding preferences. That is, when making adoption decisions, individuals include in their evaluation the privacy harm they potentially inflict on their peers.

In the following, we present the model and break down its different constituent parts. For reference, we provide a complete list of symbols used in our paper.

a_i User i's adoption choice (1 = adoption, 0 = no adoption)

c Cost of app adoption

e Individual's interdependent privacy harm resulting from her friend's adoption behavior

θ_i User i's valuation of an app (also called her *type*)

k_i User i's other-regarding preference

v_i Number of user i's friends who have already adopted the app

N Number of users in the network

n_i Number of user i's friends

p_i User i's payoff

M Number of connections per additional node in the Barabási-Albert (BA) random graph model

M_0 Number of initial seeds in the BA random graph model

SI Set of users that have already adopted the app

I Set of users that choose to adopt the app in one step

F Set of friends of users in I

In our work, we assume that individuals are rational in terms of their awareness of the privacy harm associated with app adoption. In addition, they make their adoption decisions based on the payoff of their actions. Extending Sundararajan's local network effects model, we propose that user i's payoff function is:

$$p_i = a_i[(v_i + 1)\theta_i - \frac{k_i}{v_i + 1}e \cdot n_i - c] \tag{1}$$

If the payoff from adopting the app is larger than zero, the individual will always install the app on her device; otherwise she would deny the installation offer.

There are three parts in the above payoff function: value gained from the app adoption; the perceived responsibility when inflicting privacy harm on peers (i.e., other-regarding preferences); and the cost of app adoption. We discuss each of these parts in the following subsections.

3.1   Value Gained from App Adoption

The value gained from app adoption is represented by $(v_i + 1)\theta_i$. This value can further be divided into two parts: the first part is the direct value gained from using the app; the second part refers to the positive network effects, for example, the extra enjoyment the individual will perceive when a game app can be played together with her friends rather than alone.

Since individuals have different assessments regarding an app's value, we use an individual's *type*, θ_i, to represent this heterogeneity. For example, those individuals with a higher valuation type, i.e., represented by a larger θ_i, will gain more direct value from an app compared with those who have a lower valuation type, which is represented by a smaller θ_i.

In addition, it is reasonable to consider v_i, the number of user i's friends who adopt the app, will affect the utility user i gains from installing and using the app. In particular, we assume that only the number of close friends, i.e., the neighbors in the network, will positively influence the individual's payoff. This is referred to as *positive local network effects* [30]. In practice, apps may also exhibit broader positive network effects; however, we assume that local network effects dominate the adoption decision.

3.2   Care for Privacy Harm Inflicted on Friends by Adoption Decisions

The central function of social network sites (from the user's perspective) is to find friends and interact with them. Typically, individuals will care about their close friends' well-being (however, the level of concern may differ) and try to avoid taking actions that negatively affect their friends. Experimental results provide substantial evidence of the existence of such other-regarding behaviors in group interactions [10]. Other-regarding preferences, which indicate whether and how much people tend to care about others' well-being, are described in detail in a recent review paper [9]. There are two primary types of other-regarding preferences: distributive and reciprocal. The distributive other-regarding preference is caused by people's aversion of outcome inequality [31,32]. The reciprocal aspect of the other-regarding preferences theory indicates that people tend to respond in kind to a peer's behavior [33], which means that people respond to kindness with kindness, and hostility with hostility.

Our paper focuses primarily on the reciprocal aspect of other-regarding preferences. That is, users consider the well-being of their close friends who presumably would act similarly. Under the scenario of interdependent privacy, if individual i chooses to adopt an app, she will inflict a certain amount of privacy harm, e, on her friends. More specifically, user i will incorporate partially the privacy harm she inflicts on all her n_i friends in her own payoff calculation. In our model, this other-regarding preference is represented by $e \cdot n_i$. We make the assumption that e is additive across users. In other words, if a user adopts a certain app which likely impacts her friends' privacy then her worry about this decision will increase with the number of close friends, n_i. We believe this assumption is reasonable. For example, annoyances such as spam typically affect all close friends.[3]

Studies also found that group size likely reduces the impact of other-regarding preferences due to a diffusion of responsibility [34]. When individuals know that others have taken the same potentially harmful action, they do not experience the full burden of responsibility. In our case, the guilt of inflicting privacy harm on others will be diffused with each additional close friend who has already adopted the app. Likewise, reciprocity requires an agent to respond to previous installation decisions that also impose potential privacy harm on her.[4] We use $\frac{e \cdot n_i}{v_i + 1}$ to represent the part of the remaining responsibility that user i shoulders when she calculates her payoff considering her diffused responsibility and reciprocal factors. In particular, we make the assumption that the guilt of causing privacy harm is split equally across the local peers who make the adoption decision.

In order to indicate to which degree an individual is generally concerned about privacy harm imposed on friends, we use k_i to represent agent i's other-regarding preference. A larger k indicates a higher other-regarding preference, a smaller k represents lower other-regarding preferences. Thus, $\frac{k_i}{v_i + 1} e \cdot n_i$ reflects how agent i cares about her friends' privacy harm inflicted by herself.

Please note that users apply a heuristic evaluation when they calculate the privacy harm inflicted on others with the formula stated above. For example, an exact calculation would require an assessment of the overlap between her friends, and her friends of friends. Theoretically, a user should only experience partial emotional relief for the installation decisions of her friends when not all of her own friends were affected by her friends' app adoptions. However, while in practice it is relatively easy to determine how many friends have installed a particular app; it is extremely cumbersome (if not impossible for an average user) to determine this more specific figure on most social network sites and mobile networks. In addition, user i cannot easily reciprocate in the app installation context against a specific user since her adoption decision affects the whole groups of close friends.

3.3   Cost of App Adoption

All practical costs associated with an adoption decision, except the interdependent privacy harm experienced by her choice, are included in the installation cost. For example, the installation costs contain, but are not limited to, the cost of finding and installing an app, the cost of learning how to use the app, and user's personal privacy harm when she chooses to install the app.

---

[3] For example, if Bob installs Candy Crush, a very popular third-party Facebook app, then this installation will typically trigger invitations to both his friends Eve and Trudy.

[4] Note that interdependent privacy harm user i already is suffering from cannot be influenced by herself and is therefore not part of the payoff calculation. However, it finds consideration in her other-regarding preferences.

4   Simulation Setup

Given the model we proposed above, we conduct two simulations to investigate app adoption in both its early and late phases. Based on available empirical literature on the purchasing behavior for new products, we argue that the processes for early and late adoption differ significantly. In the early phase, a pool of potential first adopters is evaluating a newly introduced product (as described in our model) while considering social and privacy factors [35]. In the later stages of adoption, users are heavily influenced by available product rankings which are interpreted as a proxy for the quality of a product [36]. (Note that early adoption decisions can be also influenced by product rankings [36]. However, social networking sites and mobile networks typically only include apps in rankings once they have reached a certain popularity threshold.)

We proceed as follows. In the first simulation, we aim to understand the percentage of users who choose to adopt an app that collects information from users' friends from the first moment the app is introduced into an app marketplace. In addition, we will show how this percentage will be affected by network size, the level of an app's interdependent privacy harm, and installation cost. In the second simulation, we simultaneously derive early adoption results for multiple apps with different interdependent privacy harms. We then proceed to rank these apps according to their associated frequency of positive early adoption decisions. Based on these rankings, we then discuss the impact of these rankings on potential later adoption by a larger pool of users.

4.1   Scale-Free Network

Evidence from measurement studies suggests that social network sites and other human-formed networks exhibit properties of scale-free networks [37,38]. We therefore conduct our simulations within the framework of a scale-free network model. The model we use to generate the network is the Barabási-Albert (BA) model [39]. The central idea of the BA model is that in a network, the more nodes a particular node connects to, the more likely the node will attract new connections. In our model, this means that the more friends a user has, the more likely others are willing to be her friends (i.e., a notion of popularity).

When using the BA model to generate a scale-free network of N people, we first randomly connect M_0 initial nodes. Then, according to the principle that the probability of connecting to an existing node is proportional to the degree of that node, each new node is connected to M existing nodes. Following this procedure, the remaining $N - M_0$ nodes are then connected to the network one by one [39].

4.2   Simulation Process

Users make their decisions according to the payoff function stated in Equation (1). Thus, before we can simulate users' behaviors, we have to decide on the parameters that appear in Equation (1). These unknowns include the number of users in the network, topology of the network, valuation type and the other-regarding preference type of each user, installation cost, and the level of an app's interdependent privacy harm. In order to decide on these unknowns, we make the assumption that given the overall number of users in the network, nature will determine how those people are connected, and what valuation type and other-regarding preference type each user has. In addition, installation cost, c, the level of an app's interdependent privacy harm, e, and the network size, N, are predefined by us (i.e., we will indicate the specific values in the following tables and figures).

Hence, before we simulate adoption rates of a newly introduced app that causes interdependent privacy harm, we need to set values for unknown parameters, i.e., e, c and N. We then use the BA model to generate a scale-free network of N users. Next, we attribute a valuation type θ_i and an other-regarding preference type k_i to each user. Although, we assume both types to follow the uniform distribution over the interval $(0, 1)$, we do not randomly attribute types to users. Instead, we assign types according to the assumption that friends tend to have similar preferences (which is motivated by social science research, e.g., [40]). This means, users are assigned to types in such a way that people who are friends tend to have similar θ_i and k_i.

After setting values for unknown parameters, we follow a fixed simulation methodology and average the percentage of individuals who install the app across 10000 simulation rounds. In our simulation procedure each individual has the opportunity to make a positive adoption decision more than once. In other words, even if a user declines to adopt an app at first inspection, she can reconsider her decision when more friends chose to adopt that app. The simulation is set up as follows:

1. For each individual in the network, set her v_i to be 0. This is reasonable since none of the N individuals has yet installed the app. In addition, we use SI to denote the set of people that have already installed the app. Here, SI is $\emptyset$.
2. Check adoption decisions of all N individuals according to the payoff function. Use set I to record the individuals that choose to adopt the app in this step. Add each person in I to SI.
3. For each individual in set I, find the friends of them and record these friends in set F. For each person in F, find her current v_i.
4. Check the adoption decision of each person in F. Change set I so that it records all the new individuals who adopted the app in this step. Add each element in I to the set SI.
5. Repeat step 3 and step 4 until there are no individuals left in set I.
6. Divide the number of individuals in SI by the total number of users in the network. Output this result, which denotes the percentage of users who have eventually decided to adopt the app.
7. Terminate this round.

The above simulation determines the adoption result for a particular app with a given combination of values e, c, and N. To help us understand how adoption

Table 2. Distribution of app adoption outcomes for various values of e and constant $c = 0$, $N = 100$

e	0.1	0.3	0.5	0.7	0.9	1.1	1.3	1.5	1.7	1.9
$\leq 10\%$	0	0.87	4.58	11.95	22.7	32.39	43.0	53.03	61.17	69.43
$10\% \sim 90\%$	0.01	0.01	0.03	0.18	0.59	1.75	2.61	3.85	4.92	4.99
$\geq 90\%$	99.99	99.12	95.39	87.87	76.71	65.86	54.39	43.11	33.91	25.58

results change with respect to each of these app dimensions, we systemically vary these parameters.

5   Simulation Results 1: Individuals' App Installation Behaviors in Early Adoption Stage

In this section, we provide simulation results that describe early stage adoption outcomes of apps with interdependent privacy consequences. The results help us understand future app adoption outcomes which we will discuss in Section 6.

In the following subsections, we focus our analysis on one particular parameter (i.e., e, c or N) to analyze its impact on app adoption.

How Is Adoption Impacted by Interdependent Privacy Harm? For this analysis, we consider changes of the level of interdependent privacy harm, e, and keep constant the values for c and N. We consider 4 different sets of (c, N) and plot graphs to show the distribution of app adoption rates for each of these 4 sets (Figure 1). The horizontal axis represents adoption rates, while the vertical axis indicates the percentage of 10,000 simulation rounds that fall into a particular range of adoption rates. Here, we consider three ranges of app adoption results: less than 10% adoption rate; adoption rates between 10% and 90%; and adoption rates above 90%.

Figure 1 shows that with increasing privacy harm the percentage of positive adoption decisions decreases. E.g., adoption rates between 90% and 100% occur much less frequently, while there is an increased possibility of falling into the lower range of adoption rates, i.e., 0% to 10%. Numeric figures are provided in Table 2.

How Is App Adoption Impacted by Installation Cost? In this subsection, we vary the installation cost, c, from 0 to 1, while keeping the parameters for privacy harm, e, and network size, N, constant. Similar to the previous analysis, we consider 4 fixed sets of (e, N). Figure 2 demonstrates that when installation costs increase, there is a higher probability that the app will suffer from a lower adoption rate. Numeric values for fixed (e, N) equaling $(0.5, 100)$ are provided in Table 3.

How is adoption impacted by network size? We consider different app network sizes from 100 to 2,000 nodes, and keep constant installation cost and

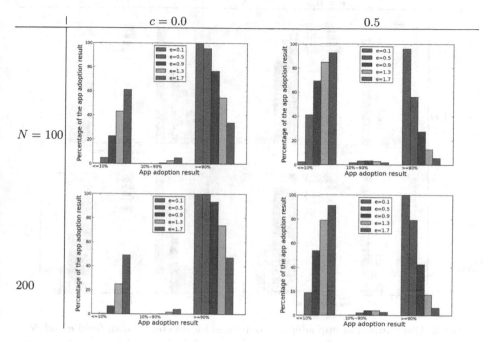

Fig. 1. Distribution of app adoption outcomes for different e with fixed c and N

Table 3. Distribution of app adoption outcomes for various values of c and constant $e = 0.5$, $N = 100$

c	0	0.1	0.2	0.3	0.4	0.5	0.6	0.7	0.8	0.9	1.0
$\leq 10\%$	4.82	8.02	12.4	18.78	28.68	41.6	55.72	72.93	87.69	96.77	99.99
$10\% \sim 90\%$	0.01	0.16	0.64	1.11	2.04	2.85	3.01	2.86	1.63	0.53	0.01
$\geq 90\%$	95.17	91.82	86.96	80.11	69.28	55.55	41.27	24.21	10.68	2.7	0

privacy harm. We consider 4 different fixed sets of (e, c), and plot the results in Figure 3. In Table 4, we provide numeric results for (e, c) equaling $(1.0, 0.5)$. From both Figure 3 and Table 4, we can observe that as the network size increases the probability of an app being adopted increases as well.

6   Simulation Results 2: App Installation Behaviors in the Late Adoption Stage

In this section, we aim to understand app installation results once rational early adopters have evaluated new apps and a ranking has become available that increases the prominence of the new apps (proportional to its ranking) to a larger user group. Rankings based on early adoption results play an important role in

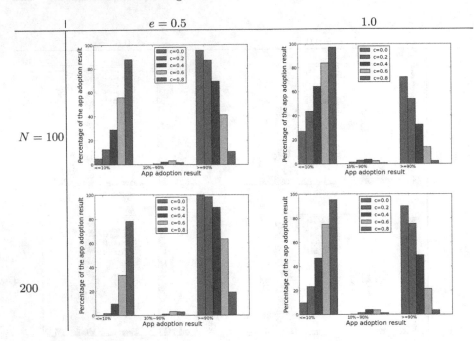

Fig. 2. Distribution of app adoption outcomes for different c with fixed e and N

Table 4. Distribution of app adoption outcomes for various values of N and constant $e = 1.0$, $c = 0.5$

N	100	200	500	1000	2000
$\leq 10\%$	74.21	61.07	37.54	19.07	5.23
$10\% \sim 90\%$	3.12	4.0	3.94	2.11	0.55
$\geq 90\%$	22.67	34.93	58.52	78.82	94.22

shaping users' adoption behavior in the later adoption phase since consumers will frequently rely on these rankings during their own adoption decisions [41,42].

For example, based on evidence from an iOS app market, Garg and Telang found that top-ranked for-pay apps generated about 150 times more downloads compared to apps ranked at about 200 [42]. Similarly, Carare showed that consumers' willingness to pay for a top-ranked app is about $4.50 greater than for the same unranked app [41]. Further direct evidence of the impact of app rankings is provided by a more recent study. Applying a data-driven approach, Ifrach and Johari studied the effect of the top-rank position on demand in the context of mobile app markets [11]. They found that the demand for an app almost doubles when its rank shifts from position 20 to position 1. Taken together, rankings can serve as an important indicator for future adoption outcomes in app markets.

In addition, previous research showed that users adopt apps with high privacy harms because they did not understand the fact that apps maliciously harvest their profile information [3,4]. Some researchers expect (e.g., [4]), it would be sufficient to protect users from undesirable apps if at least some users demonstrated

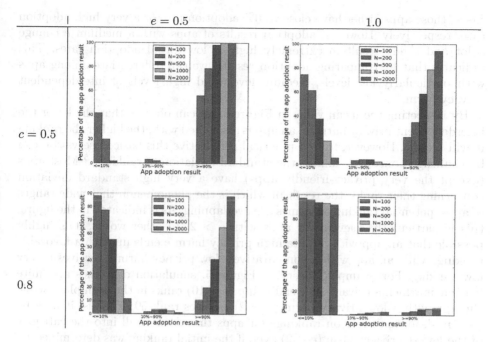

Fig. 3. Distribution of app adoption outcomes for different N with fixed e and c

awareness and understanding of permissions. However, in this section we show that even if an early adopter group rationally evaluates the different aspects of new apps, then the resulting app ranking can provide misleading signals to less savvy consumers in the later adoption phase.

We take the following approach. We first investigate early adoption results for multiple apps (by following the general methodology outlined in the previous section). More precisely, we first simulate early adoption results of 100 apps with different levels of e for 50 times. Note here, the level of privacy harm ranges from 0.1 to 10, the network size is fixed at $N = 100$, and installation cost is constant at $c = 0$. We then rank those apps based on their early adoption rates for each of the 50 simulations, i.e., we collect 50 rankings. By analyzing the variation in the early adopter rankings, we can then gain some insights about apps' likely future adoption results. Or to put it differently, we can discuss the informativeness of the app ranking as a signal to the consumer.

We show the simulation results in Figures 4 and 5. Note that in both figures, each number on the x-axis indicates a particular app with a value of e equal to that number. For example, 4 represents the app with $e = 4$. In Figure 4, the y-axis represents the number of individuals who adopt the app. The y-axis in Figure 5 represents the cardinal number of each app's rank. In both figures, the dots represent the mean value, while the ranges that include the dot in the vertical direction denote the standard deviation of the relevant value.

As we can see in Figure 4, the more interdependent privacy harm an app causes, the lower its adoption rate will become. In addition, adoption outcomes of apps with either a particularly high e or low e do not vary too much. In most cases, those apps either have close to 0% adoption rate or a very high adoption rate, respectively. However, adoption results of apps with a medium e change a lot and rarely result in extremely high or low initial adoption rates. This indicates that by comparing adoption results, we can differentiate among apps with particularly low levels, medium levels and high levels of interdependent privacy harm.

By inspecting the mean value in Figure 5, we can observe that the lower the interdependent privacy harm, e, an app is associated with, the higher the ranking it will receive. However, from a practical perspective this basic observation can be challenged once we examine the standard deviation of rankings.[5] Most apps (except the very privacy-friendly apps) have a very high standard deviation concerning their ranks; the result of which is the phenomenon that wide ranges of apps' potential ranking outcomes are overlapping. As is indicated in the figure, this is particularly relevant for apps with $e > 2$. In other words, it is highly possible that an app with a quite high privacy harm e ends up with a favorable ranking, while an app with a comparatively low privacy harm e receives a very low ranking. For example, observing Figure 5, simulation outcomes are quite feasible in which a privacy-unfriendly app ($e = 10$) ranks in the 30th place while the app with a low privacy harm ($e = 2$) receives rank 50. That is, it may be very misleading to rely on rankings for apps that do not fall into the category of the lowest privacy harm ($e < 2$) even if the initial ranking was determined by a set of rational early adopters.

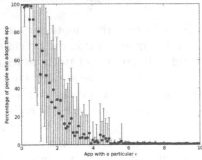

Fig. 4. Adoption results of 100 apps with e changing from 0.1 to 10   **Fig. 5.** Rankings of 100 apps with e changing from 0.1 to 10

To better illustrate that rankings cannot accurately reflect apps' interdependent privacy harm, we compare rankings between pairs of apps with one app having a medium level of privacy harm and one app with a relatively high level of privacy harm. Here, we compare four groups of apps: (1) app with $e = 2.6$

---

[5] Essentially, we investigate the variation between a low number of different ranking outcomes which in our case are 50 alternate universes of app rankings.

and app with $e = 9.5$, (2) app with $e = 3.4$ and app with $e = 9.5$, (3) app with $e = 3.8$ and app with $e = 9.8$ and (4) app with $e = 4.5$ and app with $e = 8.8$. For each group, we plot the 50 rankings of a particular pair of apps in Figure 6. As we can see in each figure, blue dots and red dots fall into the same range, and reveal no discernible pattern. We use Welch's $t-test$ to examine the relationship between the 50 rankings for each pair. The statistical results, shown in Table 5, indicate that for all of these four pairs, the potential rankings of the app with a medium level of privacy harm are not significantly different from the rankings of the app with a high level of privacy harm.

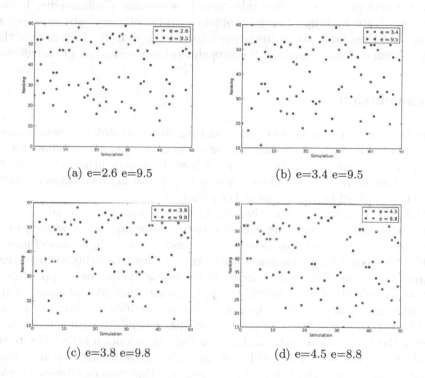

(a) e=2.6 e=9.5

(b) e=3.4 e=9.5

(c) e=3.8 e=9.8

(d) e=4.5 e=8.8

Fig. 6. Comparison of app rankings for four sample groups

Table 5. Statistical results of Welch's t-test for four sample groups

Group	F-Value	p-Value
$e = 2.6\ e = 9.5$	-1.84	0.07
$e = 3.4\ e = 9.5$	-1.42	0.16
$e = 3.8\ e = 9.8$	-1.02	0.31
$e = 4.5\ e = 8.8$	-0.63	0.53

In summary, we assume that in the early adoption period, rational users are able to identify privacy-intrusive apps; for example, partly because of the lack of market signals (i.e., a ranking) they have a higher incentive to inspect applications. However, after a sufficiently large group of early adopters has inspected the app, the platform provider will typically include the app in its rankings. Rational as well as less savvy adopters will now likely rely on the app rankings to guide their adoption behaviors. However, since the resulting ranking is not informative enough to reflect app's interdependent privacy harm level, users are likely to also fall for apps with significant privacy harm. This observation complements the findings in the behavioral literature that users adopt apps with high privacy harms mostly due to their unawareness of apps' malicious and intrusive privacy practices. However, our take-away is somewhat disillusioning. Even if we can motivate a group of early adopters to rationally evaluate apps and we assume that they understand the privacy consequences of the installation, then the long-term outcomes might still disappoint privacy and consumer advocates.

7   Conclusion

In the interconnected setting of social network sites and mobile networks, apps' practices to collect personal information of users' friends and to allow for potential misuse amplifies the importance of interdependent privacy. That is, the privacy of an individual user does not only depend on her own behavior, but it is also the result of the decisions of her friends.

Taking an economic perspective, we propose a model of the adoption behavior for social apps in a networked system where privacy consequences are interdependent. Motivated by behavioral economics research, we model users to exhibit other-regarding preferences about the privacy well-being of their peers. We present two simulation approaches to investigate individuals' app adoption behaviors: early adoption of individual apps, and later adoption of a pool of apps with different privacy harms. The simulation results indicate that in the early adoption period, either lowering the level of interdependent privacy harm or decreasing the installation cost will increase the app adoption rates. The results also show app adoption rates will increase with a growing network size. Based on the second simulation approach, we conclude that rankings based on early adoption results frequently will not accurately reflect the level of apps' interdependent privacy harm. This is especially relevant for rankings of apps that have medium and high level of privacy harm.

While further study is needed, for example, we are investigating the robustness of our results to different specifications of the model, we believe that our study can contribute to the policy discussion on app privacy [43]. Privacy advocates should cautiously reconsider the expected impact of added scrutiny by early adopters in a marketplace; that is, encouraging individuals to pay more attention to the potential privacy harm of apps may not create the anticipated ripple-effects in the marketplace. We believe that in many cases it is likely misleading to rely on such market signals when we are considering products with strong network effects and interdependent privacy harm. In addition, our work highlights the important role of the platform provider. In particular, the design and scope of rankings should be carefully tested to increase the likelihood that market signals are meaningful. For example, rankings could be limited to data with low variability of ratings.

To better understand the impact of different rankings we intend to work on actual app adoption data. Unfortunately, publicly available data usually does not provide details of app adoption dynamics. Instead, we favor an experimental approach (similar to [36]) to further calibrate our model.

References

1. Chia, P., Yamamoto, Y., Asokan, N.: Is this app safe?: A large scale study on application permissions and risk signals. In: Proceedings of the 21st International World Wide Web Conference (WWW), pp. 311–320 (April 2012)
2. Felt, A., Evans, D.: Privacy protection for social networking APIs. In: Proceedings of the 2008 Workshop on Web 2.0 Security and Privacy (W2SP) (May 2008)
3. Besmer, A., Lipford, H.: Users' (mis)conceptions of social applications. In: Proceedings of Graphics Interface (GI), pp. 63–70 (May 2010)
4. Felt, A., Ha, E., Egelman, S., Haney, A., Chin, E., Wagner, D.: Android permissions: User attention, comprehension, and behavior. In: Proceedings of the 7th Symposium on Usable Privacy and Security (SOUPS), pp. 3:1–3:14 (July 2012)
5. Biczók, G., Chia, P.H.: Interdependent privacy: Let me share your data. In: Sadeghi, A.-R. (ed.) FC 2013. LNCS, vol. 7859, pp. 338–353. Springer, Heidelberg (2013)
6. Grossklags, J., Christin, N., Chuang, J.: Secure or insure?: A game-theoretic analysis of information security games. In: Proceedings of the 17th International World Wide Web Conference (WWW), pp. 209–218 (April 2008)
7. Kunreuther, H., Heal, G.: Interdependent security. Journal of Risk and Uncertainty 26(2), 231–249 (2003)
8. Laszka, A., Felegyhazi, M., Buttyán, L.: A survey of interdependent information security games. ACM Computing Surveys (forthcoming)
9. Cooper, D., Kagel, J.: Other regarding preferences: A selective survey of experimental results (forthcoming), http://myweb.fsu.edu/djcooper/research/otherregard.pdf
10. Stahl, D., Haruvy, E.: Other-regarding preferences: Egalitarian warm glow, empathy, and group size. Journal of Economic Behavior & Organization 61(1), 20–41 (2006)
11. Ifrach, B., Johari, R.: The impact of visibility on demand in the market for mobile apps. Technical report, SSRN Working Paper (February 2014)
12. Book, T., Wallach, D.: A case of collusion: A study of the interface between ad libraries and their apps. In: Proceedings of the 3rd Annual ACM CCS Workshop on Security and Privacy in Smartphones & Mobile Devices (SPSM), pp. 79–86 (November 2013)
13. Krishnamurthy, B., Wills, C.: On the leakage of personally identifiable information via online social networks. In: Proceedings of ACM SIGCOMM Workshop on Online Social Networks (WOSN), pp. 7–12 (August 2009)
14. Steel, E., Fowler, G.: Facebook in privacy breach. The Wall Street Journal (October 2010)
15. King, J., Lampinen, A., Smolen, A.: Privacy: Is there an app for that? In: Proceedings of the 7th Symposium on Usable Privacy and Security (SOUPS), pp. 12:1–12:20 (July 2011)

16. Tam, J., Reeder, R., Schechter, S.: I'm allowing what? Disclosing the authority applications demand of users as a condition of installation. Technical Report MSR-TR-2010-54, Microsoft Research (2010)
17. Wang, N., Grossklags, J., Xu, H.: An online experiment of privacy authorization dialogues for social applications. In: Proceedings of the Conference on Computer Supported Cooperative Work (CSCW), pp. 261–272 (February 2013)
18. Wang, N., Xu, H., Grossklags, J.: Third-party apps on Facebook: Privacy and the illusion of control. In: Proceedings of the ACM Symposium on Computer Human Interaction for Management of Information Technology (CHIMIT), pp. 4:1–4:10 (December 2011)
19. Good, N., Dhamija, R., Grossklags, J., Aronovitz, S., Thaw, D., Mulligan, D., Konstan, J.: Stopping spyware at the gate: A user study of privacy, notice and spyware. In: Proceedings of the Symposium on Usable Privacy and Security (SOUPS), pp. 43–52 (July 2005)
20. Good, N., Grossklags, J., Mulligan, D., Konstan, J.: Noticing notice: A large-scale experiment on the timing of software license agreements. In: Proceedings of the ACM Conference on Human Factors in Computing Systems (CHI), pp. 607–616 (April-May 2007)
21. Shehab, M., Marouf, S., Hudel, C.S.: ROAuth: Recommendation based open authorization. In: Proceedings of the 7th Symposium on Usable Privacy and Security (SOUPS), pp. 11:1–11:12 (July 2011)
22. Wang, N.: Third-party applications' data practices on Facebook. In: Proceedings of the 2012 ACM Annual Conference on Human Factors in Computing Systems, Extended Abstracts (CHI EA), pp. 1399–1404 (May 2012)
23. Felt, A., Finifter, M., Chin, E., Hanna, S., Wagner, D.: A survey of mobile malware in the wild. In: Proceedings of the ACM Workshop on Security and Privacy in Smartphones and Mobile Devices (SPSM), pp. 3–14 (October 2011)
24. Felt, A., Greenwood, K., Wagner, D.: The effectiveness of application permissions. In: Proceedings of the 2nd USENIX Conference on Web Application Development (WebApps), p. 7 (June 2011)
25. Kelley, P., Cranor, L., Sadeh, N.: Privacy as part of the app decision-making process. In: Proceedings of the ACM Annual Conference on Human Factors in Computing Systems (CHI), pp. 3393–3402 (April 2013)
26. Beresford, A., Rice, A., Skehin, N., Sohan, R.: Mockdroid: Trading privacy for application functionality on smartphones. In: Proceedings of the 12th Workshop on Mobile Computing Systems and Applications (HotMobile), pp. 49–54 (March 2011)
27. Woollaston, V.: Is Facebook reading your TEXTS? Android update lets app access your written and picture messages. Daily Mail Online (January 2014)
28. Karambelkar, D.: Spyware: A bird's-eye view. Gulf News (February 2014)
29. Robertson, J.: Google+, 'Candy Crush' show risk of leakiest apps. Bloomberg Technology (January 2014)
30. Sundararajan, A.: Local network effects and complex network structure. The BE Journal of Theoretical Economics 7(1) (January 2007)
31. Fehr, E., Schmidt, K.: A theory of fairness, competition, and cooperation. The Quarterly Journal of Economics 114(3), 817–868 (1999)
32. Bolton, G., Ockenfels, A.: ERC: A theory of equity, reciprocity, and competition. American Economic Review 90(1), 166–193 (2000)
33. Berg, J., Dickhaut, J., McCabe, K.: Trust, reciprocity, and social history. Games and Economic Behavior 10(1), 122–142 (1995)

34. Darley, J., Latane, B.: When will people help in a crisis? In: Hochman, S. (ed.) Readings in Psychology, pp. 101–110. MSS Information Corporation (1972)
35. Fisher, R., Price, L.: An investigation into the social context of early adoption behavior. Journal of Consumer Research 19(3), 477–486 (1992)
36. Salganik, M., Dodds, P., Watts, D.: Experimental study of inequality and unpredictability in an artificial cultural market. Science 311(5762), 854–856 (2006)
37. Ahn, Y., Han, S., Kwak, H., Moon, S., Jeong, H.: Analysis of topological characteristics of huge online social networking services. In: Proceedings of the 16th International World Wide Web Conference (WWW), pp. 835–844 (May 2007)
38. Mislove, A., Marcon, M., Gummadi, K., Druschel, P., Bhattacharjee, B.: Measurement and analysis of online social networks. In: Proceedings of the 7th ACM SIGCOMM Conference on Internet Measurement (IMC), pp. 29–42 (October 2007)
39. Barabási, A., Albert, R.: Emergence of scaling in random networks. Science 286(5439), 509–512 (1999)
40. Verbrugge, L.: The structure of adult friendship choices. Social Forces 56(2), 576–597 (1977)
41. Carare, O.: The impact of bestseller rank on demand: Evidence from the app market. International Economic Review 53(3), 717–742 (2012)
42. Garg, R., Telang, R.: Inferring app demand from publicly available data. MIS Quarterly 37(4), 1253–1264 (2013)
43. Good, N., Grossklags, J., Thaw, D., Perzanowski, A., Mulligan, D., Konstan, J.: User choices and regret: Understanding users' decision process about consensually acquired spyware. I/S: A Journal of Law and Policy for the Information Society 2(2), 283–344 (2006)

Cybersecurity Games and Investments:
A Decision Support Approach

Emmanouil Panaousis[1], Andrew Fielder[2], Pasquale Malacaria[1],
Chris Hankin[2], and Fabrizio Smeraldi[1,*]

[1] Queen Mary University of London, UK
{e.panaousis,p.malacaria,f.smeraldi}@qmul.ac.uk
[2] Imperial College London, UK
{andrew.fielder,c.hankin}@imperial.ac.uk

Abstract. In this paper we investigate how to optimally invest in cybersecurity controls. We are particularly interested in examining cases where the organization suffers from an underinvestment problem or inefficient spending on cybersecurity. To this end, we first model the cybersecurity environment of an organization. We then model non-cooperative *cybersecurity control-games* between the *defender* which abstracts all defense mechanisms of the organization and the *attacker* which can exploit different vulnerabilities at different network locations. To implement our methodology we use the SANS Top 20 Critical Security Controls and the 2011 CWE/SANS top 25 most dangerous software errors. Based on the profile of an organization, which forms its preferences in terms of *indirect costs*, its concerns about different kinds of *threats* and the importance of the assets given their associated risks we derive the Nash Equilibria of a series of control-games. These game solutions are then handled by optimization techniques, in particular multi-objective, multiple choice Knapsack to determine the optimal cybersecurity investment. Our methodology provides security effective and cost efficient solutions especially against *commodity attacks*. We believe our work can be used to advise security managers on how they should spend an available cybersecurity budget given their *organization profile*.

Keywords: cybersecurity, game theory, optimization.

1 Introduction

One of the single largest concerns facing organizations today is how to protect themselves from cyber attacks whose prominence impose the need for organizations to prioritize their cybersecurity concerns with respect to their perceived threats. Organizations are then required to act in such a way so as to minimize their vulnerability to these possible threats. The report [4] published by Deloitte and NASCIO, points out that only 24% of Chief Information Security Officers

---

* The authors are supported by the Project "Games and Abstraction: The Science of Cyber Security" funded by EPSRC, Grants: EP/K005820/1, EP/K005790/1.

(CISOs) are very confident in protecting their organization's assets against external threats. Another important finding in this report is that the biggest concern CISOs face in addressing cybersecurity is a "Lack of sufficient funding" where 86% of respondents were concerned.

Most organizations will have a fixed budget for the protection of their systems. Generally this budget would not allow them to fully cover all of the vulnerabilities that their data assets are at risk from. As such an organization is interested in *how to use the limited financial budget available to best protect them* from various vulnerabilities given that the implementation of a cybersecurity control is associated with a *direct cost*.

Apart from the direct costs of controls, there are also *indirect costs* incurred by the implementation of these controls. From this point of view investing more in cybersecurity might not always be the most efficient approach that CISOs can follow. Therefore another dimension of the cybersecurity investment problem is "what is the optimal cybersecurity budget allocation given the importance that the organization places into its different assets, the system performance requirements, and the profile of employees and clients?"

1.1   Our Contributions

In this work we provide a methodology and a tool that can support security managers with decisions regarding the optimal allocation of their cybersecurity budgets.

We first motivate a method for the creation of an organization's cybersecurity strategy (Section 3). This is achieved by performing a risk analysis of the data assets that an organization has, and analyzing the effectiveness of different security controls against different vulnerabilities. We then formulate control-games (Section 4) based on these risk assessments, in order to calculate the most effective way for an organization to implement each control. In a control-game the defender aims at reducing cybersecurity risks by implementing a control in a certain way dictated by the Nash Equilibrium (NE). In this way, the defender minimizes the maximum potential damage inflicted by the attacker. The solutions of the different control-games are handled by the optimization techniques of multi-objective, multiple choice Knapsack (Section 5) to decide upon an optimal allocation of a cybersecurity budget. We also present a case study (Section 6) which includes vulnerabilities (i.e. CWE) and cybersecurity controls published by the Council on CyberSecurity. We have implemented our methodology (Section 7) for this case study by computing games solutions and investments and measure its performance in terms of cybersecurity defense for different organization profiles.

To demonstrate the effectiveness of our methodology we have implemented part of the SANS Top 20 Critical Security Controls and the 2011 CWE/SANS top 25 most dangerous software errors. We present examples of investment strategies that our tool recommends and test their optimality by looking at alternatives to show that they are the best. In this way, our work is a step towards implementing a theoretical cybersecurity investment decision-making methodology into a realistic scenario.

2 Related Work

Anderson [1] first proposed the study of security from an economics perspective putting forward the idea that cybersecurity is bounded by other non-technical incentives. Anderson highlighted with an example that although some organizations spend less money on security they spend it more effectively therefore having put in place better cyber defenses. In our work we share Anderson's view. However our approach is quite different as we focus on developing cybersecurity decision support tools to assist security managers on how to spend a cybersecurity budget in terms of different controls acquisition and implementation. Our work has been partially influenced by a recent contribution within the field of physical security [17], where the authors address the problem of finding an optimal defensive coverage. The latter is defined as the one maximizing the worst-case payoff over the targets in the potential attack set. One of the main ideas of this work we adopt here is that the more we defend the less rewards the attacker receives.

Alpcan [5] (p. 134) discusses the importance of studying the quantitative aspects of risk assessment with regards to cybersecurity in order to better inform decisions makers. This kind of approach is taken in this work where we provide an analytical method for deciding the level of risk associated from different vulnerabilities and the impact that different security controls have in mitigating these risks. By studying the incentives for risk management Alpcan [6] developed a game theoretic approach that optimizes the investment in security across different autonomous divisions of an organization, where each of the divisions is seen as a greedy entity. Furthermore Alpcan et al. examine in [14] security risk dependencies in organizations and they propose a framework which ranks the risks by considering the different complex interactions. This rank is dictated by an equilibrium derived by a Risk-Rank algorithm. Saad et al. [12] model cooperation among autonomous parts of an organization that have dependent security assets and vulnerabilities for reducing overall security risks, as a cooperative game. In [13] Bommannavar et al. capture risk management in a quantitative framework which aids decision makers upon allocation of security resources. The authors use a dynamic zero-sum game to model the interactions between attacking and defending players; A Markov model, in which states represent probabilistic risk regions and transitions, has been defined. The authors are using Q-learning to cope with scenarios when players are not aware of the different Markov model parameters. Previous work carried out by Fielder et. al. [9] considers *how to optimally allocate the time for security tasks for system administrators*. This work identifies how to allocate the limited amount of time that a system administrator has to work on the different security related tasks for an organization's data assets.

One of the initial works studying the way to model investment in cybersecurity was conducted by Gordon and Loeb [7]. The authors identify a method for determining the level of investment for the protection of individual targets, showing that the optimal level of investment should be related to the probability of a vulnerability occurring. The main message of this work is that to maximize the expected benefit from information security investment, an organization should spend only a small fraction of the expected loss due to a security breach. The work published in [8] examines the weakest target game which refers to the case where an attacker is always able to compromise the system target with the lowest level of defense and not to cause any damage to the rest of the targets. The game theoretic analysis the authors have undertaken shows that the game leads to a conflict between pure economic interests and common social norms. While the former are concerned with the minimization of cost for security investments, the latter imply that higher security levels are preferable. Cavusoglu et. al. [11] compare a decision theory based approach to game theoretic approaches for investment in cybersecurity. Their work compares a decision theory model to both simultaneous and sequential games. The results show that the expected payoff from a sequential game is better than that of the decision theoretic method, however a simultaneous game is not always better. Recent work on cybersecurity spending has been published by Smeraldi and Malacaria [10]. The authors identified the optimum manner in which investments can be made in a cybersecurity scenario given that the budget allocation problem is most fittingly represented as a multi-objective Knapsack problem. Cremonini and Nizovtsev, in [15], have developed an analytical model of the attacker's behavior by using cost-benefit analysis therefore considering rewards and costs of achieving different actions. Lastly, Demetz and Bachlechner [16] have identified, analyzed and presented a set of approaches for supporting information security investment decisions. A limitation of this paper, as highlighted by the authors, is that they assume that sufficient money is available to make an investment although in reality cybersecurity budgets are limited.

3   Cybersecurity Model

In this section we describe our cybersecurity model to illustrate an organization's network topology, systems and security components. The network architecture will determine how the different assets of an organization are interconnected. In this paper we follow the network architecture as proposed in the *SANS Critical Security Control 19-1* entitled "Secure Network Engineering" and published in [3]. This consists of three depths namely the demilitarized zone (*DMZ*), the *Middleware*, and the *Private Network*. An organization's assets that can be accessed from the Internet are placed in the *DMZ*, and they should not contain any highly sensitive data. Any asset with highly sensitive data must be located at the *Private Network*, and communicate with the outside world only through a proxy which resides on the *Middleware*.

We define the *depth* of an asset, denoted by d, as the location of this asset within an organization's network architecture. Depths are separated from each other by a set of network security software, e.g. firewalls, IDS. A depth determines (i) the level of security that needs to be breached or bypassed in order

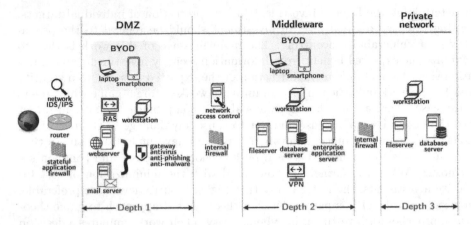

Fig. 1. Sample network architecture

for an attack to successfully exploit a vulnerability at this depth, and (ii) the importance of the data asset compromised if an attack is successful.

We denote the set of all cybersecurity targets within an organization by T and the set of all vulnerabilities threatened by commodity attacks by $\mathcal{V}$.

Definition 1. *Commodity attacks are attack methods where the attack tools can be purchased by a user, where the adversaries do not develop the attacks themselves, and only configure the tools for their own use.*

A *cybersecurity target* is defined as a *(vulnerability, depth)* pair, i.e. $t_i = (v_z, d)$. A target abstracts any *data asset*, located at depth d, that an attack threatens to compromise by exploiting the vulnerability v_z.

We define the set of all targets as $T = \{(v_z, d) | v_z \in \mathcal{V}, d \in \{1, \ldots, n\}\}$. We assume that each network architecture has its own set of targets however throughout this paper we consider the network architecture depicted in Fig. 1. In this paper, we specify that *data assets* located at the *same depth* and having the same *vulnerabilities* are abstracted by the *same target*, and they are worth the *same value* to the organization.

A *cybersecurity control* is the defensive mechanism that can be put in place to alleviate the risk from one or more attacks by reducing the probability of these attacks successfully exploiting vulnerabilities. The defender can choose to implement a control c_j at a certain level $l \in \{0, \ldots, \mathcal{L}\}$. The higher the level the greater the degree to which the control is implemented.

Definition 2. *We define a cybersecurity process as the implementation of a control at a certain level, and we denote by p_{jl} the cybersecurity process that implements the control c_j at level l.*

We define as $C = \{c_j\}$ the set of all cybersecurity controls the defender is able to implement to defend the system, and $P_j = \{p_{jl}\}$ the set of all cybersecurity processes associated with control c_j. A cybersecurity process p_{jl} has a *degree of mitigation* for each target t_i which equals the effectiveness of the cybersecurity process on this target, denoted by $e(t_i, p_{jl}) \in (0, 1]$. We also define

MITIGATION $= e(t_i, p_{jl})$. In this paper we are interested in how cybersecurity processes are combined in a proportional manner to give an implementation plan for this control. We call this *a cybersecurity plan* which allows us to examine advanced ways of mitigating vulnerabilities.

Definition 3. *A cybersecurity plan is a probability distribution over different cybersecurity processes.*

In the following we describe the notions of *Risks*, *Indirect* and *Direct Costs* resulting from the implementation and purchase of a control and the *Organization Profile* which determines the preferences of an organization in terms of risks, indirect costs and how concerned the organization is about the different threats.

Risks. The *target risks* express the damage incurred to the defender when the attacker succeeds in compromising one or more targets. The different risks we consider are *Data Loss* (DL), *Business Disruption* (BD), and *Reputation* (RE). Each risk factor depends on the depth d that the attack targets; therefore we denote by DL_d, RE_d, and BD_d the risk values associated with a depth d.

Indirect Costs. For each cybersecurity process we consider three different types of *indirect costs*. The *System Performance Cost* (SPC) is associated with anything related to system performance being affected by a cybersecurity process (e.g. processing speed affected by anti-malware scanning). The *Morale Cost* (MOC) accounts for morale issues that higher levels of security can cause to users' happiness and job satisfaction. One negative implication of high MOC is that the stricter the security measures that an organization implements, the more likely an individual will want to circumvent them if possible. In these cases the attacker is able to take advantage of the reduced security from user actions. For example, having a control about different passwords for everything, might annoy users therefore increasing MOC. This might lead to circumvention of security by the user picking weak, memorable passwords which can often be cracked by dictionary or brute force attacks. Lastly, *Re-Training Cost* (RTC) refers to the cost for re-training users, including system administration, so they can either perform the cybersecurity process in the right way or be able to continue using all systems after a security update. We express the different indirect costs of a cybersecurity process p_{jl} by SPC_{jl}, RTC_{jl}, and MOC_{jl}.

Direct Costs. Each cybersecurity process has a direct cost which refers to the budget the organization must spend to implement the control c_j at a level l. The direct cost of a cybersecurity process is split into two categories, the *Capital Cost* (CAC) and the *Labour Cost* (LAC). CAC is related to hardware or software that must be purchased for the implementation of a control at some level. LAC is the direct cost for having system administrators implementing the control such as (hours spent) × (cost/hour). When investing in cybersecurity we will be looking into the direct cost of each cybersecurity plan which is derived as a combination of the different costs of the cybersecurity processes that comprise this plan.

Vulnerability Factors. The Council on CyberSecurity has published in [2] software weaknesses (in this paper weakness and vulnerability are used interchangeably) and their factors. These factors are *Prevalence* (PR), *Attack Frequency*

(AF), *Ease of Detection* (ED), and *Attacker Awareness* (AA). For a vulnerability v_z we denote the vulnerabilities factors by PR_z, AF_z, ED_z, AA_z. The level of a factor determines its contribution towards an overall vulnerability assessment score. For a commodity attack, one can argue that AA measures whether the average adversary would know that a malicious script is for sale, and ED is a measure of the computational cost of the attack discovery process. PR indicates the number of times the weakness is found in the system (e.g. only 30% of windows systems ever downloaded a given patch), and AF dictates the number of times someone actually tries to exploit it (e.g. how many random SQL injection probes a day). We see PR and AF accounting for threats that are currently widespread (*current threats*) and ED and AA for threats that have the most potential for future attack vectors (*future potential threats*).

Organization Profile. To represent an *organization profile* we define a set $\{\mathcal{R}, \mathcal{K}, \mathcal{T}\}$ which dictates the preferences that an organization has with regards to risks, indirect costs and how concerned the organization is about well-known threats, respectively. These are given by the probability distributions $\mathcal{R} = [r_1, r_2, r_3]$, $\mathcal{K} = [k_1, k_2, k_3]$, and $\mathcal{T} = [\tau_1, \tau_2]$. The idea behind defining an organization profile is that a security manager can reason about the organization at a high level. This means whenever managers use our model they do not have to undertake some detailed security assessment, but only considers the high level needs of the organization.

The *Risk Profile*, denoted by $\mathcal{R}$, represents the importance that each of the potential areas of loss (DL,RE,BD) has to the organization. This is designed to prioritize the risk factors, such that each organization is able to identify the balance of the damage that they can expect from a successful attack. While the expectation is that data loss will be the predominant concern for most organizations, there are some that may consider that their reputation or the disruption to the operation of the business have a more significant impact. The most noticeable case for this would be organizations that predominantly deal with third party payment systems (e.g. Paypal), where the organization will hold relatively little data of value for their customers. For the *Risk Profile* weights we create the relation such that $r_1 \mapsto DL, r_2 \mapsto RE,$ and $r_3 \mapsto BD$. We implicitly assume here that the organization's risk profile remains the same at all depths. We then define $\text{RISKS} = r_1 DL_d + r_2 RE_d + r_3 BD_d$.

The *Indirect Costs Profile* $\mathcal{K}$ defines an importance for each of three different indirect cost factors SPC, RTC, and MOC. This is so that an organization can reason about the relative importance of indirect costs that it may incur when implementing a defense. The mapping of the different weights to costs are $k_1 \mapsto SPC, k_2 \mapsto RTC,$ and $k_3 \mapsto MOC$. Therefore, $\text{IND\_COSTS} = k_1 SPC_{jl} + k_2 RTC_{jl} + k_3 MOC_{jl}$.

Lastly, *Threat Concern* $\mathcal{T}$ is the level of importance that the business places on each of the threat factors. The main priority here is identifying whether the organization is concerned more about *current threats* or *future potential threats*. Therefore $\tau_1 \mapsto$ current threats and $\tau_2 \mapsto$ future potential threats. We define $\text{THREAT} = \tau_1[(PR_z + AF_z)/2] + \tau_2[(ED_z + AA_z)/2]$.

4 Cybersecurity Control-Games

In this section we use game theory to model the interactions between two players; the *defender* and the *attacker*. The defender $\mathcal{D}$ abstracts any cybersecurity decision-maker (e.g. security manager) which defends an organization's data assets by minimizing cybersecurity risks with respect to the indirect costs of the cybersecurity processes while the attacker $\mathcal{A}$ abstracts all adversaries that aim to benefit from compromising the defender's data assets. The game we model here is a two-player game where there is a negative functional correlation between the attacker and the defender payoffs; the idea is that the more an attacker gains the more the defender loses. This means that equilibria in these games are minimax in an associated zero sum game. For any control c_j we define a control-subgame as follows.

Definition 4 (Control-subgame $\mathcal{G}_{j\lambda}$). *A control-subgame $\mathcal{G}_{j\lambda}$ is a game where (i) $\mathcal{D}$'s pure strategies correspond to consecutive implementation levels of the control c_j starting always from 0 (i.e. fictitious control-game) and including all levels up to λ and, (ii) $\mathcal{A}$'s pure strategies are the different targets akin to pairs of vulnerabilities and depths.*

$\mathcal{D}$'s finite strategy space is given by the set $\mathcal{A}^D = \{p_{jl}\}$. This means that $\mathcal{D}$'s actions are the different cybersecurity processes akin to implementations of a control c_j at different levels. The attacker can choose among different targets to attack therefore $\mathcal{A}^A = \{(v_z, d)\}$. We define $\mathcal{D}$'s mixed strategy as the probability distribution $Q_{j\lambda} = [q_{j0}, \ldots, q_{j\lambda}]$. This expresses a cybersecurity plan, where q_{jl} is the probability of implementing c_j at level l in the control-subgame $\mathcal{G}_{j\lambda}$.

A mixed strategy of $\mathcal{A}$ is defined as a probability distribution over the different targets and it is denoted by $H_{j\lambda} = [h_{j1}, \ldots, h_{jn}]$, where h_{ji} is the probability of the adversary attacking target t_i when $\mathcal{D}$ has only the control c_j in their possession. $\mathcal{D}$'s aim in a control-subgame is to choose the *Nash cybersecurity plan* $Q_{j\lambda}^* = [q_{j0}^*, \ldots, q_{j\lambda}^*]$. This consists of λ cybersecurity processes chosen probabilistically as determined by the Nash Equilibrium (NE) of $\mathcal{G}_{j\lambda}$ and it minimizes cybersecurity risks and indirect costs.

Example 1. In this example we consider a security control entitled *Vulnerability Scanning and Automated Patching*, and we assume 5 different implementation levels i.e. $\{0, 1, 2, 3, 4\}$ where level 4 corresponds to *real-time scanning* while level 2 to *regular scanning*. We say that a mixed strategy $[0, 0, \frac{7}{10}, 0, \frac{3}{10}]$ determines a cybersecurity plan that dictates the following:

$\frac{3}{10} \mapsto$ real-time scanning for the 30% of the most important devices

$\frac{7}{10} \mapsto$ regular scanning for the rest 70% of devices

This mixed strategy can be realized more as advice to a security manager on how to undertake different control implementations rather than a rigorous set of instructions related only to a time factor. We claim that our model is flexible thus allowing the defender to interpret mixed strategies in different ways to satisfy their requirements.

We denote by $U_{\mathcal{D}}(p_{jl}, t_i)$ the utility of $\mathcal{D}$ when target $t_i = \langle v_z, d \rangle$ is attacked, and the cybersecurity process p_{jl} has been selected to mitigate v_z at depth d, in general:

$$U_{\mathcal{D}}(p_{jl}, \langle v_z, d \rangle) := \text{RISKS} \times \text{THREAT} \times (1 - \text{MITIGATION}) + \text{IND\_COSTS} \quad (1)$$

Theorem 1. *The zero-sum cybersecurity control-subgame $\mathcal{G}_{j\lambda}$ admits an NE in mixed strategies, $(Q^\star_{j\lambda}, H^\star_{j\lambda})$, with the property that*

$$Q^\star_{j\lambda} = \arg\max_{Q_{j\lambda}} \min_{H_{j\lambda}} U_{\mathcal{D}}(Q_{j\lambda}, H_{j\lambda}), \text{ and } H^\star_{j\lambda} = \arg\max_{H_{j\lambda}} \min_{Q_{j\lambda}} U_{\mathcal{A}}(Q_{j\lambda}, H_{j\lambda})$$

The minimax theorem states that for zero sum games NE and minimax solution coincide. Therefore in $\mathcal{G}_{j\lambda}$ any Nash cybersecurity plan mini-maximizes the attacker's payoff. If any $\mathcal{G}_{j\lambda}$ admits multiple Nash cybersecurity plans they have the ordered interchangeability property which means that $\mathcal{D}$ reaches the same level of defense independent from $\mathcal{A}$'s strategy, i.e.

$$Q^\star_{j\lambda} = \arg\min_{Q_{j\lambda}} \max_{H_{j\lambda}} U_{\mathcal{A}}(Q_{j\lambda}, H_{j\lambda})$$

Definition 5. *The non-zero sum control-subgame $\mathcal{G}'_{j\lambda} = \langle U_{\mathcal{D}}, U'_{\mathcal{A}} \rangle$ where $U'_{\mathcal{A}} = \alpha U_{\mathcal{A}} + \beta$, and α, β constants and $\alpha > 0$ is called a positive affine transformation (PAT) of the zero sum control-subgame $\mathcal{G}_{j\lambda} = \langle U_{\mathcal{D}}, U_{\mathcal{A}} \rangle$.*

Proposition 1. *If one of the game matrices of a control-subgame $\mathcal{G}_j$ is a positive affine transformations (PAT) of a zero sum control-subgame $\mathcal{G}'_j$ (and the other matrix is the same for both games) then the Nash equilibria of $\mathcal{G}_j$ are minimax strategies. These also correspond to saddle-points [5] (p. 42).*

In the rest of the paper we will restrict ourselves to control-subgames which are positive affine transformation of a zero sum control-subgame.

Definition 6 (Control-game $\mathcal{G}_j$). *For any control c_j, with $\mathcal{L}$ possible implementation levels, we define a control-game $\mathcal{G}_j$ which consists of $\mathcal{L}$ control-subgames, each of them denoted by $\mathcal{G}_{j\lambda}$, $\lambda \in \{0, 1, \ldots, \mathcal{L}\}$.*

In other words, a *control-game* is the collection of $\mathcal{L}$ control-subgames for a specific control. The solution $\mathcal{C}_j$ of a control-game for the defender is a set of Nash cybersecurity plans $\{Q^\star_{jl}\}$, $\forall l \in \{0, \lambda\}$ each of them determined by the NE of each control-subgame. The set $\{\mathcal{C}_j\}$ for all controls $c_j \in C$ contains all sets of Nash cybersecurity plans one per control.

5   Cybersecurity Investment Optimization

In the previous section we were concerned with the implementation of a cybersecurity control. Nevertheless, organizations will generally implement more than one control. In this section we identify a method for combining these controls given that an organization's budget is constrained. More specifically, we describe how the control-game solutions are handled by optimization techniques to provide investment strategies. Each cybersecurity plan imposes its own direct costs including both CAC and LAC. Given a set $\{c_j\}$ of N controls each of them being associated with a set $\{C_j\}$ of $\mathcal{L}$ Nash cybersecurity plans, and an available budget B, in this section we examine how to optimally invest in the different plans by choosing at most one plan per control.

In relation to the cybersecurity investment problem we consider a 0-1 Knapsack problem similar to the cybersecurity budget allocation problem studied by Smeraldi and Malacaria [10]. In fact, in this paper we model this cybersecurity investment optimization problem as a *0-1 Multiple-Choice Multi-Objective Knapsack Problem*.

We assume that a plan can be effective in protecting more than one target and its benefit on a target is determined by the *expected damage* caused to the target when only this plan is purchased. The benefit of an investment solution on a target is determined by the sum of the benefits of the different plans on that target where this sum never exceeds 1. Furthermore, *each investment solution has a score* determined by the maximum expected damage across all targets. When there are investment solutions with the same score we consider a *tie-break*. The question then arises, which one solution should one use? We consider that in the event of a tie-break, the solver uses the solution with the lowest cost. This tie-break makes sense as no-one would normally pay more for a defense that does no better.

The optimization method creates one objective function per target, which constraint is constrained by a common total budget B. Our goal is to derive the set of Nash cybersecurity plans (one per control) which minimizes the investment solution score. To derive the optimal investment solution, we compute the expected damage of each target for each possible set of plans. The weakest target is defined as the target that suffers the highest damage. In this way our method of evaluating the security of a system is to consider that "the security of a system is only as strong as it's weakest point". We then choose the set of plans that provides the minimum final expected damage among all highest expected damages.

Definition 7. *Defining the value of any target t_i as $\gamma_i = -\text{RISKS} \times \text{THREAT}$, considering N controls and assuming that each Nash cybersecurity plan $Q_{j\lambda}^*$ is associated with some benefit $b_{j\lambda}(t_i)$[1] upon target t_i, and it has cost $\omega_{j\lambda}$, the defender seeks a cybersecurity investment $\mathcal{I}$ such that*

$$\max_{\mathcal{I}} \min_{t_i} \{1 - \sum_{j=1}^{N} \sum_{\lambda=0}^{\mathcal{L}} b_{j\lambda}(t_i) x_{j\lambda}\} \gamma_i \tag{2}$$

subject to $\sum_{j=1}^{N} \sum_{\lambda=0}^{\mathcal{L}} \omega_{j\lambda} x_{j\lambda} \leq B$ *and* $\sum_{\lambda=0}^{\mathcal{L}} x_{j\lambda} = 1$, $x_{j\lambda} \in \{0,1\}, \forall j = 1, \ldots, N$

The objective of Definition 7 is to choose an investment solution with the lowest expected damage for the weakest target. This is subject to the condition that such an investment is within the budget B, where we consider if the control is used, given by $x_{j\lambda}$ (either 0 or 1), and the cost of implementing the control,

---

[1] We assume that $\sum_{j=1}^{N} \sum_{\lambda=0}^{\mathcal{L}} b_{j\lambda}(t_i) \leq 1$ achieved by normalized benefit values.

given by $\omega_{j\lambda}$. Additionally, we must satisfy that for each of the N controls only a single subgame solution can and must be selected. Hence although each $x_{j\lambda}$ can only take a value of 0 or 1, the sum must equal 1, ensuring that only one solution (given by a control subgame solution) is selected for each control. We denote by $\mathcal{I}$, the vector of cybersecurity plans (i.e. investment solution) purchased by solving the cybersecurity investment optimization problem for a constant number of targets.

For example in Table 5, we buy the solution (represented by a value of 1 in the knapsack) for subgame 3 of control 3, which might correspond to $[0,0,0.3,0.7,0]$, which suggests that the control is implemented at level 2, 30% of the time and at level 3, 70% of the time, with a cost of 8.2. Such that we then select 0 for all other subgame solutions for that control. For each control c_j there is a cybersecurity plan, denoted by $Q_j^\star$ that represents the optimal choice for the defender to purchase given some budget. Therefore $\mathcal{I} = [Q_1^\star, Q_2^\star, \ldots, Q_N^\star]$.

6   Case Study

In this section we describe the case study we use to implement our methodology. With regards to the organization size, we consider an SME with approximately 30 employees and we are interested in mitigating *commodity attacks*. This assumption allows us to have complete information in all control-games because the defender can be aware of the attacker's payoff when it has been disclosed online. From the 2011 CWE/SANS top 25 most dangerous software errors aka vulnerabilities published in [2], we have considered 12 of those for the purposes of this case study as described in Table 1 along with their factors, and associated levels. For each vulnerability factor different levels are defined as in [2], and summarized in Table 2. Moreover, we have chosen 6 controls out of the The SANS 20 Critical Security Controls published by the Council on Cybersecurity in [3]. These are shown in Table 4 along with the different vulnerabilities that each control mitigates. As the same vulnerability can appear at different data assets at the same depth, we assume that the implementation of a control mitigates all occurrences of this vulnerability. Otherwise, the security of the system won't increase because it is as strong as the weakest point. For a control, we assume five possible levels (i.e. 0-4) that the control can be implemented, where

Table 1. Notation of 12 examined vulnerabilities

v_z: **Vulnerability** (**CWE-code**)	PR	AF	ED	AA	**Vulnerability**	PR	AF	ED	AA
v_1: SQLi (89)	2	3	3	3	v_7: Missing encryption (311)	2	2	3	2
v_2: OS command injection (78)	1	3	3	3	v_8: Unrestricted upload (434)	1	2	2	3
v_3: Buffer overflow (120)	2	3	3	3	v_9: Unnecessary privileges (250)	1	2	2	2
v_4: XSS (79)	2	3	3	3	v_{10}: CSRF (352)	2	3	2	3
v_5: Missing authentication (306)	1	2	2	3	v_{11}: Path traversal (22)	3	3	3	1
v_6: Missing authorization (862)	2	3	2	2	v_{12}: Unchecked code (494)	1	1	2	3

Table 2. Values of vulnerabilities factors published by CWE

level	PR	AF	ED	AA
3	Widespread	Often	Easy	High
2	High	Sometimes	Moderate	Medium
1	Common	Rarely	Difficult	Low

Table 3. Indirect costs of the different cybersecurity processes

Cyber. Proc.	SPC	MOC	RTC
$p_{00}, \ldots, p_{60}$	0,0,0,0,0,0	0,0,0,0,0,0	0,0,0,0,0,0
$p_{01}, \ldots, p_{61}$	1,1,1,1,2,2	1,1,1,0,1,1	0,0,0,2,1,0
$p_{02}, \ldots, p_{62}$	2,2,1,2,2,2	2,1,1,0,2,1	0,0,0,2,1,0
$p_{03}, \ldots, p_{63}$	2,3,2,3,2,2	4,1,1,0,3,3	0,0,0,2,1,1
$p_{04}, \ldots, p_{64}$	3,3,2,4,2,2	5,2,2,0,4,3	0,0,0,2,2,2

Table 4. Vulnerabilities that each control mitigates

	v_1	v_2	v_3	v_4	v_5	v_6	v_7	v_8	v_9	v_{10}	v_{11}	v_{12}
c_1: Account Monitoring and Control	-	✓	-	-	-	✓	-	✓	✓	✓	-	-
c_2: Continuous Vulnerability Assessment and Remediation	✓	✓	✓	-	✓	-	✓	-	-	-	-	-
c_3: Malware Defenses	-	-	-	✓	-	-	-	✓	-	✓	-	✓
c_4: Penetration Tests and Red Team Exercises	✓	✓	✓	-	✓	✓	✓	-	✓	✓	✓	✓
c_5: Controlled Use of Administrative Privileges	-	-	-	✓	-	-	-	-	✓	-	✓	-
c_6: Data Loss Prevention	✓	-	-	✓	-	-	✓	✓	-	-	✓	-

level 0 corresponds to no defense against the vulnerabilities and level 4 presents the highest possible level of control implementation with no regard for system operation. In Table 3, we highlight the indirect costs for all 6 controls considered in this case study. In the following we classify the controls into different implementation methods.

Depth Based Mitigation. This refers to controls that when applied at higher levels are used to cover additional depths within a system. This form of mitigation applies a system-wide control at level 1 and then applies more advanced countermeasures at additional depths, based on the level of implementation. The different levels are $\langle c, 0 \rangle$: no implementation, $\langle c, 1 \rangle$: all depths – basic, $\langle c, 2 \rangle$: depths 1,2 – basic & depth 3 – advanced, $\langle c, 3 \rangle$: depth 1 – basic & depths 2,3 – advanced, and $\langle c, 4 \rangle$: all depths – advanced. *Associated controls:* c_1, c_3, c_6.

Frequency Based Mitigation. This type of mitigation applies a control in a system-wide manner and higher levels of implementation reduce the time between scheduled performance of the mitigation. Low levels of a frequency based control may be performed as a one-off event or very infrequently, but this is then made more frequent at higher levels, where at the highest level these actions can be performed on demand. The different levels are $\langle c, 0 \rangle$: no implementation, $\langle c, 1 \rangle$: all depths – infrequent, $\langle c, 2 \rangle$: all depths – regular, $\langle c, 3 \rangle$: all depths – frequent, $\langle c, 4 \rangle$: all depths – real-time. *Associated controls:* c_2, c_4.

Hybrid Mitigation. A *hybrid mitigation* control implements an approach to reducing the vulnerability of a system that acts with aspects of both depth based and frequency based controls. As such, these controls increase defense at lower depths as the control level increases, but additionally the frequency with which the schedule of the control at the other depths is also increased. The different levels are $\langle c, 0 \rangle$: no implementation, $\langle c, 1 \rangle$: all depths – basic & infrequent, $\langle c, 2 \rangle$: all depths – basic & regular, $\langle c, 3 \rangle$: all depths – basic & frequent, $\langle c, 4 \rangle$: all depths – advanced & real-time. *Associated control:* c_5.

Each cybersecurity plan $Q_{j\lambda} = [q_{j0}, \ldots q_{j\lambda}]$ has a benefit, denoted by $b_{j\lambda}(t_i)$, on a target t_i. This is derived by the sum of the effectiveness values of the cybersecurity processes on t_i multiplied by the corresponding probability hence $b_{j\lambda}(t_i) = \sum_{l=0}^{\lambda} e(t_i, p_{jl})q_{j\lambda}$. A cybersecurity process p_{jl} has its own direct costs denoted by y_{jl}. Therefore the direct cost of a cybersecurity plan $Q_{j\lambda}$ is given by $\omega_{j\lambda} = \sum_{l=0}^{\lambda} y_{jl}q_{jl}$. In this work here we have defined the direct costs CAC and LAC per annum. Some of the controls have a one-off cost therefore any purchase can benefit the organization's cybersecurity for the next years also. However, we examine the challenge of spending a cybersecurity budget annually assuming that in the worst case controls might need to be replaced or updated by spending an amount of money similar or even higher to the amount spent in the last year. In this paper we neither present the cybersecurity products we have chosen to implement the various controls nor their direct costs.

Lastly, the different risks values $\langle DL_d, RE_d, BD_d \rangle$ are defined as depth $1 \mapsto \langle 2, 4, 3 \rangle$, depth $2 \mapsto \langle 3, 2, 4 \rangle$, and depth $3 \mapsto \langle 4, 3, 2 \rangle$. We have chosen the value of data loss to be the highest within the *Private Network*, because this depth will generally contain the most sensitive data. We have assessed the value of reputation loss RE independently of the value of data loss DL to show the impact that only RE has to the organization. We have set the highest value of RE to the DMZ because it contains the forward facing assets of the organization. For example when the organization's website is defaced this can be seen by any potential user who visits the organization's website and harm its reputation. As most of the organization's workload is likely to be handled by devices in the *Middleware* we have assigned the highest BD value to this depth.

7   Games Solutions and Investments

This section explains the set of results we have retrieved for 3 different organization profiles (3 Cases). For each profile: (i) we solve a series of control-games therefore a set of control-subgames for each control to derive the Nash cybersecurity plans (in this section we use the terms Nash cybersecurity plans and plans interchangeably) and, (ii) we determine the optimal cybersecurity investment given a budget by using optimization techniques and the control-game solutions. In Cases 1, and 3 we consider an organization which places a high importance on its data assigning a value 0.8 to DL. RE, and BD are equally important taking the same value 0.1. We also consider here that the organization is equally interested in *current* and *potential future threats* in all cases. In Cases 1 and 2 the organization prioritizes system performance costs higher than re-training and morale costs by giving an SPC value twice that of RTC and MOC values. We have increased MOC in Case 3 making it twice as large as SPC to assess the impact of morale in cybersecurity strategies.

To derive the different Nash cybersecurity plans we have solved 24 different control-subgames (i.e. 6 control-games) for each organization profile. The game solutions were computed by using a minimax solver, implemented in Python. For simplicity reasons we have chosen the first equilibrium computed by our solver

Table 5. Nash cybersecurity plans for the Case 1 with their associated direct costs

c_j	$Q^\star_{j0}$	$Q^\star_{j1}$	$Q^\star_{j2}$	$Q^\star_{j3}$	$Q^\star_{j4}$
c_1	[1,0,0,0,0] **0**	[0,1,0,0,0] **9.7**	[0,0.7,0.3,0,0] **9.8**	[0,0.4,0.23,0.37,0] **10.7**	[0,0,0.14,0.22,0.64] **12.4**
c_2	[1,0,0,0,0] **0**	[0,1,0,0,0] **1.7**	[0,0.4,0.6,0,0] **2**	[0,0,0.5,0.5,0] **5.1**	[0,0,0.5,0.5,0] **5.1**
c_3	[1,0,0,0,0] **0**	[0,1,0,0,0] **7.1**	[0,0,1,0,0] **7.3**	[0,0,0.3,0.7,0] **8.2**	[0,0,0.3,0.7,0] **8.2**
c_4	[1,0,0,0,0] **0**	[0,1,0,0,0] **4.2**	[0,0,1,0,0] **8.3**	[0,0,0,1,0] **16.7**	[0,0,0,0,1] **33.4**
c_5	[1,0,0,0,0] **0**	[0,1,0,0,0] **4.1**	[0,0.47,0.53,0,0] **4.1**	[0,0,0.41,0.59,0] **4.1**	[0,0,0,0.33,0.67] **5.4**
c_6	[1,0,0,0,0] **0**	[0,1,0,0,0] **6**	[0,0,1,0,0] **7.4**	[0,0,0.44,0.56,0] **12**	[0,0,0.32,0.52,0.16] **13.6**

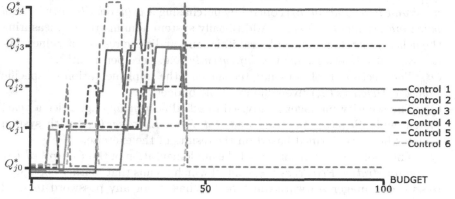

Fig. 2. Case 1: $\mathcal{R} = [0.8, 0.1, 0.1]$, $\mathcal{T} = [0.5, 0.25, 0.25]$, $\mathcal{K} = [0.5, 0.5]$

noting that all equilibria offer the same level of defense as we state in Theorem 1. In Table 5 we present the results of the control-subgames in Case 1, where the solution calculated for each control subgame taken is the strategy with the smallest support.

The graphs presented in Figs. 2 and 3 are designed to show which plans should be chosen for each possible budget level. In other words, each graph shows the optimal investment $\mathcal{I}$ which is the set of plans chosen for a certain budget. The graphs should be used to identify which are the most important plans for a given organization at an available budget. It is worth noting here that we have normalized the cost values such that the sum of the direct costs of of all the controls implemented at the highest possible level (i.e. the most expensive possible cybersecurity plans) equals 100. Our methodology uses the optimization technique presented in Section 5 to compute the investment solution $\mathcal{I}$ that has the highest score for a given budget. As we have discussed in Section 5, *each investment solution has a score* determined by the maximum expected damage across all targets. A question a security manager may ask is, how is the investment solution $\mathcal{I}$ translated in terms of controls acquisition and how can someone describe that it is better, in terms of cyber defense, than alternative solutions where $\mathcal{I}' \neq \mathcal{I}$?

Example 2. From Fig. 2 we consider an available budget of 17. In this example our decision support methodology advices the security manager to implement $\mathcal{I} = [Q^\star_{10}, Q^\star_{21}, Q^\star_{30}, Q^\star_{42}, Q^\star_{50}, Q^\star_{61}]$ with a cost of 16.102. The above solution determines a set of plans to be selected for the implementation of the 6 SANS controls as defined in our case study. To be able to translate the solution into the implementation of the different available controls in Table 5 we present the Nash cybersecurity plans of Case 1. According to $\mathcal{I}$ the controls that should be implemented and the manner in which they are implemented is listed as follows

- $Q_{10}^\star$: With the given budget, *Account Monitoring and Control* (c_1) software should not be purchased, nor should system administrators spend time on activities to this control.
- $Q_{21}^\star$: The organization must implement the *Continuous Vulnerability Assessment and Remediation* (c_2) control by purchasing a *vulnerability scanner and patch management* software. Additionally system administrators measuring the delay in patching new vulnerabilities and audit the results of vulnerability scans at all network depths *infrequently* (e.g, once per month).
- $Q_{30}^\star$: The decision tool does not recommend the implementation of specific *Malware Defenses* (c_3) given the available budget.
- $Q_{42}^\star$: The security manager is advised to schedule *regular* (e.g. twice a year) system-wide *Penetration Tests and Red Team Exercises* (c_4), with system updates being performed based on the results of the exercise.
- $Q_{50}^\star$: The tool does not recommend the implementation of the *Controlled Use of Administrative Privileges* (c_5) control which means that neither enterprise password manager software must be purchased nor any password renewal policy must be enforced.
- $Q_{61}^\star$: The tool recommends the implementation of the *Data Loss Prevention* (c_6) control system-wide and at a basic level (e.g. integrated services router with security, VPN).

By using Table 4 we see that with these controls all targets are covered to some degree. In the following we consider alternative cases to highlight the optimality of the solution. If we implement system-wide *Penetration Tests and Red Team Exercises* infrequently (e.g. once per year) $(Q_{41}^\star)$ instead of regularly $(Q_{42}^\star)$ then we release a budget of 4.174 therefore we can implement *Controlled Use of Administrative Privileges* by using an *enterprise password manager* software and *renew passwords* of all systems infrequently (e.g. annually) $(Q_{51}^\star$ with cost 4.153). This gives another investment $\mathcal{I}' = [Q_{10}^\star, Q_{21}^\star, Q_{30}^\star, Q_{41}^\star, Q_{51}^\star, Q_{61}^\star]$ with cost 16.081. Under $\mathcal{I}'$ the *Controlled Use of Administrative Privileges* control improves the defense on targets associated with the following vulnerabilities; XSS (v_4), Unnecessary privileges (v_9), and Path traversal (v_{11}). But it then leaves worse off, due to the less frequent *Penetration Tests and Red Team Exercises*, 8 vulnerabilities namely; SQLi (v_1), OS command injection (v_2), Buffer overflow (v_3), Missing authentication (v_5), Missing authorization (v_6), Missing encryption (v_7), CSRF (v_{10}), and Unchecked code (v_{12}). Due to $\mathcal{I}$ being the choice of the optimization the score achieved by $\mathcal{I}$ is higher than $\mathcal{I}'$ therefore the weakest target in $\mathcal{I}'$ must appear in these 8 vulnerabilities and it must be weaker than the weakest target under $\mathcal{I}$. By saying weakest target we refer to the target with the maximum expected damage. Therefore our methodology advices the security manager to undertake *Penetration Tests and Red Team*

Exercises regularly (e.g. twice a year, $Q_{42}^\star$) without implementing *Controlled Use of Administrative Privileges* at all.

If we do not spend any money on *Penetration Tests and Red Team Exercises* we then have an available budget of 8.347 which can be spent in implementing *Malware Defenses* by installing a free anti-malware software with manual scheduled scans and database updates in all devices of the organization ($Q_{31}^\star$, 7.095). Therefore another investment is $\mathcal{I}' = [Q_{10}^\star, Q_{21}^\star, Q_{31}^\star, Q_{40}^\star, Q_{50}^\star, Q_{61}^\star]$. Under $\mathcal{I}'$ targets associated with `Missing authorization` (v_6) and `Unnecessary privileges` (v_9) are not covered by any control thus one of these becomes the weakest target under $\mathcal{I}'$. Due to $\mathcal{I}$ being the optimal investment solution provided by our tool, the weakest target (not covered at all) under $\mathcal{I}'$ must be weaker than the weakest (partially covered) target under $\mathcal{I}$. Therefore our solution recommends not to ignore (even at some basic level) the implementation of *Penetration Tests and Red Team Exercises* which can actually identifies if a user can access a given resource, despite not being authorized for that (v_6) and it can also mitigate v_9 by identifying cybersecurity processes that run with extra privileges, such as root or Administrator, and they can disable the normal security checks.

Finally, if we assume a slightly higher budget of 17.145 we can choose the investment strategy $\mathcal{I}' = [Q_{10}^\star, Q_{21}^\star, Q_{31}^\star, Q_{42}^\star, Q_{50}^\star, Q_{60}^\star]$ which does not implement the *Data Loss Prevention* control but it installs free anti-malware with manual scheduled scans and database updates system-wide ($Q_{31}^\star$). This is not a better investment than $\mathcal{I}$ despite being more expensive. Both *Data Loss Prevention* (in $\mathcal{I}$) and *Malware Defenses* (in $\mathcal{I}'$) mitigate `XSS` (v_4) and `Unrestricted upload` (v_8) which *Penetration Tests and Red Team Exercises* does not. Thus from this point of view the replacement of *Data Loss Prevention* by *Malware Defenses* does not affect the effectiveness of the targets associated with `XSS` and `Unrestricted upload`. However due to $\mathcal{I}$ being the choice of the optimization the target associated with `Path traversal` (v_{11}), which is mitigated only by *Penetration Tests and Red Team Exercises* under $\mathcal{I}'$, is weaker than a target associated with `CSRF` (v_{10}) or `Unchecked code` (v_{12}) mitigated only by *Penetration Tests and Red Team Exercises* under $\mathcal{I}$. In other words, according to the effectiveness values we have provided in our case study, `Path traversal` is not mitigated as much as `CSRF` and `Unchecked code`, which are both mitigated by *Penetration Tests and Red Team Exercises*, therefore $\mathcal{I}$ is better than $\mathcal{I}'$.

Example 3. According to Fig. 2 for a budget of 28 our methodology gives the investment solution $\mathcal{I} = [Q_{13}^\star, Q_{21}^\star, Q_{31}^\star, Q_{41}^\star, Q_{52}^\star, Q_{60}^\star]$ with a total direct cost 27.80. This solution provides the following list of recommendations.

– $Q_{13}^\star$: Implementation of *Account Monitoring and Control* (c_1) at a basic level (e.g. control built into OS and manually review all accounts or set files/folders auditing properties) in all devices in *DMZ*; in 63% of the devices in *Middleware*; and in 40% of the devices in *Private Network*. The control must be also implemented at an advanced level (e.g. vulnerability scanner and patch management software) in 37% of the devices in *Middleware* and 60% of the devices in *Private Network*.

- $Q_{21}^\star$: System-wide *Continuous Vulnerability Assessment and Remediation* (c_2) must be implemented infrequently (e.g. once per month).
- $Q_{31}^\star$: System-wide *Malware Defenses* (c_3) must be implemented at a basic level (e.g. free anti-malware with manual scheduled scans and database updates).
- $Q_{41}^\star$: *Penetration Tests and Red Team Exercises* (c_4) to be undertaken infrequently (e.g. once per year).
- $Q_{52}^\star$: *Controlled Use of Administrative Privileges* (c_5) to be implemented at a basic level (e.g. using an enterprise password manager software) with 47% of the devices to change passwords infrequently (e.g. once per year) and 53% regularly (e.g. every 4 months).
- $Q_{60}^\star$: The purchase of a *Data Loss Prevention* control is not recommended.

To see how $\mathcal{I}$ outperforms other investments we have considered some alternative investments for a budget of 28. As the first alternative investment $\mathcal{I}'$, we decide not to follow $Q_{13}^\star$ therefore saving 10.68. By doing that the targets associated with $v_2, v_6, v_8, v_9, v_{10}$ are now defended in a lower degree than in $\mathcal{I}$ as the effectiveness of $Q_{13}^\star$ does not count in the sum of the benefits for these targets. Also under $\mathcal{I}'$ the targets associated with v_2, v_9, and v_{10} are defended by two controls while the targets with v_6, and v_8 are only defended by one control. With a budget of 10.68 available we can purchase c_6 and implement it according to plan $Q_{62}^\star$ with cost 7.408 or according to $Q_{61}^\star$ with cost 6.052. If we choose the former the control is implemented at an advanced level (e.g. drive encryption, system recovery) in the *Private network*, and at a basic level (e.g. integrated services router with security, VPN) in *DMZ* and *Middleware*. *Data Loss Prevention* improves the defense of targets with v_1, v_4, v_7, v_8, and v_{11}. When implementing *Data Loss Prevention*, v_6 (**Missing authorization**) is mitigated only by one control therefore making any target with this vulnerability likely to be the weakest among all in $\mathcal{I}'$ and weaker than the weakest target under $\mathcal{I}$. Our solution $\mathcal{I}$ dictates that it is preferable for the security manager to purchase the *Account and Monitoring Control* as opposed to *Data Loss Prevention* to prevent unauthorized users accessing resources or data of the organization in the first case rather than allowing such access and hoping that data encryption and system recovery capabilities can discourage an adversary from attacking.

Next, we assume another variation $\mathcal{I}'$ of our investment where *Malware Defenses* (c_3) is removed and 7.095 budget is available for spending in other controls. By not purchasing c_3 vulnerabilities v_4, v_8 and v_{10} are mitigated by one less control. With an available budget of 7.095 we can purchase c_6 and implementing it according to $Q_{61}^\star$. The difference between $\mathcal{I}$ and $\mathcal{I}'$ is that v_{10} is mitigated by one more control in $\mathcal{I}$. Since the latter has been the result of optimization, any target with v_{10} (**CSRF**) is the weakest target and weaker than the weakest target under $\mathcal{I}$. In other words $\mathcal{I}$ advises the security manager to purchase *Malware Defenses* rather than *Data Loss Prevention* to detect malware that can be installed when a **CSRF** attack is launched. Again here $\mathcal{I}$ dictates that stopping the attack at a first infection stage is more important than guaranteeing that stolen data are encrypted thus unreadable. Besides the attacker's motivation might be to just corrupt or delete data which is something *Data Loss Prevention* can address only at high levels of implementation which require a higher budget.

Graph Trends. From the graphs in Fig. 2 we see the results level off (at around a budget of 45), when the perceived benefit from a combination of plans brings the expected damage down to a minimum, this is such that adding a new plan or a plan at a higher level won't improve the defense of the system. This is as a result of us capping the sum of improvements to 1, but would exist in any form of interdependent control methodology and only the point at which it levels off would change. Furthermore, this observation dictates that *cybersecurity does not get improved by investing more in cybersecurity plans*. With higher budgets it is much more feasible to reduce the damage of not just the weakest target, but other targets as well. A spike exists when there is a small budget range that opens up a number of new cybersecurity plans. In reality a number of the solutions in that range will have similar expected damage values, but we only see the best solution for that particular budget. For budgets 1-19 the progression is the same regardless of the plans. The reason for this is that at these levels only certain plan combinations are available and we want to ensure that as low an expected damage as possible is achieved. At these levels it is seen as most important to cover all of the targets with some form of plan, bringing down the system-wide expected damage.

While there is a consistent strategy for investment with budget levels for solutions up to a budget of 20, after this budget we see that different investment profiles are suggested by our methodology across the different organization profiles. From budgets 22 to 26 and from 36 to 38, there is no change in the solution. While alternative solutions may become available in these ranges, none of these solutions will improve on the security of the weakest target, which means that as with very low budgets there is no incentive to implement a more expensive plan combination that does not improve the effectiveness of defense on the weakest target. Between budgets 30 and 36 as the budget increases more, there are new combinations of plans that become available at each of these levels that will improve the overall defense of the system. However it can also be seen that in order to implement a different solution some components of the previous solution need to be removed in order to reduce the cost to fit within the budgetary constraint.

Sensitivity to Organizations' Profile Perturbations. One question that arises is how robust is the proposed approach to informing the way an organization should invest in cyber security? We have focused on the importance of the decisions made by the organization with regards to their profile. In this way we have looked at how small perturbations in a single case affect the allocation of investment. We consider the two cases of $[0.75, 0.125, 0.125]$ and $[0.85, 0.075, 0.075]$ for $\mathcal{R}$. Both alternative profiles have minor deviations from the original solution. Each of the deviations found would cause the solution to differ for up to a maximum of 3 consecutive budget levels, before the proposed controls would realign. Using the values of $[0.55, 0.45]$ and $[0.45, 0.55]$ for $\mathcal{T}$ we find that, in the case of $[0.55, 0.45]$ there is a different investment strategy that is proposed between controls 5 and 6 for budgets between 13 and 17. This is the only case we have seen

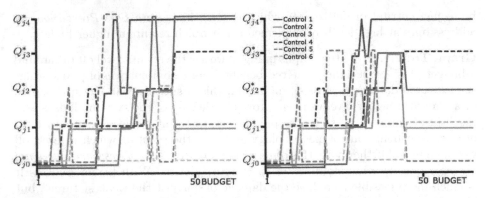

Fig. 3. (i) Case 2: $\mathcal{R} = [0.6, 0.4, 0], \mathcal{T} = [0.5, 0.25, 0.25], \mathcal{K} = [0.5, 0.5]$, (ii) Case 3: $\mathcal{R} = [0.8, 0.1, 0.1], \mathcal{T} = [0.3, 0.1, 0.6], \mathcal{K} = [0.5, 0.5]$

where there is a difference in the low budget strategies across all the cases tested for this work. With a value for $\mathcal{T}$ of $[0.45, 0.55]$, we find that there is no change to the proposed investment plans. In the case of $\mathcal{K}$ we consider $[0.45, 0.275, 0.275]$ and $[0.55, 0.225, 0.225]$, which for both values give us no change in the proposed investment from the original case. Importantly in all of the cases tested we have seen that the stable investment solution for all of the results is the same as the case presented in Fig. 2.

8   Conclusions

This paper presents a cybersecurity decision support methodology for calculating the optimal security investment for an organization. This is formulated as a multiple choice and multi-objective Knapsack problem which handles the solutions of cybersecurity control-games. Our methodology creates strategies for each control at different levels of implementation and enforcement, where the combination of the most effective controls within a budget are suggested for implementation. The model supports the movement of human decision making from trying to analyze the explicit security requirements of the system to deciding upon an organization's priorities. The feature of the model that helps to create this movement is the *organization profile*, where a profile allows the model to reflect the individual nature of different organizations in the proposed investment. One of the most important factors that this work highlights is that it is important for an organization to know how to appropriately generate their profile. This is crucial because it influences the way an organization should invest in their cybersecurity defenses. From the results we have noticed that for similar organizations the best protection will be similar if not the same, because the results of the control-games will favor certain targets or controls.

In this paper we have assumed additive benefits for the different plans and the same target. One important future aim is to better understand the steps of the attacks, as such identifying steps in the chain will better inform the way in which different security controls interact in order to better cover different targets. This will better inform the way in which the subgames results are combined in the investment problem and reflect in a more realistic way how cyber defenses work.

At the moment our data is generated with the advice of a limited set of experts. In the future we aim to increase the number of experts involved to better understand their cyber environment needs. This will allow us to implement our methodology in a realistic environment. Additional limitations of our work that we wish to address in future work is to consider a higher number of available controls and continuous values for the levels of controls implementation. Moreover, at the moment our control-subgames are games of complete information. In the future we will examine incomplete information games where the defender is not aware of the attacker's payoff therefore any investment solution has to respect this uncertainty which highlights a situation very close to realistic environments that are prone to 0-days attacks and Advanced Persistent Threats (APTs). Finally, we do not see a strong case for using Stackelberg games in the case of commodity attacks where both players have publicly available information about attack types. The case would have been stronger if we were considering sophisticated cyber criminals or nation states where surveillance of the defender's actions prior to the attack would be important for the recognition of the defending mechanisms and the exploitation of one or more weak targets.

References

1. Anderson, R.: Why Information Security is Hard. In: Proc. of the 17th Annual Computer Security Applications Conference (2001)
2. CWE.: 2011 CWE/SANS Top 25 Most Dangerous Software Errors, http://cwe.mitre.org/top25/ (accessed, May 2014)
3. Council on Cybersecurity: The critical security controls for effective cyber defense (version 5.0), http://www.counciloncybersecurity.org/attachments/article/12/CSC-MASTER-VER50-2-27-2014.pdf (accessed, May 2014)
4. 2012 Deloitte-NASCIO Cybersecurity Study State governments at risk: A call for collaboration and compliance, https://www.deloitte.com/assets/Dcom-UnitedStates/Local%20Assets/Documents/AERS/us_aers_nascio%20Cybersecurity%20Study_10192012.pdf (accessed, May 2014)
5. Alpcan, T., Basar, T.: Network Security: A Decision and Game-Theoretic Approach. Cambridge University Press (2010)
6. Alpcan, T.: Dynamic incentives for risk management. In: Proc. of the 5th IEEE International Conference on New Technologies, Mobility and Security, NTMS (2012)
7. Gordon, L.A., Loeb, M.P.: The economics of information security investment. In: ACM Transactions on Information and System Security, TISSEC (2002)
8. Johnson, B., Grossklags, J., Christin, N., Chuang, J.: Nash equilibria for weakest target security games with heterogeneous agents. In: Jain, R., Kannan, R. (eds.) Gamenets 2011. LNICST, vol. 75, pp. 444–458. Springer, Heidelberg (2012)
9. Fielder, A., Panaousis, E., Malacaria, P., Hankin, C., Smeraldi, F.: Game theory meets information security management. In: Cuppens-Boulahia, N., Cuppens, F., Jajodia, S., Abou El Kalam, A., Sans, T. (eds.) SEC 2014. IFIP AICT, vol. 428, pp. 15–29. Springer, Heidelberg (2014)

10. Smeraldi, F., Malacaria, P.: How to Spend it: Optimal Investment for Cyber Security. In: Proc. of the 1st International Workshop on Agents and CyberSecurity, ACySe (2014)
11. Cavusoglu, H., Srinivasan, R., Wei, T.Y.: Decision-theoretic and game-theoretic approaches to IT security investment. Journal of Management Information Systems(ACySe) 25(2), 281–304 (2008)
12. Saad, W., Alpcan, T., Basar, T., Hjorungnes, A.: Coalitional game theory for security risk management. In: Proc. of the 5th International Conference on Internet Monitoring and Protection (ICIMP), pp. 35–40 (2010)
13. Bommannavar, P., Alpcan, T., Bambos, N.: Security risk management via dynamic games with learning. In: Proc. of the 2011 IEEE International Conference on Communications (ICC), pp. 1–6 (2011)
14. Alpcan, T., Bambos, N.: Modeling dependencies in security risk management. In: Proc. of the Fourth International Conference on Risks and Security of Internet and Systems (CRiSIS), pp. 113–116 (2009)
15. Cremonini, M., Nizovtsev, D.: Understanding and influencing attackers' decisions: Implications for security investment strategies
16. Demetz, L., Bachlechner, D.: To Invest or Not to Invest? Assessing the Economic Viability of a Policy and Security Configuration Management Tool. In: The Economics of Information Security and Privacy, pp. 25–47. Springer, Heidelberg (2013)
17. Kiekintveld, C., Islam, T., Kreinovich, V.: Security games with interval uncertainty. In: Proc. of the 12th International Conference on Autonomous Agents and Multiagent Systems (AAMAS 2013), pp. 231–238. International Foundation for Autonomous Agents and Multiagent Systems, Richland (2013)

Data Integrity and Availability Verification Game in Untrusted Cloud Storage

Brahim Djebaili[1], Christophe Kiennert[1], Jean Leneutre[1], and Lin Chen[2]

[1] Télécom ParisTech, 46 rue Barrault, 75013 Paris, France
[2] Université Paris Sud, 15 rue Georges Clémenceau, 91400 Orsay, France

Abstract. The recent trends towards outsourcing data to the Cloud as well as various concerns regarding data integrity and availability created an increasing interest in enabling secure Cloud data-centers. Many schemes addressing data integrity issues and complying with various requirements came to place: high scheme efficiency, stateless verification, unbounded use of queries and retrievability of data. Yet, a critical question remains: how to use these schemes efficiently, i.e. how often should data be verified. Constantly checking is a clear waste of resources but only checking at times increases risks. This paper attempts to resolve this thorny issue by formulating the data integrity check problem as a non-cooperative game and by performing an in-depth analysis on the Nash Equilibrium and the engineering implications behind. Based on our game theoretical analysis, the course of action was to anticipate the Cloud provider's behavior; we then derive the minimum verification resource requirement, and the optimal strategy of the verifier. Finally, our game theoretical model is validated by showing correctness of the analytical results via simulation on a case study.

Keywords: Cloud computing, Game theory, Data integrity, Data availability, Nash equilibrium.

1 Introduction

Cloud computing is a model for enabling ubiquitous, convenient, on-demand network access to a shared pool of configurable computing resources (e.g., networks, servers, storage, applications, and services) that can be rapidly provisioned and released with minimal management effort or service provider interaction [11].

However, all the benefits brought by the cloud, such as lower costs and ease of use, come with a tradeoff. Users will have to entrust their data to a potentially untrustworthy cloud provider (CP). As a result, cloud security has become an important issue for both industry and academia [2].

One important security problem with cloud data storage is data integrity and availability, since the client lacks control over his data, entailing difficulties in ensuring that data stored in the Cloud are indeed left intact. Moreover, the storage service provider, which experiences Byzantine failures occasionally, may decide to hide data errors from the clients for his own benefit. On top of that,

R. Poovendran and W. Saad (Eds.): GameSec 2014, LNCS 8840, pp. 287–306, 2014.

for both money and storage space saving purposes, the service provider might deliberately delete rarely accessed data files that belong to an ordinary client.

In order to solve these problems, many verification schemes are provided in the literature [10]. In all these works, it has taken major efforts to design solutions that meet various requirements: low time complexity, stateless verification, unbounded use of queries and retrievability of data, etc. In spite of these numerous features, knowing how to use these schemes efficiently remains a major issue. Indeed, it would be a waste of time and resources if the verifier checks the data all the time while the CP is being honest. On the other hand, it would be risky if the verifier checks the data just a few times while the CP is being dishonest. The best approach for the verifier is to find the right frequency of verification for the minimum cost, while maintaining accuracy and consistency of data. The natural way to achieve this last condition is to use *game theory*, by modeling the process of data verification as a game that contains two players, the defender (verifier) and the attacker (CP).

Considering the role of the verifier, all the proposed schemes fall into two categories: private verification, in which the client performs the auditing operation himself, and public verification, that consists in using a third party auditor (TPA). In this paper, we focus on the latter, because in many cases, clients do not know how to check data integrity, nor do they know which protocol they should use. Moreover, a client who owns a considerable amount of outsourced data (like a company) will have no incentive to check his data, as this process requires considerable resources and time.

In such an environment, the major questions are: What is the expected behavior of a rational attacker (CP)? What is the optimal strategy of the defender (TPA)?

In this paper, we answer these questions by developing a non-cooperative game model of Cloud storage verification problem, analyzing the resulting equilibria, investigating the engineering implications behind the analytical results, and then deriving the optimal strategy for the defender. It is worth noting that the different cases taken into account in this work represent realistic situations, in which a client expects a specific service level from the TPA as stated in his contract with the TPA, which can be seen as an *Audit Level Agreement*.

Our main contributions can be summarized as follows:

1) We provide a game theoretical framework of cloud storage verification, by analyzing as a first model the case of deterministic verification. Then, as extensions, we study the case of the Leader/Follower game (Stackelberg game) in the second model, and probabilistic verification in the third one.

2) For each model, we derive the expected behavior of a rational attacker, the minimum verification resource requirements of the defender, as well as his optimal strategy in terms of resource allocation.

The remainder of the paper is organized as follows: In Section 2, we describe the technical background on which our work is based on. In Section 3, we study the Nash equilibrium (NE) of the Cloud storage game for deterministic verification. In Section 4, we explore several variants and extensions of the game, by analyzing the case of the Stackelberg game, and the case of probabilistic verification. Section 5 provides numerical results of the game theoretical framework. Finally, our concluding remarks are given in Section 6.

2   Technical Background

2.1   Integrity Verification Schemes

In recent years, a considerable amount of data integrity schemes were proposed by different researchers, and have been gradually adapted to specific use cases such as outsourced databases and Cloud Computing. Among these schemes, Provable Data Possession (PDP) for ensuring possession of data, and Proof of Retrievability (POR) for data possession and retrievability are the two main directions explored by researchers.

The main idea of PDP is that a data owner generates some metadata information for a data file to be used later for verification purposes. Many extensions of this scheme managed to decrease the communication cost and complexity, as well as to allow dynamic operations on data such as insertion, modification, or deletion. Moreover, [18] and [16] proposed PDP schemes fitting requirements specific to Cloud Computing.

The POR scheme is considered as a complementary approach to PDP. [9] was among the first papers to consider formal models for POR schemes. In this scheme, disguised blocks (called sentinels) are embedded into the data before outsourcing. The verifier checks randomly picked sentinels which would be influenced with a certain probability if data are corrupted. [10] gives a detailed survey of the contributions of numerous extensions of the PDP and POR schemes.

The aforementioned schemes primarily focus on a single data file copy. Yet, other schemes, such as [6], allow the verifier to check multiple copies of a data file over multiple Cloud servers.

2.2   Approaches Related to Game Theory

Several works handle cloud-related problems using game theory. Most focus on solutions such as resource allocation and management [8] or Cloud service negotiation [17], while few papers addressed the problem of Cloud security [12,13]. [12] addressed Cloud integrity issues by proposing a model where a client checks the correctness of calculations made on the data by the CP. They considered the case where for two CPs, the client sends a query to one of the two servers chosen randomly, and with a fixed probability, he sends the query to the other server as well.

Nix and Kantarcioglu also proposed in [13] to study the case of querying one single cloud provider, since checking data at multiple CPs is prohibitively expensive. [12,13] focused on checking that the queries sent to the CP are being computed correctly, under the condition that the stored data is intact. On a side note, they did not mention which type of verification protocol (deterministic or probabilistic) they used. Besides the Cloud, game theory has already been applied to study network security [7] [1], intrusion detection [5], Botnet defense [4], etc. The work presented in this paper was actually strongly inspired by [5].

3 Untrusted Cloud Storage Game for Deterministic Verification

As a first step, we considered a basic model in which the data integrity verification protocol is deterministic and always returns correct information. The main problem of deterministic verification schemes is the fact that they are computationally expensive, since the TPA performs the verification process on the entire data. After solving this game and finding its Nash Equilibrium (NE), which describes the optimal strategies of both players from which neither of them has incentive to deviate unilaterally, we will progressively refine this model by taking more realistic hypotheses into account.

3.1 Game Features

- Players: The game features two players, the auditor (TPA: third party auditor) and the outsourced server (CP: Cloud provider).

   - Information: The CP stores the client's data $D = \{D_1, D_2, ..., D_N\}$, with different importances and sizes. We consider that the TPA checks the data by using a deterministic scheme guaranteeing a probability of detecting data modification or deletion equal to 1.

   - Actions: We consider *mixed strategies* where a probability is assigned to each strategy of each player. Thus, for each data D_i, the auditor may choose to check its integrity and availability with probability t_i that stems from a probability distribution $t = \{t_1, t_2, ..., t_N\}$. On the other side, the CP can modify or delete data D_i with probability p_i steming from a probability distribution $p = \{p_1, p_2, ..., p_N\}$. Both TPA and CP have resource constraints respectively designated by $T \leq 1$ and $P \leq 1$.

   - Payoffs: The two TPA possible actions are *Check* and *Not Check*. Meanwhile, the CP may *Modify/Delete* a data or not, hence possibly leading to *Corrupted/Unavailable data*.

   If the corrupted or unavailable data D_i is not checked, then the CP gains S_i, which represents the size of the data, with $S_1 \geq S_2 \geq ... \geq S_N$, while the TPA loses data value and importance designated by F_i. If the TPA decides not to verify, and the CP has the correct data, then both players will neither lose nor gain anything.

Table 1. Cloud Storage Game with Deterministic Verification

CP \ TPA	Check	Not check
Correct/Available data	0 , $-C^t S_i - C^s S_i$	0 , 0
Corrupted/Unavailable data	$-C^s S_i - S_i$, $-C^t S_i + F_i$	S_i , $-F_i$

Let C^tS_i be the cost of the verification process by the TPA, and C^sS_i be the cost of executing the verification query by the Cloud Provider. Both costs are proportional to the size of data D_i.

If the TPA verifies the data whereas the CP has the correct data, we then consider that the TPA should pay the cost of CP verification process C^sS_i, since the data are intact. However, when the CP chose to modify or delete the data, the TPA will gain F_i, which is the the importance of data D_i, minus the verification cost C^tS_i, while the CP will lose S_i, minus the cost of verification C^sS_i. Table 1 illustrates the matrix payoff of both players (CP/TPA) in the strategic form.

The overall payoffs of the TPA (U_t) and the CP (U_p) are defined as follows:

$$U_t(t,p) = \sum_{i=1}^{N} t_i[p_i(2F_i + C^sS_i) - (C^tS_i + C^sS_i)] - \sum_{i=1}^{N} p_iF_i$$

$$U_p(t,p) = \sum_{i=1}^{N} p_iS_i[1 - t_i(2 + C^s)]$$

We finally define the Cloud storage verification game G.
Definition 1: the two players Cloud storage verification game G is defined as:

Players: Attacker (CP), Verifier (TPA).
Strategy type: Mixed strategy.
Strategy set: Attacker:

$$W_P = \left\{ p \cdot p \in [0, P]^N, \sum_{i=1}^{N} p_i \leq P \right\}$$

Verifier:

$$W_T = \left\{ t : t \in [0, T]^N, \sum_{i=1}^{N} t_i \leq T \right\}$$

Payoff: U_p for attacker, U_t for verifier.
Game rule: The attacker/verifier selects his strategy
$p/t \in W_P/W_T$ to maximize U_p/U_t.

3.2 Solving the Game

For non-cooperative games like ours, the most essential solution concept is the Nash Equilibrium (NE), which can be considered as the optimal agreement between the players, i.e. an equilibrium in which no player has any incentive to unilaterally deviate from his current strategy in order to maximize his payoff.

1) Data Distribution

Since the attacker has limited attack resources, a relevant approach consists in determining if a rational attacker will target any data, or if he will tend to focus on specific data only. This question will be studied before starting the NE analysis.

First, we introduce two sets that will be of use to clarify data distribution: the attractive set D_A, and the unattractive set D_U. In order to do so, we will introduce the notations $\mathcal{N} = \{1, ..., N\}$, $\mathcal{N}_A = \{i \in \mathcal{N}/D_i \in D_A\}$, and $\mathcal{N}_U = \{i \in \mathcal{N}/D_i \in D_U\}$.

Definition 2: The two sets D_A and D_U are defined as follows:

We set $C = \dfrac{|\mathcal{N}_A|\left(\frac{1}{2+C^s}\right) - T}{\sum_{j \in \mathcal{N}_A}\left(\frac{1}{2S_j + C^s S_j}\right)}$

$$\begin{cases} S_i > C, & \forall i \in \mathcal{N}_A \\ S_i < C, & \forall i \in \mathcal{N}_U \end{cases}$$

where $|\mathcal{N}_A|$ is the number of data contained in $\mathcal{N}_A$. The case where $S_i = C$ does not need to be taken into account, since it happens with very low probability and since these values rely on estimations. Therefore, should this case happen, replacing S_i with a slightly different estimation $S_i + \epsilon$ or $S_i - \epsilon$ would be enough to solve the situation.

Lemma 1: Given a Cloud provider that stores N Data, $\mathcal{N}_A$ is uniquely determined and consists of N_S data with the biggest sizes, such that:

1) if $S_N > \dfrac{N\left(\frac{1}{2+C^s}\right) - T}{\sum_{j=1}^{N}\left(\frac{1}{2S_j + C^s S_j}\right)}$, then $N_S = N$.

2) if $S_N < \dfrac{N\left(\frac{1}{2+C^s}\right) - T}{\sum_{j=1}^{N}\left(\frac{1}{2S_j + C^s S_j}\right)}$, N_S is determined as follows:

$$\begin{cases} S_{N_S} > \dfrac{N_S\left(\frac{1}{2+C^s}\right) - T}{\sum_{j=1}^{N_S}\left(\frac{1}{2S_j + C^s S_j}\right)} \\[6mm] S_{N_S+1} < \dfrac{N_S\left(\frac{1}{2+C^s}\right) - T}{\sum_{j=1}^{N_S}\left(\frac{1}{2S_j + C^s S_j}\right)} \end{cases}$$

Proof: See Appendix I.

Now we will study the implication of data distribution on the players' decisions.

Theorem 1: A rational attacker has no incentive to attack any data $D_i \in D_U$.

Proof: See Appendix II.

The theorem shows that the attacker only needs to attack data that belong to D_A in order to maximize his payoff. From this point, the defender has no incentive to verify data that will not be attacked. The meaning of the theorem is to assert the existence of data that are too small to be worth attacking to free significant space. As a consequence, it would be a waste of resource for the TPA to verify the integrity of such data.

Guideline 1: A rational defender has only to verify the integrity and the availability of data in D_A.

2) NE Analysis

Definition 3: A strategy profile (p^*, q^*) is a Nash Equilibrium of the Cloud storage verification game G, when both players (CP and TPA) cannot improve their payoff by unilaterally deviating from their current strategy.

As G is a two-player game with mixed strategies, it admits at least one NE, according to Theorem 1 in [14]. Let (t^*, p^*) denote the NE, it holds that:

$$0 \leq p_i^*(2F_i+C^sS_i)-(C^tS_i+C^sS_i)=p_j^*(2F_j+C^sS_i)-(C^tS_i+C^sS_i) \geq$$
$$p_k^*(2F_k+C^sS_i)-(C^tS_i+C^sS_i) \ \forall i,j,k \in \mathcal{N}, t_i^*, t_j^* > 0, t_k^* = 0 \qquad (1)$$

Equation (1) can be shown by noticing the TPA payoff function. Indeed, if the TPA gain when verifying D_k is lower than when verifying D_i, then in order to maximize his payoff, the TPA will not have incentive to verify D_k and will set $t_k = 0$. The same thing remains valid for the CP, and by noticing his payoff function, it holds that:

$$0 \leq S_i(1 - 2t_i^*) - t_i^*C^aS_i = S_j(1 - 2t_j^*) - t_j^*C^sS_i \geq$$
$$S_k(1 - 2t_k) - t_k C^sS_i \ \forall i,j,k \subset \mathcal{N}, p_i^*, p_j^* > 0, p_k^* \quad 0 \qquad (2)$$

These two equations allow us to find the NE, which we study in two different cases according to the players resource constraints. The NE is hence defined in the following cases:

Case 1: $\sum_{i \in \mathcal{N}} t_i^* = T$ and $\sum_{i \in \mathcal{N}} p_i^* = P$:

In this case, both TPA and CP use all their resources in order to verify/attack data. The game can be seen as a resource allocation problem, in which each player seeks to choose the most profitable strategy.

By combining (1) and (2), we get the NE displayed hereby:

$$t_i^* = \begin{cases} \dfrac{T - \dfrac{N_S}{2 + C^s} + S_i \sum_{j=1}^{N_S}\left(\dfrac{1}{2S_j + C^sS_j}\right)}{(2S_i + C^sS_i)\sum_{j=1}^{N_S}\left(\dfrac{1}{2S_j + C^sS_j}\right)}, & i \in \mathcal{N}_A \\[4mm] 0, & i \in \mathcal{N}_U \end{cases}$$

$$p_i^* = \begin{cases} \dfrac{P - \sum_{j=1}^{N_S}\left(\dfrac{(C^t + C^s)(S_j - S_i)}{2F_j + C^sS_j}\right)}{(2F_i + C^sS_i)\sum_{j=1}^{N_S}\left(\dfrac{1}{2F_j + C^sS_j}\right)}, & i \in \mathcal{N}_A \\[4mm] 0, & i \in \mathcal{N}_U \end{cases}$$

The necessary condition for the obtained result to be a NE is:

$$\begin{cases} p_i^*(2F_i + C^sS_i) - (C^tS_i + C^sS_i) \geq 0, \\ S_i[1 - t_i^*(2 + C^s)] \geq 0 \end{cases} \qquad i \in \mathcal{N}_A$$

$$\Longrightarrow \begin{cases} \dfrac{P}{C^t + C^s} \geq \sum_{i=1}^{N_S} \left(\dfrac{1}{\dfrac{2F_i}{S_i} + C^S} \right) \\ N_S \geq T(2 + C^s) \end{cases}$$

It is worth noting that $U_t(t^*, p^*)/U_p(t^*, p^*)$ is monotonously increasing in T/P, which means that the more resources are available to both players, the more payoff they will get.

This case is actually the most realistic situation to be considered, for both the TPA and the CP. The number of data that are usually outsourced in the Cloud is high enough to prevent both the attacker and the verifier from targeting every data. Actions, both in attack and verification, are therefore limited to the attractive data set D_A.

Case 2: $\displaystyle\sum_{i \in \mathcal{N}} t_i^* < T$ and $\displaystyle\sum_{i \in \mathcal{N}} p_i^* < P$:

In this case, both the CP and the TPA have sufficient resources, so they do not use up all their resources to respectively attack and verify data. Noticing U_t and U_p, we have:

$$\begin{cases} S_i(1 - 2t_i^* - t_i^*C^s) = 0 \\ p_i^*(2F_i + C^sS_i) - (C^tS_i + C^sS_i) = 0, \end{cases} \qquad i \in \mathcal{N}$$

$$\Longrightarrow NE = \begin{cases} t_i^* = \dfrac{1}{2 + C^s}, & i \in \mathcal{N} \\ p_i^* = \dfrac{C^tS_i + C^sS_i}{2F_i + C^sS_i}, & i \in \mathcal{N} \end{cases}$$

At the NE, we have:

$$\begin{cases} U_t(p^*, t^*) = -\displaystyle\sum_{i=1}^{N} \left(\dfrac{F_i(C^tS_i + C^sS_i)}{2F_i + C^sS_i} \right) \\ U_p(p^*, t^*) = 0 \end{cases}$$

In this case, the necessary condition for this result to be a NE is $N < T(2+C^s)$. Lemma 1 then states that $N_S = N$, which means that $D_U = \emptyset$. This is an expected result since both players have enough resources to target any data.

Moreover, from the above utility, it appears that having sufficient resources drags the utility of the attacker to zero, and leads the defender to be able to face greater risks by verifying more valuable data. The fact that the NE does not depend on the available resources is therefore consistent. Finally, the NE values show that the TPA will spend the necessary amount in order to prevent the CP from gaining anything. In other words, the CP cannot expect to gain anything when the TPA has enough resources to verify all the outsourced data.

However, for medium and large companies, it is very unlikely that this case could actually occur given the amount and the wide diversity of data that are usually outsourced.

In the previous analysis, we identified the specific amount of resources that both the TPA and the CP should allocate for respectively verifying and attacking the attractive data set, in two different cases. A numerical analysis of this model is provided in section V.

However, this model obviously lacks some more realistic hypotheses, such as taking into account the fact that both players are more likely to act one after the other rather than at the same time, or taking into account a probabilistic integrity checking protocol instead of a deterministic one. The next section therefore considers such extensions of our primary model.

4   Extensions

4.1   Cloud Storage for Stackelberg Game

In the previous model, we considered that the two players take their decisions locally and simultaneously. However, a player can follow a certain strategy taking into account his opponent's decision (meaning that the follower makes his choice only after knowing the other's strategy). In this extended model, we address this case by modeling the interaction between TPA and CP as a Stackelberg game. The leader begins by choosing his best strategy, then the follower, after being informed about the leader's choice, chooses his own strategy which will maximize his payoff. We define the Stackelberg game for the Cloud storage verification like this: In this definition, the TPA is assumed as a leader, and the CP as a follower.

Players: Leader : verifier side;
                    Follower : attacker side;
Strategy type: Mixed strategy.
Strategy: $t \in W_T$ and $p \in W_P$
Payoff: U_T for leader and U_P for follower
Game rule: the leader decides t first, the follower
                    decides p after knowing t.

Follower's problem:
According to the leader's chosen strategy, the follower chooses the strategy that maximizes his payoff (best response). Formally, for any chosen strategy t, the follower solves the following optimization problem:

$$p(t) = \arg \max_{p \in W_P} U_p(p, t)$$

Leader's problem:
The leader chooses his strategy which will maximize his payoff, given the follower will subsequently choose his best strategy. In other words, the leader

Table 2. Payoff matrix of the lead-or-follow game in extensive form

TPA/CP	Lead (p^L)	Follow (p^F)
Lead (t^L)	$U_p = -\delta \sum_{i \in \mathcal{N}} \left(\dfrac{S_i(C^t + C^s)(2 + C^s)}{\frac{2F_i}{S_i} + C^s} \right)$ $U_t = -\epsilon \sum_{i \in \mathcal{N}} \left(\dfrac{(2F_i + C^s S_i)}{2 + C^s} \right)$ $- \sum_{i \in \mathcal{N}} F_i \left(\dfrac{C^t S_i + C^s S_i}{2F_i + C^s S_i} - \epsilon \right)$	$U_p = 0$ $U_t = - \sum_{i \in \mathcal{N}} \left(\dfrac{C^t S_i + C^s S_i}{2 + C^s} \right)$ $-\delta \sum_{i \in \mathcal{N}} (C^t S_i + C^s S_i)$
Follow (t^F)	$U_p = \sum_{i \in \mathcal{N}} S_i \left(\dfrac{C^t S_i + C^s S_i}{2F_i + C^s S_i} - \epsilon \right)$ $U_t = - \sum_{i \in \mathcal{N}} F_i \left(\dfrac{C^t S_i + C^s S_i}{2F_i + C^s S_i} - \epsilon \right)$	$U_p = 0$ $U_t = 0$

chooses his strategy that gives the maximum gain in the worst case scenario. Formally, the leader solves the following optimization problem:

$$t(p) = \arg \max_{t \in W_T} U_t(p(t), t)$$

In most cases, Stackelberg games are solved by the backward induction technique. The solution consists of taking the follower's best response strategy as a function of the leader's strategy. Then, giving follower's best chosen response, the leader chooses his best strategy. The obtained equilibrium is referred to as a Stackelberg equilibrium (SE) or Stackelberg– Nash equilibrium (SNE).

Next, we address all possible cases, starting by considering the attacker as a leader and the verifier as a follower, then the verifier as a leader and the attacker as a follower, then we lastly examinate with the case when a player decides to be a leader or a follower without knowing the adversary's choice. In our study, we focus on the scenario where the attacker and the verifier have sufficient resources.

1) *Leader:* Attacker side; *Follower:* Verifier side

As the attacker will choose his strategy before the verifier, we have to find his best strategy subject to the constraint that the verifier makes a decision according to his best response function. We first start solving the verifier's best response by performing backward induction as follows:

$$t_i(t) = \begin{cases} = 0, & p_i < H_i, & i \in \mathcal{N} \\ \in [0, 1], & p_i = H_i, & i \in \mathcal{N} \\ = 1, & p_i > H_i, & i \in \mathcal{N} \end{cases}$$

Where $H_i = \dfrac{C^t S_i + C^s S_i}{2F_i + C^s S_i}$.

By noticing the leader's utility function $\sum_{i \in \mathcal{N}} p_i S_i [1 - t_i(2 + C^s)]$, we obtain the following SNE :

$$\begin{cases} t_i^S = 0, & i \in \mathcal{N} \\ p_i^S = H_i, & i \in \mathcal{N} \end{cases}$$

The corresponding payoff of both TPA and CP is as follows:

$$\begin{cases} U_t(t^S, p^S) = -\sum_{i \in D} F_i \, H_i, & i \in \mathcal{N} \\ U_p(t^S, p^S) = \sum_{i \in D} S_i \, H_i, & i \in \mathcal{N} \end{cases}$$

The fact that $U_t(t^S, p^S) = U_t(t', p^S), \forall t' \in W_T$ makes the above solution a weak Stackelberg equilibrium. Hence, the leader risks getting a negative payoff $(U_p(t^S, p^S) = -\sum_{i \in \mathcal{N}} H_i(S_i(1 + C^s)))$, since the follower can set $t_i = 1$ for all targets instead of t^S. This is clearly not acceptable for the attacker while his payoff is 0 when doing nothing.

As a solution, the attacker has to decrease his strategy a little bit by setting $p_i = p_i^S - \epsilon = H_i - \epsilon$, where ϵ is a small positive number, in order to guarantee that TPA will operate on t^S. As a result, the payoff will be $\sum_{i \in \mathcal{N}} S_i H_i - \epsilon \sum_{i \in \mathcal{N}} S_i$, which is slightly less than his desired payoff, since ϵ is sufficiently small.

2) *Leader:* Verifier side; *Follower:* Attacker side:

In this case, as the verifier plays the role of the leader, we will try to find the maximum value of his minimum payoff. Following the same analysis of the first case, The SNE is:

$$\begin{cases} t_i^S = \dfrac{1}{2 + C^s}, & i \in \mathcal{N} \\ p_i^S = 0, & i \in \mathcal{N} \end{cases}$$

In order to make sure that the attacker will operate on p^s, the verifier needs to increase his strategy a little bit by setting $t_i = t_i^S + \delta = \left(1/(2 + C^s)\right) + \delta$, where δ is a small positive number. In such a situation, the TPA payoff will be $-\sum_{i \in \mathcal{N}}(C^t S_i + C^s S_i/(2 + C^a)) - \delta \sum_{i \in \mathcal{N}}(C^t S_i + C^s S_i)$, which is a slightly less than his desired payoff at the SNE.

3) *Lead or Follow:*

Here we look at an interesting scenario where each player decides to choose the leader or the follower strategy, without knowing his adversary's choice. In this case, we aim to address the following questions: Is being a leader a better strategy than being a follower? Does the leader always control the behavior of the follower?

We formulate the (lead or follow) Cloud storage verification game as follows: the players are the verifier and the attacker; each player seeks to maximize his payoff by operating either on the leader strategy that we denote by t^L and p^L, respectively, or the follower strategy denoted by t^F and p^F, respectively. $\forall i \in \mathcal{N}$, we have:

$$t_i^L = \frac{1}{2 + C^s} + \delta, \quad t_i^F = 0, \quad p_i^L = H_i - \epsilon, \quad p_i^F = 0$$

Table 2 shows the payoff of both the attacker and the verifier. We ignored the terms that contain $\epsilon\delta$ due to their small value.

For the verifier, we can notice from Table 2 that the first row is strictly dominated by the second row, which means that it is better off for the verifier to be the follower. Hence, (p^L, t^F) is the NE of the game; the case when the attacker plays the role of the leader and the verifier follows.

From the above result, we can notice that the NE of the game is more favorable to the CP than the TPA, since the leader can control the behavior of the follower and pushes him to keep silent. Nevertheless, the TPA (follower) can influence the attacker's strategy, since both the strategy and the payoff of the attacker at the NE depends on the verification cost of the verifier. That being said, if $C^t \ll F_i$; both p_i and U_p are very small at the NE.

For the TPA, we would like to mention that his strategy at the NE $t_i^F = 0$ does not mean that no defender is needed, since before reaching the equilibrium, both players may try different strategies before choosing the one that maximizes their payoff.

Guideline 2: The TPA should choose the follower strategy in order to maximize his payoff, while leader is the best strategy for the CP.

4.2 Cloud Storage Game for Probabilistic Verification

Unlike the previous models, in which we consider that the TPA uses a deterministic verification protocol that guarantees a probability of detecting data modification or deletion equal to 1, in this extended model, we analyze the case of a probabilistic verification protocol that guarantees a detection probability inferior to one ($a < 1$) such as [3,9,15], since the TPA only performs verification on some parts of the data, in order to alleviate the verification cost. This means that there is a possibility that the TPA could not detect the incorrectness of the data with probability $(1 - a > 0)$. On top of that, we now consider that the CP loses some storage cost when he does not attack the data while the TPA does not verify it.

Table 3 shows the matrix payoff of both players (CP/TPA) in the following extensive form: when the CP does not attack the data while the TPA does not verify it, the CP loses a payoff proportional to the size of the data, denoted by BS_i, where $B \in [0,1]$. If the TPA verifies the data when it happens to be corrupted, then the TPA will gain ($-C^sS_i+F_i$) while CP gets ($-C^sS_i-S_i$), with probability a. With probability $(1 - a)$, the TPA has to pay the cost of the verification that is executed in both parts and also loses the data size, which means ($-C^sS_i-C^tS_i-F_i$) while CP gains ($-C^sS_i+C^sS_i+S_i$) = S_i.

The utility functions of CP and TPA are defined as follows:

$$U_t(t,p) = \sum_{i \in \mathcal{N}} t_i \Big[p_i a \big(2F_i + C^s S_i \big) - \big(C^t S_i + C^s S_i \big) \Big] - \sum_{i \in \mathcal{N}} p_i F_i$$

Table 3. Cloud storage game for probabilistic verification

CP \ TPA	Check		Not check
Available/	$U_p = 0$		$U_p = -BS_i$
Correct data	$U_t = -C^tS_i - C^sS_i$		$U_t = 0$
Unavailable/	$U_p = (1 - 2a)S_i - aC^sS_i$		$U_p = S_i$
Corrupted data	$U_t = -(1 - 2a)F_i - (1 - a)C^sS_i-C^tS_i$		$U_t = -F_i$

$$U_p(t,p) = \sum_{i \in \mathcal{N}} p_i \left[t_i \left(-(B + 2a)S_i - aC^s S_i \right) + (1 + B)S_i \right] - \sum_{i \in \mathcal{N}} (1 - t_i) B S_i$$

For data distribution, we keep the same characteristics as in the first model, in which data are distributed in two sets: the attractive set D_A, and the unattractive set D_U. The sets $\mathcal{N}_A$ and $\mathcal{N}_U$ are defined as in section III as well.

Now, we will investigate the NE of the game, according to players resource constraints. In this model, D_A and D_U are defined as follows.

Let $W = \dfrac{(1 + B)|\mathcal{N}_A| - T(B + a(2 + C^s))}{(1 + B)\sum_{j \in \mathcal{N}_A} \frac{1}{S_j}}$.

Then :

$$\begin{cases} S_i > W, & \forall i \in \mathcal{N}_A \\ S_i < W, & \forall i \in \mathcal{N}_U \end{cases}$$

It is interesting to note that the detection rate a has a real influence on the constitution of the data sets D_A and D_U, since it follows from the preceding definition that D_A grows as a increases. This remark can be interpreted as follows: when the detection rate is low, the CP can target the most interesting data to corrupt without being detected, whereas with a high detection rate, the CP will have to take more targets into consideration in order to mitigate the risk of being detected.

As in section 3, the NE can be analyzed following two different cases, depending on the players resource constraints.

Case 1: $\sum_{i \in \mathcal{N}} t_i^* = T$ and $\sum_{i \in \mathcal{N}} p_i^* = P$:

This case represents the most frequent situation, encountered when both players do not have enough resources to attack or defend every target.

The NE, obtained by a reasoning similar to section 3, is as follows:

$$t_i^* = \begin{cases} \dfrac{T + \dfrac{1 + B}{B + 2a + aC^s} \sum_{j=1}^{N_S} \left(\dfrac{S_i - S_j}{S_j} \right)}{S_i \sum_{j=1}^{N_S} \left(\dfrac{1}{S_j} \right)}, & i \in \mathcal{N}_A \\[4ex] 0, & i \in \mathcal{N}_U \end{cases}$$

$$p_i^* = \begin{cases} \dfrac{P + \sum_{j=1}^{N_S}\left(\dfrac{(C^t + C^s)(S_i - S_j)}{a(2F_j + C^sS_j)} \right)}{a(2F_i + C^sS_i) \sum_{j=1}^{N_S}\left(\dfrac{1}{2F_j + C^sS_j} \right)}, & i \in \mathcal{N}_A \\[4mm] 0, & i \in \mathcal{N}_U \end{cases}$$

The necessary condition for the solution to be a NE is:

$$\begin{cases} \dfrac{P}{C^t + C^s} \geq \sum_{i=1}^{N_S}\left(\dfrac{1}{a\left(\dfrac{2F_i}{S_i} + C^S \right)} \right) \\[4mm] N_S(1 + B) \geq T(B + a(2 + C^s)) \end{cases}$$

In this case, as in the deterministic verification model, both players try to use the maximum of their resources in order to maximize their payoff. Moreover, calculating $U_t(t^*, p^*)$ shows, as expected, that improving the detection rate of the protocol used by the TPA (i.e., increasing a) can increase his utility and alleviate the attack intensity.

Case 2: $\sum_{i \in \mathcal{N}} t_i^* < T$ and $\sum_{i \in \mathcal{N}} p_i^* < P$:

Both players have enough resources to attack and verify every data. The NE is then:

$$\begin{cases} t_i^* = \dfrac{1 + B}{B + 2a + aC^s}, & i \in \mathcal{N} \\[4mm] p_i^* = \dfrac{C^tS_i + C^sS_i}{a(2F_i + C^sS_i)}, & i \in \mathcal{N} \end{cases}$$

Where the necessary condition is $N(1 + B) < T(B + a(2 + C^s))$.

As shown in the payoff values at the NE given below, having sufficient resources for both players is not suitable for the CP, who gets a negative payoff due to the fact that he loses some storage cost even when he does not attack. Since the TPA can target every data for verification, the CP has overall no chance to gain anything when attacking a data, and also suffers some loss, at least in this model, when doing nothing.

At the NE, the corresponding payoffs are indeed:

$$\begin{cases} U_t(t^*, p^*) = -\sum_{j=1}^{N}\left(\dfrac{F_i(C^tS_i + C^sS_i)}{a(2F_i + C^s)} \right) \\[4mm] U_p(t^*, p^*) = -\sum_{j=1}^{N}\left[BS_i\left(1 - \dfrac{(1 + B)}{B + a(2 + C^s)} \right) \right] \end{cases}$$

It is also interesting to note that when the detection rate a increases, the TPA payoff increases, and the CP payoff decreases, which is a consistent result since a higher detection rate means that the TPA will have less failed verification attempts, while it will be harder for the CP to behave fraudulently without being detected.

From this analysis, we conclude that this theoretical model is realistic and consistent, and we were able to deduce the optimal strategies for both players in the two preceding cases, while putting into relief the importance of the detection rate a in the data distribution as well as in the players payoffs. A numerical study will now allow us to confirm these theoretical results.

5 Numerical Study

In this section, we validate the analytical results of the previous models by performing a numerical study.

In order to simplify the analysis, we consider that a client stores 20 data in the Cloud provider's data center with different sizes and sensibilities. We therefore consider that each data D_i has a size S_i and an importance F_i equal to $(21 - i) * 0.05$, $(i = 1, 2, ..., 20)$. As we mentioned earlier, the client delegates the check process to a special third party auditor TPA, that is equipped with high-performance verification modules and powerful processing capabilities. Thus, we set $C_t = C_s = 0.1$ for the case of deterministic verification schemes, and $C_t = C_s = 0.01$ for probabilistic schemes, since these schemes are much lighter, in terms of complexity, than deterministic ones.

For the deterministic verification model, according to Definition 2, our data are distributed into two sets: the first nine data belong to the attractive set D_A, whereas the remaining data are unattractive.

In the third model, where the verification process is probabilistic, we set $B = 0.001$. As expected, the data distribution is influenced by the probability of detecting data tampering a. In the case where $a = 0.9$ the attractive data are almost identical to the first model, since a is not so far from 1, while for $a = 0.5$, the number of attractive data decreases to 5, until reaches 3 for $a = 0.1$. This observation confirms our remark made in the previous section about the effect of a on the size of the data sets D_A and D_U. To further evaluate our analytical results, we investigate the case where TPA deviates from the NE. We thus simulate 10000 random strategies for the TPA under the condition that the CP chooses always his best response for each random strategy, in order to maximize his payoff.

For the deterministic model, Table 5 shows the strategies and the utility functions for both players at the NE, while Table 6 shows the payoffs of the TPA when he deviates from the NE. $U_t(t^r, p')_B$ is the best and the maximum payoff that the TPA can gain, where t^r is the random strategy for TPA, and p' is the CP's best response. $U_t(t^r, p')_W$ is the worst and minimal gain for TPA, while $U_t(t^r, p')_A$ is the average of all 10000 random strategies.

Table 5 and 6 clearly show that the best strategy for the TPA that maximizes his payoff is the NE, since $U_t(t^r, p')_B < U_t(t^*, p^*)$.

Fig.1 shows the utility functions of the TPA and the CP in the probabilistic verification model, under different values of the detection rate a. The valuable information that can be drawn here is that the TPA loss increases every time a decreases, while the CP gains more payoff every time a decreases, due to the fact

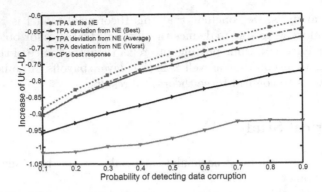

Fig. 1. Influence of the detection rate a on the TPA payoff in the probabilistic model

Table 4. Deterministic Verification Nash Equilibrium

The Defender (TPA)	The Attacker (CP)
$t_1^* = 0.19189$	$p_1^* = 0.10759$
$t_2^* = 0.17692$	$p_2^* = 0.10824$
$t_3^* = 0.16030$	$p_3^* = 0.10897$
$t_4^* = 0.14172$	$p_4^* = 0.10978$
$t_5^* = 0.12081$	$p_5^* = 0.11068$
$t_6^* = 0.09712$	$p_6^* = 0.11171$
$t_7^* = 0.07004$	$p_7^* = 0.11289$
$t_8^* = 0.03880$	$p_8^* = 0.11425$
$t_9^* = 0.00235$	$p_9^* = 0.11583$
$t_{10}^* - t_{20}^* = 0$	$p_{10}^* - p_{20}^* = 0$
$U_t(t^*, p^*) = -0.77100$	$U_p(t^*, p^*) = 0.59702$

that the more resources the CP uses to attack the first data in the attractive set, the more space he gains. Moreover, it appears that the TPA gets less payoff when he deviates from the NE.

These numerical results therefore corroborate our analysis of these theoretical models, and prove the consistency of the NE concept as the optimal strategy from which no player has any incentive to deviate in order to maximize his payoff.

6   Conclusion

In this paper, we focused on the problem of verifying data integrity in the case of data outsourced to an untrusted Cloud provider. We formulated the interaction between the verifier and the Cloud provider as a noncooperative game with mixed strategies, before performing an in-depth analysis on a deterministic model and on two extensions, namely the Stackelberg game for deterministic verification model, and a probabilistic verification model. Based on our analytical

Table 5. TPA Deviation From NE in Deterministic Verification Model

$U_t(t^r,p')_B$	$U_t(t^r,p')_A$	$U_t(t^r,p')_W$
-0.79884	-0.89058	-1.01050

results, we presented the expected behavior of a rational attacker, then derived the minimum verification resource requirement and the optimal strategy of the defender. We were also able to validate our analytical results by performing simulations.

However, the usual hypothesis of perfectly rational players limit the results of this work to very experienced attackers and verifiers who had a thoughtful approach of their actions. While not being unrealistic, given the fact that the CP and TPA entities are both very rational players by nature, this hypothesis remains a potential limitation to the superposition of this model with the objective behaviour of such entities in the reality.

Moreover, this work does not take into account several variants of the situation, such as the introduction of a penalty symbolizing the reputation loss in case of fraud from the CP, possibility to outsource numerous versions of a data to a CP, or the possibility for a CP to store multiple copies of each data with replication. Also, both the TPA and the CP can target more than one data at a time, which can be represented by a multiple-shot game. These variants will be the subject of future works that will aim at deepening this study in order to refine the model and integrate the hypotheses that are closer to reality.

References

1. Alpcan, T., Basar, T.: Network Security: A Decision and Game-Theoretic Approach. Cambridge University Press (2010)
2. Armbrust, M., Fox, A., Griffith, R., Joseph, A.D., Katz, R., Konwinski, A., Lee, G., Patterson, D., Rabkin, A., Stoica, I., et al.: A view of cloud computing. Communications of the ACM 53(4), 50–58 (2010)
3. Ateniese, G., Di Pietro, R., Mancini, L.V., Tsudik, G.: Scalable and efficient provable data possession. In: Proceedings of the 4th International Conference on Security and Privacy in Communication Netowrks, p. 9. ACM (2008)
4. Bensoussan, A., Kantarcioglu, M., Hoe, S(C.): A game-theoretical approach for finding optimal strategies in a botnet defense model. In: Alpcan, T., Buttyán, L., Baras, J.S. (eds.) GameSec 2010. LNCS, vol. 6442, pp. 135–148. Springer, Heidelberg (2010)
5. Chen, L., Leneutre, J.: A game theoretical framework on intrusion detection in heterogeneous networks. IEEE Transactions on Information Forensics and Security 4(2), 165–178 (2009)
6. Curtmola, R., Khan, O., Burns, R., Ateniese, G.: Mr-pdp: Multiple-replica provable data possession. In: The 28th International Conference on Distributed Computing Systems, ICDCS 2008, pp. 411–420. IEEE (2008)
7. Gueye, A., Marbukh, V.: A game-theoretic framework for network security vulnerability assessment and mitigation. In: Grossklags, J., Walrand, J. (eds.) GameSec 2012. LNCS, vol. 7638, pp. 186–200. Springer, Heidelberg (2012)

8. Hassan, M.M., Song, B., Huh, E.-N.: Distributed resource allocation games in horizontal dynamic cloud federation platform. In: 2011 IEEE 13th International Conference on High Performance Computing and Communications (HPCC), pp. 822–827. IEEE (2011)

9. Juels, A., Kaliski Jr., B.S.: Pors: Proofs of retrievability for large files. In: Proceedings of the 14th ACM Conference on Computer and Communications Security, pp. 584–597. ACM (2007)

10. Kochumol, A., Win, M.J.: Proving possession and retrievability within a cloud environment: A comparative survey. International Journal of Computer Science and Information Technologies 5(1), 478–485 (2014)

11. Mell, P., Grance, T.: The NIST definition of cloud computing (draft). NIST Special Publication 800(145), 7 (2011)

12. Nix, R., Kantarcioglu, M.: Contractual agreement design for enforcing honesty in cloud outsourcing. In: Grossklags, J., Walrand, J. (eds.) GameSec 2012. LNCS, vol. 7638, pp. 296–308. Springer, Heidelberg (2012)

13. Nix, R., Kantarcioglu, M.: Efficient query verification on outsourced data: A game-theoretic approach. arXiv preprint arXiv:1202.1567 (2012)

14. Ben Rosen, J.: Existence and uniqueness of equilibrium points for concave n-person games. Econometrica: Journal of the Econometric Society, 520–534 (1965)

15. Sebé, F., Domingo-Ferrer, J., Martinez-Balleste, A., Deswarte, Y., Quisquater, J.: Efficient remote data possession checking in critical information infrastructures. IEEE Transactions on Knowledge and Data Engineering 20(8), 1034–1038 (2008)

16. Yang, J., Wang, H., Wang, J., Tan, C., Yu, D.: Provable data possession of resource-constrained mobile devices in cloud computing. JNW 6(7), 1033–1040 (2011)

17. Zheng, X., Martin, P., Powley, W., Brohman, K.: Applying bargaining game theory to web services negotiation. In: 2010 IEEE International Conference on Services Computing (SCC), pp. 218–225. IEEE (2010)

18. Zhu, Y., Wang, H., Hu, Z., Ahn, G.-J., Hu, H., Yau, S.S.: Efficient provable data possession for hybrid clouds. In: Proceedings of the 17th ACM Conference on Computer and Communications Security, CCS 2010, pp. 756–758. ACM (2010)

Appendix I : Proof of Lemma 1

Here, we will prove that $\mathcal{N}_A$ contains d data with the biggest sizes, and $d = N_S$ by showing that neither $d < N_S$ nor $d > N_S$ is achieved.

In this proof, We need to only focus on the second case of the lemma, since the first case is straightforwardly evident. Before delving into the proof that $\mathcal{N}_A$ is unique, we should mention that it clearly appears that the N_S data with the biggest sizes that satisfy the second case of the lemma constitute the attractive data set $\mathcal{N}_A$, since the very definition of $\mathcal{N}_A$ given in Definition 2 is satisfied.

We first show that if $i \in \mathcal{N}_A$, then $\forall j < i$ $(S_j \geq S_i)$, it holds that $j \in \mathcal{N}_A$. Suppose this is not the case. Then, there exist $j_0 < i$ $(S_{j_0} \geq S_i)$ such that $j_0 \in \mathcal{N} - \mathcal{N}_A$. It follows that $S_{j_0} \leq C$. On the other hand, from Definition 2, we have $S_i > C$. It follows that $S_i > S_{j_0}$, which contradicts with $S_{j_0} \geq S_i$. Hence, $\mathcal{N}_A$ consist of the d data with the biggest sizes.

Now, we have to prove that $d = N_S$. Suppose first that $d < N_S$. From case 2 of the Lemma, we have:

$$S_{N_S} > \frac{N_S\left(\frac{1}{2+C^s}\right) - T}{\sum_{j=1}^{N_S}\left(\frac{1}{2S_j+C^sS_j}\right)} \implies S_{N_S}\sum_{j=1}^{N_S}\left(\frac{1}{2S_j+C^sS_j}\right) > N_S\left(\frac{1}{2+C^s}\right) - T$$

$$\implies S_{N_S}\sum_{j=1}^{N_S}\left(\frac{1}{2S_j+C^sS_j}\right) - \frac{1}{2+C^s}(N_S - d) > d\frac{1}{2+C^s} - T.$$

Noticing that $S_{N_S} \leq S_i, \forall i \leq N_S$ and $d < N_S$ (i.e. $S_{d+1} \geq S_{N_S}$), we have:

$$S_{d+1}\sum_{j=1}^{d}\left(\frac{1}{2S_j+C^sS_j}\right) \geq S_{N_S}\sum_{j=1}^{d}\left(\frac{1}{2S_j+C^sS_j}\right)$$

$$\geq S_{N_S}\sum_{j=1}^{N_S}\left(\frac{1}{2S_j+C^sS_j}\right) - S_{N_S}\sum_{j=d+1}^{N_S}\left(\frac{1}{2S_j+C^sS_j}\right)$$

$$> S_{N_S}\sum_{j=1}^{N_S}\left(\frac{1}{2S_j+C^sS_j}\right) - \frac{1}{2+C^s}(N_S - d) > d\frac{1}{2+C^s} - T$$

Hence, $S_{d+1} > \dfrac{d\left(\frac{1}{2+C^s}\right) - T}{\sum_{j=1}^{d}\left(\frac{1}{2S_j+C^sS_j}\right)}$. On the other hand, from Definition 2, we have

$S_{d+1} \leq \left(d\left((1/(2+C^s))-T\right)/\left(\sum_{j-1}^{d}\left((1/2S_j+C^sS_j)\right)\right)\right)$. This contradiction shows that it is impossible that $d < N_S$. Similarly, we can show that it is impossible that $d > N_S$. Hence, $d = N_S$ is uniquely determined, and so is $\mathcal{N}_A$. It follows obviously that $\mathcal{N}_U$ is also uniquely determined.

Appendix II : Proof of Theorem 1

The proof consists of showing that regardless of the verifier's strategy t, for any $p \in W_P$ such that $\exists i \in \mathcal{N}_U, p_i > 0$, we can construct another strategy p' such that $p'_i = 0, \forall i \in \mathcal{N}_U$ and $U_p(t, p) < U_p(t, p')$.

If $S_N \geq C$, then $\mathcal{N}_U = \emptyset$; the theorem holds evidently. We focus in our proof in the case where $S_N < C$, in other words, $\mathcal{N}_U \neq \emptyset$.

We consider a vector $t^0 = (t_1^0, t_2^0, ..., t_N^0)$ where:

$$t_i^0 = \begin{cases} \dfrac{T - \dfrac{N_S}{2+C^s} + S_i\sum_{j=1}^{N_S}\left(\dfrac{1}{2S_j+C^sS_j}\right)}{(2S_i + C^sS_i)\sum_{j=1}^{N_S}\left(\dfrac{1}{2S_j+C^sS_j}\right)}, & i \in \mathcal{N}_A \\[20pt] 0, & i \in \mathcal{N} - \mathcal{N}_A \end{cases}$$

It holds that $t_i^0 \geq 0$ and $\sum_{i=1}^{N_S} t_i^0 = T$. Let $t = (t_1, t_2, ..., t_N)$ denote the verification probability distribution of the verifier, with $\sum_{i=1}^{N_S} t_i \leq T$. By the Pigeon Hole Principle, it holds that $\exists m \in \mathcal{N}_A$ such that $t_m \leq t_m^0$.

We now consider any attacker strategy $p = (p_1, p_2, ..., p_N) \in W_P$ satisfying $\sum_{i\in\mathcal{N}_U} p_i > 0$, i.e; the attacker attacks at least one target outside the attractive data set with nonzero probability. We construct another attacker strategy profile p' based on p such that:

$$p'_i = \begin{cases} p_i, & i \in \mathcal{N}_A \text{ and } i \neq m \\ p_m + \sum_{j \in \mathcal{N}_U} p_j, & i = m \\ 0, & i \in \mathcal{N}_U \end{cases}$$

By comparing the attacker payoff at p and p', noticing that $\forall i \in \mathcal{N}_U$,

$$S_i < \frac{N_s\big((1/(2+C^s))-T\big)}{\Big(\sum_{j=1}^{N_S}\big((1/2S_j+C^sS_j)\big)\Big)}, \text{ we obtain:}$$

$$U_P(p) - U_P(p') = \sum_{i \in \mathcal{N}} p_i S_i \Big(1 - t_i(2+C^s)\Big) - \sum_{i \in \mathcal{N}} p'_i S_i \Big(1 - t_i(2+C^s)\Big)$$

$$= \sum_{i \in \mathcal{N}} p_i S_i \Big(1 - t_i(2+C^s)\Big)$$

$$- \left(\sum_{i \in \mathcal{N}_A, i \neq m} p_i S_i \Big(1 - t_i(2+C^s)\Big) + \Big(p_m + \sum_{i \in \mathcal{N}-\mathcal{N}_A} p_i\Big) S_m \Big(1 - t_m(2+C^s)\Big) \right)$$

$$= \sum_{i \in \mathcal{N}-\mathcal{N}_A} p_i S_i \Big(1 - t_i(2+C^s)\Big) - \sum_{i \in \mathcal{N}-\mathcal{N}_A} p_i S_m \Big(1 - t_m(2+C^s)\Big)$$

$$\leq \sum_{i \in \mathcal{N}-\mathcal{N}_A} p_i S_i \Big(1 - t_i(2+C^s)\Big) - \sum_{i \in \mathcal{N}-\mathcal{N}_A} p_i S_m \Big(1 - t_m^0(2+C^s)\Big)$$

$$= \sum_{i \in \mathcal{N}-\mathcal{N}_A} p_i S_i \Big(1 - t_i(2+C^s)\Big) - \sum_{i \in \mathcal{N}-\mathcal{N}_A} p_i \left(\frac{N_S \frac{1}{2+C^s} - T}{\sum_{j=1}^{N_S} \big(\frac{1}{2S_j+C^sS_j}\big)} \right)$$

$$\leq \sum_{i \in \mathcal{N}-\mathcal{N}_A} p_i S_i - \sum_{i \in \mathcal{N}-\mathcal{N}_A} p_i \left(\frac{N_S \frac{1}{2+C^s} - T}{\sum_{j=1}^{N_S} \big(\frac{1}{2S_j+C^sS_j}\big)} \right) < 0$$

Hence, the strategy p' gives more payoff to the CP than the strategy p. A rational CP therefore has no incentive to attack any data $D_i \in D_U$.

Short Papers

Empirical Comparisons of Descriptive Multi-objective Adversary Models in Stackelberg Security Games

Jinshu Cui and Richard S. John

University of Southern California, Department of Psychology
Los Angeles, CA, USA

Abstract. Stackelberg Security Games (SSG) have been used to model defender-attacker relationships for analyzing real-world security resource allocation problems. Research has focused on generating algorithms that are optimal and efficient for defenders, based on a presumed model of adversary choices. However, relatively less has been done descriptively to investigate how well those models capture adversary choices and psychological assumptions about adversary decision making. Using data from three experiments, including over 1000 human subjects playing over 25000 games, this study evaluates adversary choices by comparing 9 adversary models both nomothetically and ideographically in a SSG setting. We found that participants tended to be consistent with utility maximization and avoid a target with high probability of being protected even if the reward or expected value of that target is high. It was also found in two experiments that adversary choices were dependent on the defender's payoffs, even after accounting for attacker's own payoffs.

Keywords: adversary modeling, Stackelberg Security Game, utility function.

1 Introduction

Relationships between attackers and defenders have been modeled as Stackelberg Security Games (SSG). In SSG, a defender moves first as a leader, an attacker then observes the defender's strategy and choose a target to attack. Security resource allocation research has focused on identifying defenders' optimal strategy. One approach is to generate a robust method that is independent of adversaries' strategies[1]. Another approach to determine a defender's optimal strategy is to model adversaries' strategies and construct an optimal defense in response[2, 3]. The approach that considerably models adversaries' choices has been proved to be more effective.

However, relatively less has been done descriptively to investigate how well the adversary-based defenders' algorithms capture adversary decision making and the psychological assumptions of adversaries' choice behavior. This study aims to explore adversaries' choices by comparing different adversary models in a SSG setting. Using data from three experiments, including over 1000 human subjects playing over 25000 games, nine models were evaluated nomothetically and ideographically.

The models compared in this paper all measure adversaries' choices as probabilistic choices, that is, if the probability of choosing one target is higher than that of

R. Poovendran and W. Saad (Eds.): GameSec 2014, LNCS 8840, pp. 309–318, 2014.

choosing an alternative, the adversary will choose that target to attack. In decision making research, Luce's Choice Axiom (LCA)[4] assumes that choice behavior is probabilistic instead of deterministic. McFadden[5, 6] applied LCA to preferential choice in economic analysis. His model was able to exaggerate the differences between different alternatives by exponentiating utilities and the optimal choice is consistent with utility maximization. McKelvey and Palfrey[7] later developed Quantal Response Equilibrium (QRE) in economics, which assumes that the chance of selecting a non-optimal strategy increases as the level of error increases, in which λ captures the rational level (absence of errors) of a player. Since expected utility maximizing is the baseline of a rational decision maker and it is easier to measure a parameter close to 0, we adjusted the quantal response model by reversing the parameter and let λ represent the level of error (softmax): a player chooses randomly when $\lambda \to \infty$ and maximizes expected utility when $\lambda \to 0$. Let $q_i \in [0, 1]$ represent the probability that target t_i will be attacked:

$$q_i(x_i) = \frac{e^{U_{A_i}(x_i)/\lambda}}{\sum_{t_k \in T} e^{U_{A_k}(x_k)/\lambda}}, \lambda \geq 0 \qquad (1)$$

Using the softmax function, we evaluated adversary decision making by assessing four different aspects of the proposed choice models:

(1) Consistency level with utility function maximizing. As suggested by bounded rationality[8], inconsistency with utility function maximization could result when an attacker has limited time and resources to contemplate the optimal choice. The actual choice could deviate from optimal choice and magnitude of deviation is represented by the inconsistency level (λ).

(2) Attention to probability of success. It has been assumed that adversaries pay more attention to probability rather than consequences such that they tend to choose targets with higher probability of success ("soft targets")[9]. We hypothesized that an attacker would pay extra attention to the probability of sucess.

(3) Dependence on defender's utility. Given that adversaries may be driven by emotion, it is reasonable to assume an attacker could "sacrifice" part of their own reward to "hurt" the enemy. We anticipated that terrorists would choose a target that could create more damage to the targeted population, even though that choice could have a lower expected return.

(4) Risk attitude. Past research indicates that emotions can influence risk attitude, such that fear can lead to risk-aversion and anger can lead to risk-seeking[10]. There is little basis to assume adversaries are risk neutral or risk averse[11], especially for an attacker who could experience strong emotions.

2   Models

For each of the proposed models, we aim to capture an adversary's consistency with utility maximization, attention to probability of success, dependence on defender's utility, and risk attitude. The various proposed utility models all utilize the softmax function to calculate the attacker's probability of choosing a particular target. The nine utility functions can be partitioned into five categories: (1) attacker's expected value (EV), (2) attacker's expected utility (EU) accounting for risk attitude, (3) lens model[12, 13] with a weighted average of p(success), attacker's reward and penalty and defender's reward and penalty, (4) lens model accounting for risk attitude, and (5) multi-attribute utility (MAU) model with a weighted average of p(success), attacker's EV and defender's EV[14].

A summary of the nine models grouped in five categories is presented in the Table 1. All nine models capture the inconsistency level (λ). EV is the baseline model. The five lens models and the MAU model capture an attacker's trade-offs among competing cues (or objectives). The EU–α, lens–3–α and lens–5–α allow risk attitude to be accounted for; lens–4, lens–5, lens–5–α, and MAU model take defender's utility into account for attacker's utility function.

2.1 Attacker's Expected Value

The basic utility function of an adversary only captured the expected utility of an attacker who is risk neutral (expected value). The model was first introduced by Yang and colleagues[2] in the name of Quantal Response model. If the attacked target t_i (i = 1,2,..,8) is covered by the defender, the attacker receives penalty P_{A_i} and the defender receives reward R_{D_i}; if the attacked target is not covered by the defender, the attacker received reward R_{A_i} and the defender receives penalty P_{D_i}. Let x_i denotes the probability of a guard at t_i, attacker's expected utility at t_i is

$$U_{A_i}(x_i) = x_i P_{A_i} + (1 - x_i)R_{A_i} \qquad (2)$$

Yang et al. [2] further modified the model by adding an extra weight (λ_s, $\lambda_s \geq 0$) to the target that is least protected by the defender, that is, the least defended target is given a bonus in the SOFTMAX calculation. This assumption is consistent with "soft target" hypothesis. Let $S_i(x_i)$ denote whether a target is covered by the least resource:

$$S_i(x_i) = \begin{cases} 1, if\, x_i \leq x_{i'} \\ 0, otherwise \end{cases} \qquad (3)$$

2.2 Attacker's Expected Utility Accounting for Risk Attitude

A simple power utility function was constructed by adding a parameter α to capture risk attitude where $\alpha>1$ indicates risk seeking and $0<\alpha<1$ indicates risk aversion. Assuming the same risk attitude for gain and loss, expected utility of target t_i is:

$$U_{A_i}(x) = x_i P_{A_i}^{\alpha} + (1 - x_i)R_{A_i}^{\alpha} \qquad (4)$$

2.3 Lens Model

The lens model suggests that attacker judgments depend on a linear combination of multiple observable cues. Therefore, the expected utility function of an attacker can

Table 1. A Summary of the Nine Models Grouped in Five Categories

Category	Model	Abbreviation	Equation
Attacker's expected utility models	Attacker's expected utility model	EU	$q_i(x_i) = \dfrac{e^{[x_iP_{A_i}+(1-x_i)R_{A_i}]\lambda}}{\sum_{t_k\in T} e^{[x_kP_{A_k}+(1-x_k)R_{A_k}]/\lambda}}$
	Attacker's expected utility model accounting for soft target	EU–soft target	$q_i(x_i) = \dfrac{e^{[x_iP_{A_i}+(1-x_i)R_{A_i}]/\lambda+\lambda_sS_i(x_i)}}{\sum_{t_k\in T} e^{[x_kP_{A_k}+(1-x_k)R_{A_k}]/\lambda+\lambda_sS_k(x_k)}}$
Attacker's expected utility model accounting for risk attitude	Attacker's expected utility model accounting for risk attitude	EU–α	$q_i(x_i) = \dfrac{e^{[x_iP_{A_i}^\alpha+(1-x_i)R_{A_i}^\alpha]/\lambda}}{\sum_{t_k\in T} e^{[x_kP_{A_k}^\alpha+(1-x_k)R_{A_k}^\alpha]/\lambda}}$
Lens models	Lens model – three parameters	Lens–3	$q_i(x_i) = \dfrac{e^{(w_1x_i+w_2P_{A_i}+w_3R_{A_i})/\lambda}}{\sum_{t_k\in T} e^{(w_1x_k+w_2P_{A_k}+w_3R_{A_k})/\lambda}}$
	Lens model – four parameters	Lens–4	$q_i(x_i) = \dfrac{e^{(w_1x_i+w_2P_{A_i}+w_3R_{A_i}+w_4(P_{D_i}+R_{D_i}))/\lambda}}{\sum_{t_k\in T} e^{(w_1x_k+w_2P_{A_k}+w_3R_{A_k}+w_4(P_{D_k}+R_{D_k}))/\lambda}}$
	Lens model – five parameters	Lens–5	$q_i(x_i) = \dfrac{e^{(w_1x_i+w_2P_{A_i}+w_3R_{A_i}+w_4P_{D_i}+w_5R_{D_i})/\lambda}}{\sum_{t_k\in T} e^{(w_1x_k+w_2P_{A_k}+w_3R_{A_k}+w_4P_{D_k}+w_5R_{D_k})/\lambda}}$
Lens models accounting for risk attitude	Lens model – three attributes accounting for risk attitude	Lens–3–α	$q_i(x_i) = \dfrac{e^{(w_1x_i+w_2P_{A_i}^\alpha+w_3R_{A_i}^\alpha)/\lambda}}{\sum_{t_k\in T} e^{(w_1x_k+w_2P_{A_k}^\alpha+w_3R_{A_k}^\alpha)/\lambda}}$
	Lens model – five attributes accounting for risk attitude	Lens–5–α	$q_i(x_i) = \dfrac{e^{[w_1x_i+w_2(P_{A_i}^\alpha+R_{A_i}^\alpha)+w_3(P_{D_i}+R_{D_i})]/\lambda}}{\sum_{t_k\in T} e^{[w_1x_k+w_2(P_{A_k}^\alpha+R_{A_k}^\alpha)+w_3(P_{D_k}+R_{D_k})]/\lambda}}$
Multi-attribute utility model	Multi-attribute utility model	MAU	$q_i(x_i) = \dfrac{e^{w_1x_i+w_2[x_iP_{A_i}+(1-x_i)R_{A_i}]+w_3[x_iP_{D_i}+(1-x_i)R_{D_i}]}}{\sum_{t_k\in T} e^{w_1x_k+w_2[x_kP_{A_k}+(1-x_k)R_{A_k}]+w_3[x_kP_{D_k}+(1-x_k)R_{D_k}]}}$

be a linear combination of three attributes that are important to the decision (x_i, R_{A_i}, and P_{A_i}). The model, labeled the Subjective Utility Quantal Response (SUQR), was first proposed by Nguyen and colleagues [3]. The utility function was defined as:

$$U_{A_i}(x_i) = w_1x_i + w_2P_{A_i} + w_3R_{A_i} \tag{5}$$

We then extended this utility function to a linear combination of five cues with four weighting parameters (x_i, R_{A_i}, P_{A_i}, R_{D_i}, and P_{D_i}) with a common weight for the sum of defender's penalty and reward. We also extended this model to a linear combination of all five cues with separate weighting parameter for each cue:

$$U_{A_i}(x_i) = w_1x_i + w_2P_{A_i} + w_3R_{A_i} + w_4(P_{D_i} + R_{D_i}) \tag{6}$$

$$U_{A_i}(x_i) = w_1x_i + w_2P_{A_i} + w_3R_{A_i} + w_4P_{D_i} + w_5R_{D_i} \tag{7}$$

2.4 Lens Model Accounting for Risk Attitude (lens-α)

Risk attitude can be captured by introducing the parameter α to the lens model:

$$U_{A_i}(x_i) = w_1x_i + w_2P_{A_i}^\alpha + w_3R_{A_i}^\alpha \tag{8}$$

Risk attitude was also captured in the lens model with five cues. To reduce the number of parameters, we assumed a common weight on attacker's reward and penalty and another common weight on defender's reward and penalty. The evaluation of choosing target t_i then is:

$$U_{A_i}(x_i) = w_1 x_i + w_2 (P_{A_i}^\alpha + R_{A_i}^\alpha) + w_3 (P_{D_i} + R_{D_i}) \tag{9}$$

2.5   Multi-Attribute Utility Model

Inspired from the lens model which assumed expected utility as a linear combination of different attributes, we developed a new model of multi-attribute utility assuming that the adversary had multiple objectives. We assumed adversaries had three objectives: (1) maximize the probability of success, (2) maximize their expected utility and (3) minimize defender's expected utility. The probability of choosing target t_i is:

$$U_{A_i}(x_i) = w_1 x_i + w_2 EU_{A_i} + w_3 EU_{D_i}$$

$$= w_1 x_i + w_2 [x_i P_{A_i} + (1 - x_i) R_{A_i}] + w_3 [x_i P_{D_i} + (1 - x_i) R_{D_i}] \tag{10}$$

3   Experiment

3.1   Method

The three experiments used the same game paradigm called "The Guards and The Treasure" written in PHP. Each participant was asked to play as an attacker and choose one out of eight gates to attack given x_i, R_{A_i}, P_{A_i}, R_{D_i}, and P_{D_i} for each alternative. The three experiments differ in attacker and defender payoff matrixes, defender's guarding strategies and experiment procedures[1]. The published work [1-3] focused on evaluating algorithms for defender strategy in terms of defender EV. This paper reports new analyses of data from the three experiments, focusing on evaluating attackers' choices.

Amazon Mechanical Turk (AMT) was used to collect data. In experiment I, 102 participants, each played 40 rounds, and completed 4080 rounds in total. Forty of the 102 participants were from the US and 48 were from India. Thirty-six (35%) were female. In experiment II, a total of 653 US participants, each played 25 rounds and completed 16325 rounds in total. Two-hundred and seventy-two (42%) were female. In experiment III, a total of 294 US participants, each played 25 to 33 rounds and completed 8538 rounds in total. Eighty-nine (30%) were female.

---

[1] Please refer to the published worksfor the game procedures, payoff matrices and algorithms.

Table 2. Estimates of Parameters and AIC for Experiments I, II and III

Model	AIC	Experiment I Parameters estimation	AIC	Experiment II Parameters estimation	AIC	Experiment III Parameters estimation
EU	15036	λ=.09	60674	λ=.08	33334	λ=.20
EU–soft target	14820	λ=(.09,.59)	50548	λ=(.07,1.89)	27802	λ=(.41, 1.79)
EU–α	15012	λ=.08, α=.86	59169	λ=.06, α =.7	31065	λ=.08, α=.33
Lens–3	14670	λ=.05, w=(-.32,.44,.24)	52014	λ=.04, w=(-.42,.35,.23)	25445	λ=.07, w=(-.16,.18,.67)
Lens–4	14656	λ=.02, w=(-.31,.44,.23,.02)	48218	λ=.01, w=(-.36,.30,.20,.14)	22937	λ=.02, w=(-.47,.03,.19,.31)
Lens–5	14658	λ=.05, w=(-.30,.43,.23,.02,.02)	43265	λ=.02, w=(-.31,.26,.17,.04,.20)	22592	λ=.04, w=(-.44,-.01,.04,.30,.21)
Lens–3–α	14645	λ=.05, w=(-.32,.35,.34), α=1.47	51929	λ=.04, w=(-.42,.30,.28), α =1.25	25159	λ=.07, w=(-.09,.11,.80), α =1.86
Lens–5–α	14624	λ=.08, w=(-.46,.5,.04), α=1.51	48121	λ=.04, w=(-.45,.32,.18), α =1.32	23228	λ=.05, w=(-.58,.07,.35), α =.47
MAU	14973	λ=.08, w=(-.06,.84,.10)	45335	λ=.03, w=(-.32,.39,.29)	26540	λ=.07, w=(-.66,-.01,.33)

3.2 Results

Nomothetic Analysis. Maximum Likelihood Estimation (MLE) [15] was employed to fit the data over all the games played in each of the three experiments and estimate parameters for all nine models. The likelihood function for each model is:

$$L = \prod_{i=1,2,...,N} q_i(x_i) \tag{11}$$

The Akaike Information Criterion (AIC) [16] was calculated using equation 12 for each model in the three experiments, where k is the number of parameters of a model. AIC is an estimate of the expected, relative distance between the fitted model and the unknown true mechanism that generated the observed data [17]. The model with the minimum AIC is the best among the alternatives.

$$AIC = -2 \ln L + 2k \tag{12}$$

The estimates of the parameters and AICs for the nine models tested in experiments I, II and III are summarized in Table 2. In experiment I, AIC results indicate that models EV and EU–α were similar in terms of fit; model EU–soft target was slightly better than EU and EU–α. The lens models fit better than model EU–soft target; among lens models, lens–5–α was the best. The MAU model did not fit as well as the linear utility models. Parameter estimates indicate that participants were consistent with maximization of the various evaluation functions ($\lambda < 0.1$) for all nine models. Both the lens models and the MAU model resulted in a negative weight on the probability of being caught, which suggests that participants tended to give a bonus to targets that are less likely to be guarded. Parameter estimates for the four models that captured the weight participants put on defender's rewards and penalties (Lens–4, lens–5, lens–5–α and MAU) suggest that the weight on defender's side was much lower than that put on attacker's rewards and penalties (about 1/10). Finally, model EU–α indicated that participants were risk-averse, while lens–3–α and lens–5–α indicated that attackers were risk-seeking.

In experiment II, the AIC fit indices indicated consistency with experiment I; model EV was the worst model among the nine. Model EU–α was slightly better than EV but was worse than EU–soft target. The lens models and MAU were again better than EU–soft target. The MAU model was not as good as the lens models. Among the five lens models, lens–4 was better than lens–3 and lens–5 was better than lens–4. Adding a parameter for risk attitude on lens–3 (lens–3–α) improved the model slightly. Adding a parameter for risk attitude on lens–5 and combining attacker's side and defender's side (lens–5–α) did not improve the model. Parameter estimates indicated that participants were rational ($\lambda<0.1$ for EU–α, lens models, and MAU while $\lambda<0.5$ for EV and EU–soft target). Again, lens models and the MAU model indicated that a negative weight was put on the probability of being caught. Results of the four models that capture the weight attackers place on the defender's rewards and penalties (lens–4, lens–5, lens–5–α and MAU) suggested that the weight on defender's rewards and penalties was as high as the weight on attacker's side. Finally, EU–α and EU–5–α indicated that participants were risk-averse while EU–3–α indicated that participants were risk-seeking.

In experiment III, AIC results were consistent with those from experiments I and II in that model EV was the worst model among the nine. EU–α was slightly better than EV but was worse than EU–soft target. EU–soft target was better than lens–3 and lens–3–α, and was worse than lens–4, lens–5, lens–5–$\alpha$and MAU (all models accounted for defender's rewards and penalties). Among the five lens models, lens–4 was better than lens–3 and lens–5 was better than lens–4. Adding a parameter for risk attitude on lens–3 (lens–3–α) improved the model slightly. Adding a parameter for risk attitude on lens–5and combining attacker's rewards and penalties and defender's rewards and penalties (lens–5–α) did not improve the model. The MAU model did not fit as well as lens–5 but was better than the other seven models. Parameter estimates indicated that participants were rational ($\lambda<0.1$) for all models. Again, lens models and MAU indicated that a negative weight was put on the probability of being caught. Results of the four models that captured the weight participants put on defender's rewards and penalties (lens–4, lens–5, lens–5–α and MAU), suggested that the weight put on defender's rewards and penalties was as high as that put on attacker's rewards and penalties. Finally, EU–α indicated that attackers were risk-averse while lens–3–$\alpha$and lens–5–α indicatedthat attackers were risk-seeking.

Ideographical Analysis. We expected there were individual differences in utility function parameters. For instance, some attackers may have multiple objectives of maximizing expected utility, minimizing the chance of being caught and minimizing their enemies' (defenders) expected utility at the same time (captured in MAU). Some attackers may only maximize their own expected value (captured in EV). It is impossible to differentiate different types of "adversaries" with the nomothetic analysis alone. An ideographical analysis allows us to evaluate how each individual attacker made the decision and how that person is different from others. Again, parameters were estimated using MLE. Since the sample size (N) is small with respect to the number of parameters (k) (N/k < 40 using the k from the most complex model), AICc was calculated for comparisons over different models[17]:

$$AICc = -2 \ln L + 2k \left(\frac{N}{N-k-1}\right) \tag{13}$$

The number of times each model has a minimum AICc is summarized in Table 3. Out of 102 attackers (each playing 40 games) in Experiment I, results indicated that lens–3 scored the minimum AICc most often. MAU model, EU–α and lens–4 also scored the minAICc more often than the other models. In Experiment II, out of 653 attackers (playing 25 games each), results indicated that lens–5 scored the minimum AICc most often; MAU and lens–4 also scored the minimum AICc more often than other models. In Experiment III, out of 294 attackers (each playing 25-33 games), results indicated that lens–5 scored the minimum AICc most often, and lens–4 more often scored the minimum AICc compared to other models. EU and EU–α never scored the minimum AICc across all 294 attackers.

Table 3. Number of Times Model i has Minimum AICc for Experiments I, II and III

	EU	EU – soft target	EU–α	Lens – 3	Lens – 4	Lens – 5	Lens–3–α	Lens–5–α	MAU	Total
Experiment I	7	3	14	28	11	3	7	8	21	102
Experiment II	7	13	6	32	105	282	14	26	168	653
Experiment III	0	29	0	34	56	96	11	33	35	294

4   Discussion and Conclusion

We found that attackers in all three experiments tended to behave consistently with the proposed evaluation functions ($\lambda \to 0^+$). This suggests that in general attackers select targets based on maximizing one of the proposed evaluation functions. The EV model never provided as good a fit as the other eight models, suggesting that the traditional expected value model for an attacker cannot account for adversary choice. Moreover, while model EU–α was superior to model EV, it did not perform as well as the other seven models, suggesting that risk attitude alone does not fully explain adversaries' deviations from EV.

In addition to maximizing attackers' own expected utility, it was found that another predictor of adversaries' choices is defender's payoffs and rewards. In the nomothetic analysis, Experiment I demonstrated that evaluation functions with more parameters (e.g., lens–5) did not fit any better than evaluation functions with fewer parameters (e.g., lens–3). However, in both experiments II and III the evaluation functions with more parameters were better. Model lens–5–α was the best model in experiment I, and lens–5 was the best model in experiments II and III. Both models indicate that attackers take defender's rewards and penalties into account when selecting a target. Additionally, results from experiments II and III indicated a comparable weight of defender's payoffs with the weight of attacker's own payoffs, which implied that attackers gave as much weight to the defenders' rewards and penalties as they did to their own payoffs. The idiographic analysis revealed substantial variability among attackers; however, model lens–5 was found to provide the best fits for the most attackers, consistent with findings from the nomothetic analysis.

We also found that another determinant of adversaries' target selection is the likelihood of success. Participants tended to overvalue the target that was less likely to be guarded. For instance, in the MAU model, which double-counts the probability of success (or probability of being caught) both directly and in the EU calculation, was found to be a competitive model in both the nomothetic analysis and the idiographic analysis. We also found consistently in all three experiments that models accounting directly for success probability (lens models and MAU model) are better than models that account for success probability only in the calculation of EV or EU.

Results from the idiographic analysis indicated that there is no best model among the nine that generally accounts for most of the attackers' choices. Our results suggest that attackers used different evaluation functions to compute the "best" choice in a game. Therefore, individual differences in adversaries (diversity) should be taken into consideration when predicting attacker behavior. It is necessary to identify different types of adversaries in order to predict their choices and to compute optimal strategies for defenders.

References

1. Pita, J., John, R., Maheswaran, R., Tambe, M., Yang, R., Kraus, S.: A robust approach to addressing human adversaries in security games. In: Proceedings of the 11th International Conference on Autonomous Agents and Multiagent Systems, vol. 3 (2012)
2. Yang, R., Kiekintveld, C., Ordóñez, F., Tambe, M., John, R.: Improving resource allocation strategies against human adversaries in security games: An extended study. Artificial Intelligence 195, 440–469 (2012)
3. Nguyen, T.H., Yang, R., Azaria, A., Kraus, S., Tambe, M.: Analyzing the effectiveness of adversary modeling in security games. In: Conference on Artificial Intelligence (2013)
4. Luce, R.D.: Individual choice behavior. John Wiley & Sons, Inc., New York (1959)
5. McFadden, D.L.: Quantal choice analaysis: A survey. Annals of Economic and Social Measurement 5(4), 363–390 (1976)
6. McFadden, D.: Economic choices. American Economic Review, 351–378 (2001)
7. McKelvey, R.D., Palfrey, T.R.: Quantal response equilibria for normal form games. Games and Economic Behavior 10(1), 6–38 (1995)
8. Simon, H.A.: A behavioral model of rational choice. The Quarterly Journal of Economics 69(1), 99–118 (1955)
9. Asal, V.H., Rethemeyer, R.K., Anderson, I., Stein, A., Rizzo, J., Rozea, M.: The softest of targets: A study on terrorist target selection. Journal of Applied Security Research 4(3), 258–278 (2009)
10. Lerner, J.S., Keltner, D.: Beyond valence: Toward a model of emotion-specific influences on judgement and choice. Cognition & Emotion 14(4), 473–493 (2000)
11. Stott, H.P.: Cumulative prospect theory's functional menagerie. Journal of Risk and Uncertainty 32(2), 101–130 (2006)
12. Brunswik, E.: The conceptual framework of psychology, vol. 1. University of Chicago Press (1952)
13. Hammond, K.R.: Probabilistic functioning and the clinical method. Psychological Review 62(4), 255 (1955)

14. Keeney, R.L., Raiffa, H.: Decisions with multiple objectives: Preferences and value trade-offs (1976)
15. Scholz, F.: Maximum likelihood estimation. Encyclopedia of Statistical Sciences (1985)
16. Akaike, H.: A new look at the statistical model identification. IEEE Transactions on Automatic Control 19(6), 716–723 (1974)
17. Burnham, K.P., Anderson, D.R.: Model selection and multimodel inference: A practical information-theoretic approach. Springer (2002)

A Dynamic Bayesian Security Game Framework for Strategic Defense Mechanism Design

Sadegh Farhang[1,*], Mohammad Hossein Manshaei[2],
Milad Nasr Esfahani[2], and Quanyan Zhu[1]

[1] Department of Electrical and Computer Engineering,
Polytechnic School of Engineering, New York University, New York, USA
[2] Department of Electrical and Computer Engineering,
Isfahan University of Technology, Isfahan, Iran
{farhang,quanyan.zhu}@nyu.edu,
{manshaei,m.nasresfahani}@cc.iut.ac.ir

Abstract. In many security problems, service providers are basically unaware of the type of their clients. The client can potentially be an attacker who will launch an attack at any time during their connections to service providers. Our main goal is to provide a general framework for modeling security problems subject to different types of clients connected to service providers. We develop an *incomplete information two-player* game, to capture the interaction between the service provider (i.e., the server) and an unknown client. In particular, we consider two types of clients, i.e., attacker and benign clients. We analyze the game using *perfect Bayesian Nash equilibrium* (PBNE) with different conditions. We finally design an algorithm using the computed PBNE strategy profiles to find the best defense strategy.

1 Introduction

With the rapid deployment of new computing and networking technologies and services, we are witnessing different types of clients having access to service providers via different communication infrastructures, such as the Internet. Service providers (e.g., servers in the Internet) are generally unaware of the type of their clients. These clients could be benign (legitimate) or attacker (malicious). Moreover, there exists different malicious clients with different goals and abilities. This includes but not limited to hackers, crackers, malicious insiders, industrial spy, cybercriminals, hacktivist, and cyber terrorist. In summary, in many security problems the identity of a client is unknown to the server.

Note that if a server only considers legitimate clients (i.e., optimistic point of view) to design its defense mechanism, attackers would breach to the system easily. But the server can provide a good quality of services to benign clients in such cases. On the other hand, if the server assumes that each client is potentially an attacker (i.e., pessimistic point of view), it would degrade the quality of services for the connected benign clients. Therefore, to design optimal defense

---

* S. Farhang was with IUT during part of this research.

R. Poovendran and W. Saad (Eds.): GameSec 2014, LNCS 8840, pp. 319–328, 2014.
© Springer International Publishing Switzerland 2014

mechanism that prevents malicious activities and provides good quality of services to benign clients, we should consider both types of clients simultaneously. Game theory is an appropriate tool that can be used to deal with such problems.

Game theory has been used widely to tackle security issues in computer and communication networks [7,4]. Most security problems are usually modeled between a defender (i.e., server) and an attacker (i.e., client), where the identity of the players is clearly distinguished. However, it is not always possible to assume that the identity of client (i.e., benign or malicious) is known to the server [6,12,10,5]. Game theory enables the server to model its interaction with clients whose identities are unknown to the server [2,13,8]. The main goal of this paper is to propose a new class of security games that can be used to model the interactions between a server and its client which can be either an attacker or a benign client. By using *multi-stage games with observed action and incomplete information*, we capture uncertainties that are dynamically evolving in this type of security problems. This leads to the definition of *perfect Bayesian Nash equilibrium* concept. We apply the computed PBNE to identify server's uncertainty about its clients. Furthermore, we propose the mechanism for the server to prevent the malicious activities of the attacker client as well as provide good quality service to the benign client.

Bayesian games have been used to model the uncertainties of one player about its opponent. In [10], Parunchuri et al. consider Bayesian Stackelberg games to model airport security problem, in which the leader is uncertain about the types of adversary. In this model, leader assigns prior probability to each type of adversary, i.e., follower. However, during the game, these prior probabilities will remain constant. Our model modifies the belief of defender based on the clients' behavior during the game. As an example in network security, Liu et al. [3] use Bayesian games to model the interaction between defender and the connected node that can be malicious or regular in wireless ad hoc network. In this model, the best strategy of defender is computed by only considering malicious nodes. In our model, we consider both types of clients in finding the server's best defense strategy.

This paper is organized as follows. In Section 2, we propose our system model. We analyze the game in Section 3, followed by protocol design in Section 4. Finally, we conclude the paper in Section 5.

2   System Model

In this section, we propose our model for security games when a server is uncertain about type of a client. We name this security game as G^S. Game theory enables us to deal with the lack of knowledge about the identity of players [9]. As shown in Fig. 1, security game G^S is a *two-player* game with a server S and client C as players. Each player $i \in \{S, C\}$ has a type θ_i in a finite set Θ_i.

In the security game G^S, we consider two types ($\theta_C = 0$ denotes benign client and $\theta_C = 1$ denotes attacker one) for clients of our server. Indeed, in our game the server type is always $\theta_S = 0$. The nature of communications between server

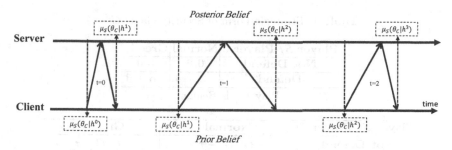

Fig. 1. G^S is a two-player repeated game between server and client. Client could be either benign or attacker. At each stage (with various duration), the server updates its belief about the type of its client based on the client's current action and the history.

and client is repeated and played at stages $t = 0, 1, 2, \ldots, T$. Different packets are sent by the client to the server and vice versa. Each packet and the corresponding response can be considered as one stage of the security game G^S.

We model security game G^S using *multi-stage games with observed actions and incomplete information* [1] to deal with incomplete information about client C's type. Our security game G^S is perfect information since each player can observe the action of another player. Moreover, the server does not know the identity of client. So, security game G^S is incomplete information. Note that different G^S games are played in parallel when different clients are connected to the server at the same time.

In G^S game, h^t is the history at the beginning of stage t. History denotes actions of all players in all previous stages until stage t. h^0 is the history at the beginning of the game. h^1 is the history at the end of stage zero and denotes actions of both players at stage zero. Similarly, h^2 represents the actions of both players at stages zero and one.

The server has a prior knowledge about the type of its clients, i.e., $\mu_S(\theta_C|h^0)$. In other words, this prior knowledge is the belief of server about type of its clients at the beginning of the game. In general, $\mu_S(\theta_C|h^t)$ is the belief of server about the type of player C at time t. When $t > 0$, this belief is also called posterior probability. This belief could be potentially updated at each stage. Players consider history, action of other players at this stage and the belief at the previous stage to update their beliefs about their opponents. Bayes' rule is used to update belief at the end of each stage. In stage zero, server uses its prior knowledge besides client's action to update its belief at the end of the stage. In stage 1, server updates its belief using Bayes' rule, $\mu_S(\theta_C|h^0)$, h^1, and client's current action. In other stages, for example stage t, server updates its belief similar to stage 1, i.e., using Bayes' rule, $\mu_S(\theta_C|h^{t-1})$, h^t, and client's current action. In the rest of this section, we first define both players' strategies and then calculate players' payoffs.

In our security game G^S presnted in Table 1, the strategy set of player C, i.e., client, is limited to *Greedy* and *Normal*, i.e., $s_C = \{Greedy, Normal\}$, and the server should only select between two strategies of *Defend* or *Not Defend*, i.e., $s_S = \{Defend, Not D - efend\}$. Let's explain how strategies can be defined with two security examples.

Table 1. Strategic Form of Security Game G^S

Player S/ Player C	Normal	Greedy
Not Defend	$0, 0$	$0, 0$
Defend	$-\alpha, -\beta$	$-\alpha, -\beta$

Player C is Benign

Player S/ Player C	Normal	Greedy
Not Defend	$-G', G' - \tau'$	$-G, G - \tau$
Defend	$g' - G' - \alpha, G' - g' - \tau'$	$g - G - \alpha, G - g - \tau$

Player C is Attacker

The first example is Intrusion Detection Systems (IDS) in mobile ad hoc network which has been presented in [6]. In this case, *Greedy* strategy means that client sends more packets than a certain threshold to the server. From server's point of view, *Defend* strategy could be interpreted as *monitoring* the client. The Second example is password reset which is presented in [11]. In this example, *Defend* strategy is *moving* into a new state. In other words, server changes the password because the attacker might have penetrated to its system. Moreover, *Greedy* strategy means that client tries to penetrate server's system by examining different passwords.

Our security game G^S is a two-player repeated game between a server and a client as players. The server is uncertain about type of its client which could be either benign or attacker. We assume that the players' identities remain consistent throughout the game. To calculate server's payoff, we consider a constant cost for its *Defend* strategy, i.e., α, regardless of the type of player C. On the other hand, we do not consider a cost for server's strategy of *Not Defend*.

Benign type of player C, i.e., $\theta_C = 0$, might play *Greedy*. If the server is smart enough, it will not play *Defend*. Server sometimes plays *Defend* against benign type of player C due to lack of sufficient information. This server's action leads to degradation in service or problem in communication to the benign client. We quantify this degradation or problem by β. Note that there does not exist any difference between *Greedy* strategy and *Normal* strategy of benign type of player C. Because, the goal of server is to provide good quality service to benign clients.

Following our discussion in payoff calculation, now we consider attacker type of player C, i.e., $\theta_C = 1$. We represent attacker's cost of playing *Greedy* and *Normal* by τ and τ', respectively. In some cases, playing *Normal* leads to spend more time to launch successful attack. Therefore, alarming tools, such as IDS, will be suspicious about type of this client. So, we do not consider $\tau' = 0$ and we assume that $\tau' \geq \tau$.

If every step of game is done completely, the attack will be successful. Some steps of attack are necessary for successful attack. Let us assume that each step of attack gives attacker fraction of information for successful attack. We define G as the information that is gained by attacker when attacker plays *Greedy*. Similarly, we define G' as the attacker's gain of information when it plays *Normal*. One can simplify the model by not considering the attacker's gain of information when playing *Normal* strategy. Note that, attacker's gain of information when playing *Greedy* is not lower than attacker's gain when playing *Normal*, i.e., $G \geq G'$.

In our security game G^S, playing *Defend* has a cost for server as well as prevents leakage of information to the attacker. This prevention must be different with respect to attacker's strategy. So, we define g and g' as prevention from information's leakage when attacker's strategies are *Greedy* or *Normal*, respectively. We assume that $g \geq g'$.

3   Security Game Analysis

In this section, we analyze the security game G^S to propose optimum probability of playing *Defend* strategy for the server where it is uncertain about type of its client. We find the best responses of players in Lemmas 1, 2, and 3. Furthermore, Conjecture 1 shows that how the server can distinguish between the attacker or the benign type of player C by using Bayes' rule.

First, we show that four requirements in the definition of PBNE, i.e., B(i)-B(iv), are satisfied for our security game G^S (Please see [1] for the definition of PBNE and its four requirements). B(i) is satisfied, because server has one type. B(ii) is satisfied, since we use Bayes' rule to update server's belief. The action of the server does not have any impact on the belief of the server about the type of its client. In other words, $\mu_S(\theta_C|h^t)$ is just affected by the action of client C. Therefore, B(iii) is also satisfied. Finally, B(iv) is satisfied, because this game is a two-player game.

Let us define the following parameters to simplify the representations of the actions' probabilities given players' actions as well as the type of clients:

$$
\begin{aligned}
r_0 &:= \sigma_S(a_S^t = Defend|h^t, \theta_C = 0) \\
r_1 &:= \sigma_S(a_S^t = Defend|h^t, \theta_C = 1) \\
q &:= \sigma_C(a_C^t = Greedy|h^t, \theta_C = 0) \\
p &:= \sigma_C(a_C^t = Greedy|h^t, \theta_C = 1) \\
r &:= \sigma_S(a_S^t = Defend|h^t)
\end{aligned}
\tag{1}
$$

Let's first assume that the server knows the type of its client. Lemma 1 shows the best strategy of the server in such cases (All proofs can be found in Appendix A.).

Lemma 1 *In our security game G^S, if the server knows that its client is an attacker, then it defends with probability equal to r_1^* given in Table 2. Otherwise, it does not defend, i.e., $r_0^* = 0$.*

Lemma 1 identifies five different cases, given that the server knows that its client is attacker. In each case, it shows that how the server plays *Defend*.

• **Cheap Defense:** in this state, server plays *Defend* in all stages against an attacker. Because the cost of playing *Defend* is lower than what server acquires when playing *Defend*, i.e., $\alpha < g' \leq g$.

Table 2. The best *Defend* strategy, given different power of client and the cost of defend, when the server knows that its client is an attacker

Defense State	Condition	r_1^*
Cheap Defense	$\alpha < g'$	1
Expensive Defense	$\alpha > g$	0
Greedy	$G - g - \tau > G' - g' - \tau'$ & $g' \leq \alpha \leq g$	1
Uncertain	$G - g - \tau \leq G' - g' - \tau'$ & $g' \leq \alpha \leq g$ & $g > g'$	$\frac{(G-\tau)-(G'-\tau')}{g-g'}$
Baffled	$G - g - \tau \leq G' - g' - \tau'$ & $g' = \alpha = g$	Any Probability

• **Expensive Defense:** in this state, cost of playing *Defend* is greater than what server acquires when playing *Defend*, i.e., $\alpha > g \geq g'$. Therefore, server does not play *Defend* at all.

• **Greedy:** in this state, attacker always plays *Greedy* and consequently, the server plays *Defend* in all stages.

• **Uncertain:** in this state, server plays *Defend* by certain probability. In this condition, if attacker plays *Greedy*, server will play *Defend*. Moreover, server plays *Not Defend* when attacker plays *Normal*.

• **Baffled:** in this state, there is no difference between *Normal* and *Greedy* strategy of attacker. Similarly, there is no difference between *Defend* and *Not Defend* strategy of server. Hence, server can play *Defend* by any probability.

Lemma 2 *In the security game G^S, the benign type of player C plays Greedy by any probability, i.e., q^*.*

Lemma 2 states that the behavior of benign type is independent from the belief of the server, as there is no difference between *Normal* and *Greedy* strategy of benign client. The server does not know the type of its client. The server has a belief about the type of its client. Attacker uses this belief to find its best response. Lemma 3 represents the best response of the attacker in different conditions.

Lemma 3 *In our security game G^S, the attacker plays Greedy with probability p^*:*

$$p^* = \begin{cases} median \{0, \frac{\alpha - \mu_S(\theta_C = 1|h^t)g'}{\mu_S(\theta_C = 1|h^t)(g-g')}, 1\} & if \ g > g' \\ 1 & if \ g = g' \ \& \ G > G' \\ 1 & if \ g = g' \ \& \ \tau < \tau' \\ any \ probability & if \ g = g' \ \& \ \tau = \tau' \ \& \ G = G'. \end{cases} \quad (2)$$

Note that the attacker knows that the server is uncertain about type of its clients. Server only has a belief about the type of its client, i.e., $\mu_S(\theta_C = 1|h^t)$. Therefore, attacker uses this belief to calculate its best response, i.e., p^*. The higher the belief is, the lower the p^* is. In other words, attacker tries to decrease belief of server in the next stage by playing *Greedy* with lower probability.

Server updates its belief about the type of client at the end of each stage by using Bayes' rule. Note that server uses p^* and q^* according to Lemma 2 and Lemma 3, respectively, in Bayes' rule to update its belief.

Finally, Conjecture 1 shows the optimum strategy of server in which it is uncertain about type of its client.

Conjecture 1 *In our security game* G^S, *server must play Defends with probability* r^*:

$$r^* = r_1^* \mu_S(\theta_C = 1|h^t) + r_0^* \mu_S(\theta_C = 0|h^t) \tag{3}$$

Where r_1^* *and* r_0^* *are calculated according to Lemma 1. The belief of server about the type of its client is calculated based on Bayes' rule.*

Note that in Conjecture 1 contrary to Lemma 1, we consider both types of player C in calculating probability of playing *Defend*. In Equation (3), the server's belief weights the probability of playing *Defend* given that the server knows the identity of its client. For example, when $\mu_S(\theta_C = 1|h^t)$ is high (low), i.e., more probable that the client is attacker (benign), server plays *Defend* (r) with higher (lower) probability. In other words, the higher the $\mu_S(\theta_C = 1|h^t)$ is, the higher the r is.

4   Protocol Design

The above results provides guidelines for designing a defense mechanism named *SmartTypeDetector*, enabling the server to prevent malicious activities of the attacker while providing service to the benign clients. In other words, we employ our results for optimal *Defend* strategy presented in Conjecture 1 to compute the probability of *Defend* strategy, i.e., r. Note that, one G^S game is played for one client and different clients are independent from each other. In summary, the server finds p^* and q^* at each stage when the client plays *Greedy*. When the client plays *Normal*, the server calculates $1 - p^*$ and $1 - q^*$. The server uses these probabilities, Bayes' rule, and its belief in the previous stage to update its belief (update μ). The server calculates r according to Conjecture 1 where the server's belief has important influence in r.

Let's consider a situation of the game G^S between server and the attacker in which $p^* = 1$. Rational attacker will always play *Greedy*. But, irrational attacker may play *Normal* strategy in some stages. If the attacker plays *Normal* in this situation, the server's belief will be equal to zero and remain constant for the rest of the game. To avoid irrational behavior of the attacker, we also apply upper and lower bounds on p^*.

Algorithm 1. SmartTypeDetector

1: run this algorithm for each stage
2: **if** the client plays Greedy **then**
3:     find q^* (i.e., Lemma 2)
4:     find p^* (i.e., Lemma 3)
5: **else**
6:     find $1 - q^*$ (i.e., Lemma 2)
7:     find $1 - p^*$ (i.e., Lemma 3)
8: update μ
9: calculate r (according to Conjecture 1)
10: $A = rand$ (random number with uniform distribution in [0,1])
11: **if** $r \geq A$ **then**
12:     Defend
13: **else**
14:     Not Defend

---

5   Conclusion

In this paper, we have proposed a Bayesian security game framework to tackle with lack of knowledge about the type of the server's client. In our game-theoretic model, the game is between server and its client which could be either benign or attacker. We analyzed the game using *perfect Bayesian Nash equilibrium* concept and proposed SmartTypeDetector algorithm, based on our PBNE calculation. In this algorithm, server uses its belief about the identity of its client to determine which client is connected to the server. This framework can be applied in many security problems, such as OS fingerprinting attack and IDS. We believe that the framework is an efficient tool to model security problems in real life, where defender does not have enough information about the type of attackers.

References

1. Fudenberg, D., Tirole, J.: Game theory. MIT Press (1991)
2. Jain, M., An, B., Tambe, M.: Security games applied to real-world: Research contributions and challenges. In: Moving Target Defense II, pp. 15–39. Springer (2013)
3. Jin, X., Pissinou, N., Pumpichet, S., Kamhoua, C.A., Kwiat, K.: Modeling cooperative, selfish and malicious behaviors for trajectory privacy preservation using bayesian game theory. In: Local Computer Networks (LCN), pp. 835–842. IEEE (2013)
4. Liang, X., Xiao, Y.: Game theory for network security. IEEE Communications Surveys & Tutorials 15(1), 472–486 (2013)
5. Lin, J., Liu, P., Jing, J.: Using signaling games to model the multi-step attack-defense scenarios on confidentiality. In: Grossklags, J., Walrand, J. (eds.) GameSec 2012. LNCS, vol. 7638, pp. 118–137. Springer, Heidelberg (2012)
6. Liu, Y., Comaniciu, C., Man, H.: A bayesian game approach for intrusion detection in wireless ad hoc networks. In: Proceeding from the 2006 Workshop on Game Theory for Communications and Networks, p. 4. ACM (2006)
7. Manshaei, M.H., Zhu, Q., Alpcan, T., Bacşar, T., Hubaux, J.-P.: Game theory meets network security and privacy. ACM Comput. Surv. 45(3), 1–39 (2013)

8. Nguyen, K.C., Alpcan, T., Basar, T.: Security games with incomplete information. In: International Conference on Communications (ICC), pp. 1–6. IEEE (2009)
9. Osborne, M.J.: An introduction to game theory, vol. 3. Oxford University Press, New York (2004)
10. Paruchuri, P., Pearce, J.P., Marecki, J., Tambe, M., Ordonez, F., Kraus, S.: Playing games for security: An efficient exact algorithm for solving bayesian stackelberg games. In: Proceedings of AAMAS 2008, pp. 895–902 (2008)
11. Pham, V., Cid, C.: Are we compromised? Modelling security assessment games. In: Grossklags, J., Walrand, J. (eds.) GameSec 2012. LNCS, vol. 7638, pp. 234–247. Springer, Heidelberg (2012)
12. Rahman, M.A., Manshaei, M.H., Al-Shaer, E.: A game-theoretic approach for deceiving remote operating system fingerprinting. In: IEEE CNS, pp. 73–81 (2013)
13. Tsai, J., Kiekintveld, C., Ordonez, F., Tambe, M., Rathi, S.: Iris-a tool for strategic security allocation in transportation networks (2009)

A Proof of Lemmas

Proof of Lemma 1: If the server knows that its client is benign, *Defend* strategy is strictly dominated by *Not Defend* strategy. So, probability of *Defend* given that the server knows that dealing with benign client is equal to zero, i.e., $r_0^* = 0$. On the other hand, when the server knows that its client is attacker, r_1^* is calculated as follows:

• **Cheap Defense** ($\alpha < g' \leq g$): the server's dominant strategy is to play *Defend*, i.e., $r_1^* = 1$.

• **Expensive Defense** ($\alpha > g \geq g'$): the *Defend* strategy of the server is strictly dominated by *Not Defend* strategy, i.e, $r_1^* = 0$.

• **Greedy** ($G - g - \tau > G' - g' - \tau'$ and $g' \leq \alpha \leq g$): attacker's dominant strategy is to play *Greedy*. So, the server will play *Defend*, i.e., $r_1^* = 0$.

• **Uncertain** ($G - g - \tau \leq G' - g' - \tau'$ and $g' \leq \alpha \leq g$ and $g > g'$): in this condition, there does not exist any dominant or dominated strategy. To determine r_1^*, first we calculate attacker's expected payoff by playing *Normal* when the server plays its mixed strategy:

$$Eu_C[((r_1, 1 - r_1), Normal|\theta_C = 1)] = r_1(G' - g' - \tau') + (1 - r_1)(G' - \tau') \quad (4)$$

Then, attacker's expected payoff by playing *Greedy* when the server plays it mixed strategy is:

$$Eu_C[((r_1, 1 - r_1), Greedy|\theta_C = 1)] = r_1(G - g - \tau) + (1 - r_1)(G - \tau) \quad (5)$$

We derived r_1^* by setting the Equations (4) and (5) equal:

$$r_1^* = \frac{(G - \tau) - (G' - \tau')}{g - g'} \quad (6)$$

• **Baffled** ($G - g - \tau \leq G' - g' - \tau'$ and $g' \leq \alpha \leq g$ and $g = g'$): as presented in Section 2, we assume that $G - G' \geq 0$ and $\tau - \tau' \leq 0$. The condition $G - g - \tau \leq$

$G' - g' - \tau'$ could be written as $G - G' \leq \tau - \tau'$. The left side of this inequality is nonnegative, while the right side is nonpositive. So, both sides must be equal to zero, i.e., $G = G'$ and $\tau = \tau'$. In Baffled, there is no difference between server's strategies as well as attacker's. Hence, the server could play *Defend* by any probability.

Proof of Lemma 2: There is no difference between *Normal* and *Greedy* strategy of the benign client. So, the benign client could play *Greedy* by any probability.

Proof of Lemma 3: To calculate probability of playing given that the client is attacker, we consider following conditions:

- $g > g'$: The server's expected payoff for playing *Defend* given that both types of its client playing their mixed strategy is calculated as:

$$Eu_S(Defend) = \mu_S(\theta_C = 1|h^t)(p(g - G - \alpha) + (1 - p)(g' - G' - \alpha)) \\ + \mu_S(\theta_C = 0|h^t)(-\alpha) \tag{7}$$

And the server's expected payoff for playing *Not Defend* when both types of its client playing their mixed strategy is:

$$Eu_S(Not\,Defend) = \mu_S(\theta_C = 1|h^t)(p(-G) + (1 - p)(-G')) \tag{8}$$

Note that in Equations (7) and (8), these expected payoffs are not function of q. Since, there is no difference between *Normal* and *Greedy* strategy of the benign client regardless of the server's actions.

The attacker chooses p^* to keep the server indifferent between *Defend* and *Not Defend* strategy. p^* is derived by setting Equations (7) and (8) equal, i.e.,

$$p^* = \frac{\alpha - \mu_S(\theta_C = 1|h^t)g'}{\mu_S(\theta_C = 1|h^t)(g - g')} \tag{9}$$

In Equation (9), p^* is function of $\mu_S(\theta_C = 1|h^t)$. If $\mu_S(\theta_C = 1|h^t) < \frac{\alpha}{g}$, p^* is bigger than 1. In this situation, we use $p^* = 1$. Moreover, p^* is less than 0 when $\mu_S(\theta_C = 1|h^t) > \frac{\alpha}{g}$. In this situation, we use $p^* = 0$. Hence, we have *median* $\{0, \frac{\alpha - \mu_S(\theta_C=1|h^t)g'}{\mu_S(\theta_C=1|h^t)(g-g')}, 1\}$. Where, the median of a finite list of numbers can be found by arranging all the numbers from lowest value to highest value and picking the middle one.

- $g = g'$ and $G > G'$: *Normal* strategy of the attacker is strictly dominated by *Greedy*, i.e., $p^* = 1$.
- $g = g'$ and $\tau < \tau'$: *Normal* strategy of the attacker is strictly dominated by *Greedy*, i.e., $p^* = 1$.
- $g = g'$ and $\tau = \tau'$ and $G = G'$: there is no difference between *Normal* and *Greedy* strategy of the attacker, i.e., $p^* = Any\,probability$.

Can Less Be More? A Game-Theoretic Analysis of Filtering vs. Investment

Armin Sarabi, Parinaz Naghizadeh, and Mingyan Liu

University of Michigan, Ann Arbor, USA
{arsarabi,naghizad,mingyan}@umich.edu

Abstract. In this paper we consider a single resource constrained strategic adversary, who can arbitrarily distribute his resources over a set of nodes controlled by a single defender. The defender can (1) instruct nodes to filter incoming traffic from another node to reduce the chances of being compromised due to malicious traffic originating from that node, or (2) choose an amount of investment in security for each node in order to directly reduce loss, regardless of the origin of malicious traffic; leading to a *filtering* and an *investment* game, respectively. We shall derive and compare the Nash equilibria of both games for different resource constraints on the attacker. Our analysis and simulation results show that from either the attacker or the defender's point of view, none of the games perform uniformly better than the other, as utilities drawn at the equilibria are dependent on the costs associated with each action and the amount of resources available to the attacker. More interestingly, in games with highly resourceful attackers, not only the defender sustains higher loss, but the adversary is also at a disadvantage compared to less resourceful attackers.

1 Introduction

The continuous attempts by malicious entities to discover and exploit security vulnerabilities in networks and the ensuing efforts of network administrators at patching up such exploits have evolved into a cat and mouse game between attackers and defenders. In addition to research on mitigating security flaws and building more robust networks by analyzing specific hardware and software involved in a network, the problem has also been addressed by game theorists. Game theory provides a broad framework to model the behavior of rational parties involved in a competitive setting, where each party seeks to maximize their own net worth. For instance, the interdependent nature of cyber-security leads to numerous studies on games describing the behavior of multiple interdependent agents protecting their assets in a network [8].

In this paper we study the strategic interaction between an attacker and a defender[1], both taking actions over a set of interconnected nodes or entities. The

---

[1] For the remainder of the paper, to eliminate confusion, we will use *he/him* to refer to the attacker, and *she/her* to refer to the defender.

R. Poovendran and W. Saad (Eds.): GameSec 2014, LNCS 8840, pp. 329–339, 2014.
© Springer International Publishing Switzerland 2014

attacker is resource limited but can arbitrarily spread his resources or effort over this set of nodes. The amount of effort exerted over a node determines the attacker's likelihood of infiltrating the node and inflicting a certain amount of loss; a compromised node can go on to contaminate other connected nodes to inflict further loss. From the defender's point of view, the interactions between nodes present possible security risks, but also value derived from the communication.

We consider two types of actions the defender can take. The first is inbound filtering, whereby a certain amount of traffic from another node is blocked. This is routinely done in practice, through devices such as firewalls and spam filters, based on information provided by sources such as host reputation blacklists (RBLs) [7, 9], where traffic originating from IP addresses suspected of malicious activities (listed by the RBLs) are deemed unsafe and blocked. Ideally, if the defender could distinguish between malicious and innocuous traffic, she could block all malicious traffic and achieve perfect security. However, blocking traffic comes at a price, since no detection mechanism is without false alarms. Thus, filtering decisions leads to tradeoffs between balancing security risks and communication values. The second type of action is self-protection through investing in security. In this case the defender foregoes filtering, but instead focuses on improving its ability to resist malicious effort in the presence of tainted traffic. Self-protection is more costly than inbound filtering, but it does not put legitimate communication at risk since it carries no false alarms.

These two types of actions result in a *filtering game* and an *investment game*, respectively, which we analyze in this paper. Specifically, we derive Nash equilibria in both scenarios. We shall see that for both games, more powerful attackers, or those with larger amounts of resources, do not necessarily draw more utility at the equilibrium. By contrast, a defender always prefers to face less powerful attackers. In addition, we will compare these two games, and conclude that highly resourceful attackers favor facing a defender that invests, while less resourceful attackers' preference depends on the cost of security investments.

Most of the existing literature on interdependent security games focus on a collection of agents responding to a constant exogenous attempt to breach their systems and inflict damage, while fewer publications have addressed games with a strategic adversary. In reality, malicious sources have shown highly strategic behavior. For instance, in November 2008 the McColo ISP was effectively blocked by the rest of the Internet due to its massive operation in spam, and its takedown was estimated to have contributed to a two-thirds reduction in global spam traffic in the immediate aftermath [2]. However, by the second half of March, the seven-day average spam volume was back at the same volume seen prior to the blocking of McColo ISP [1]. In other words, if a defender decides to completely secure her assets, then the attacker will likely respond strategically by redirecting resources.

Studies on strategic attackers and most relevant to the present paper include [3–6, 10]. Specifically, in [3] Fultz and Grossklags propose a complete information game consisting of a single attacker and N defenders, where defenders can decide to protect their systems through security measures and/or self-insurance. The attacker is assumed to decide only on the number of targeted nodes, with the intensity being equal among all. In [10] Nochenson and Heimann consider a single attacker competing with a single defender in a game of incomplete information, where the players can only choose from a set of action classes (e.g. protecting the highest value node, protecting proportional to nodes' values, etc). In [4–6], Hausken considers one-shot and sequential attacker-defender games, under different assumptions on independent and interdependent security models, attacker income and substitution effects, and so on.

Compared to the above references, the present paper examines a network with a large number of nodes, where the attacker can spread his efforts over the network arbitrarily. Moreover, the utility models studied herein differ from those in [4–6]. Perhaps most importantly, our study complements existing literature by considering filtering actions, in addition to the security investment actions, in order to evaluate and compare the effectiveness of security measures and blacklisting against strategic attackers from a game theoretical point of view.

In the remainder of the paper we present our model, provide intuition on how it relates to current cyber-security practices, and derive the Nash equilibria of games under discussion. We will then simulate, discuss and compare the games and their respective equilibria. The proof of the theorems are omitted for brevity.

2 Filtering

We consider a network consisting of N inter-connected nodes. There is a single attacker and a single defender both acting over these nodes. The attacker has a fixed amount of resources he can use toward compromising any subset of the N nodes. A compromised node sustains a certain amount of (direct) loss; a compromised node is also assumed to inflict further (indirect) losses on nodes it communicates with, thus modeling interdependence. On the defender's side, one mitigating option is inbound/outbound filtering over these nodes. Filtering traffic can effectively reduce the amount of malicious traffic received by a node, thereby reducing its probability of being compromised, or the incurred losses. The extreme form of filtering is *takedown*, whereby traffic from a node is completely blocked, effectively isolating this node from the rest of the network. An advantage of filtering is low cost; it takes relatively little to perform inbound filtering, and we will assume its cost is zero in our analysis. The downside of filtering is false positives, which reduce the *value* represented by communication between two nodes; this aspect is explicitly modeled in this case.

Following the discussion above, the defender's actions can be modeled by a vector $f \in [0,1]^N$, where f_i is the percentage of node i's outgoing traffic that is being dropped. We assume this filtering is performed uniformly, either by outbound filtering across all egress points, or inbound filtering done by all other nodes which have agreed upon the same filtering level. In reality, this corresponds to the observation that filtering decisions are often source-based rather that destination-based. The attacker's actions are modeled by a vector $r \in \mathbb{R}_+^N$, where r_i is the amount of effort spent by the attacker to breach node i. Increased effort exerted over a node leads to increased losses (e.g., through increased probability of compromising a node). The total amount of loss inflicted constitutes the attacker's profit. We further assume that the attacker has a limited amount of resources r, so that $\sum_{i=1}^{N} r_i \leq r$.

We will adopt the simplification of only considering indirect losses. The justification is that in large networks, the amount of direct loss sustained by a node is negligible compared to the total indirect losses it can inflict on the network. In a sense, the attacker's main objective is to contaminate a large set of nodes through network effects, rather than drawing utility from compromising selected nodes. Let L_{ij} denote the maximum loss per unit of effort that can be inflicted on node j through a breached node i, when node i's traffic is unfiltered. Note that filtering the traffic leaving a breached node does not protect the node itself against losses; it protects to some degree the rest of the network from indirect losses from that node. Thus, the attacker's utility is given by:

$$u_a^F(\boldsymbol{r}, \boldsymbol{f}) = \sum_{i=1}^{N} r_i \sum_{\substack{j=1 \\ j \neq i}}^{N} L_{ij} g^F(f_i) \ , \ \text{s.t.} \ \sum_{i=1}^{N} r_i \leq r \ . \tag{1}$$

Here, $g^F : [0,1] \to [0,1]$ is a risk function with respect to the filtering policy, which we will take to be linear ($g^F(f_i) = 1 - f_i$). To further illustrate, it is more natural to view a node as a network (a collection of individual machines or IP addresses); in this case the single defender becomes a convenient way to model consistent actions taken by different networks against other networks of known malicious activities. For instance, benign networks may adopt similar inbound filtering policies against a network known to send out large quantities of malicious traffic (e.g., given by the reputation blacklists). More specifically, a network may decide that all traffic from another network with a certain presence on the RBLs (percentage of its IPs listed) shall be filtered at a certain level (with some probability). In this case, the filtering level leads to linear reduction in risk and loss in value for the node. Alternatively, a network may decide that all traffic from listed IPs shall be blocked, in which case the amount of filtering is equal to the fraction of blacklisted IP addresses. However, with this interpretation, the reduction in risk and loss in value are no longer linear with respect to filtering levels. This is because targeted filtering is presumably more accurate, leading to higher risk reduction. We will revisit this case after deriving the equilibrium of our game, and explain how the results might also hold for the nonlinear case.

Define $L_i := \sum_{j \neq i} L_{ij}$ as the total indirect loss incurred by node i. We assume without loss of generality that users are indexed such that L_i is a decreasing sequence. Equation (1) can then be re-written as:

$$u_a^F(\boldsymbol{r}, \boldsymbol{f}) = \sum_{i=1}^{N} r_i L_i (1 - f_i) \ , \ \text{s.t.} \ \sum_{i=1}^{N} r_i \leq r \ . \tag{2}$$

From the defender's viewpoint, let V_i be the value associated with node i's traffic. Similar to the definition of L_i, $V_i = \sum_{j \neq i} V_{ij}$ is the value of traffic from node i to the rest of the nodes. Note that by filtering inbound traffic, the defender is inevitably losing a portion of a node's value, as she is filtering parts of the legitimate traffic as well. The defender's utility is thus given by:

$$u_d^F(\boldsymbol{r}, \boldsymbol{f}) = -u_a^F(\boldsymbol{r}, \boldsymbol{f}) + \sum_{i=1}^{N} V_i(1 - f_i) . \qquad (3)$$

Together, $\mathcal{G}^F := \langle(\text{attacker, defender}), (\boldsymbol{r}, \boldsymbol{x}), (u_a^F, u_d^F)\rangle$ defines a one-shot simultaneous move filtering game with perfect information between an attacker and a defender. Theorem 1 characterizes the Nash equilibrium of the game $\mathcal{G}^F$.

Theorem 1. *Assume $r \leq \sum V_i/L_i$. Define k to be the smallest integer such that $r \leq \sum_{i=1}^{k} V_i/L_i$. Define vectors $\boldsymbol{r}^*, \boldsymbol{f}^*$ as follows:*

$$(r_i^*, f_i^*) = \begin{cases} \left(\frac{V_i}{L_i}, 1 - \frac{L_k}{L_i}\right) & i < k , \\ \left(r - \sum_{j<k} r_j^*, 0\right) & i = k , \\ (0, 0) & i > k . \end{cases} \qquad (4)$$

Then $(\boldsymbol{r}^*, \boldsymbol{f}^*)$ forms a Nash equilibrium for $\mathcal{G}^F$. Also if $L_i \neq L_j$ for $i \neq j$, and $\sum_{i=1}^{k} V_i/L_i \neq r$, then this Nash equilibrium is unique.

For $r > \sum V_i/L_i$, any $\boldsymbol{r}$ such that $r_i \geq V_i/L_i$ can constitute an NE. The defender's response in such equilibria is $f_i = 1$ for all i.

Note that at the Nash equilibrium, $u_a^F(\boldsymbol{r}^*, \boldsymbol{f}^*) = rL_k$. Therefore, the efficiency of the attacker is equal to L_k, where k is the strongest node under attack. It is also worth noting that the attacker will only dedicate a maximum of V_i/L_i of resources to a node i; since beyond this point, the defender would filter that node completely. Consequently, V_i/L_i can be viewed as the capacity, or saturation point, of each node, while $\sum V_i/L_i$ is the capacity of the network. When direct losses are not negligible, but still less that the total indirect losses, L_i can be redefined to include direct losses, and the results of Theorem 1 would still hold.

The game presented in this section can be viewed as a probabilistic filtering game. In other words, f_i represents the probability of blocking each unit of node i's outgoing traffic. It is also possible to consider the non-probabilistic, or binary, version of this game, where the defender's action space is $\{0, 1\}^N$. However, such games do not generally have a pure strategy Nash equilibrium. Another interesting observation is that at the NE, no nodes are being completely blocked. In fact, the maximum filtering level is $f_1 = 1 - L_k/L_1$. If this maximum is sufficiently small, then our assumption on the linearity of g^F is justified.

While our model does not restrict the type of malicious activities the attacker engages in, it helps to interpret the model in a more specific application context. We will use spam as an example. In this case the "single" attacker more aptly models a single spam campaign orchestrated by certain entity or entities. The attacker's effort translates into attempts toward acquiring bandwidth or processing power from a machine, either by purchasing or hijacking it. The indirect loss inflicted on other machines by an infected machine includes resources spent in processing or acting on spam traffic (e.g., from running the spam filter, storage, reading spams, to possible economic losses when taken in by spams).

3 Investment

In this section, we consider a similar strategic game between the defender and the attacker. However, the defender's action here is to choose a level of protection, effort, or investment in security, for each node, in order to mitigate the attacks. More precisely, the defender can choose to invest an amount $x_i \in [0,1]$ on node i's security. This investment in turn decreases the effectiveness of the attacker's effort. The defender incurs a cost of $c_i > 0$ per unit of investment. Investing at level $x_i = 1$ is assumed to provide node i with perfect protection. The attacker's utility when the defender invests in security measures is given by:

$$u_a^I(\boldsymbol{r}, \boldsymbol{x}) = \sum_{i=1}^{N} r_i(1 - x_i) \sum_{\substack{j=1 \\ j \neq i}}^{N} L_{ij}(1 - x_j), \text{ s.t. } \sum_{i=1}^{N} r_i \leq r . \tag{5}$$

Here, L_{ij} is the loss inflicted on node j per unit of attack on node i, when both are unprotected. The utility of the defender is given by:

$$u_d^I(\boldsymbol{r}, \boldsymbol{x}) = -u_a^I(\boldsymbol{r}, \boldsymbol{x}) - \sum_{i=1}^{N} c_i x_i . \tag{6}$$

We refer to the game $\mathcal{G}^I := \langle (\text{attacker, defender}), (\boldsymbol{r}, \boldsymbol{x}), (u_a^I, u_d^I) \rangle$ as the one-shot investment game with perfect information between an attacker and a defender.

In order to choose an optimal action, each player solves the KKT conditions for their respective optimization problem, assuming the other player's action is given. Therefore, at an NE, the following sets of conditions have to be satisfied:

$$
\begin{cases}
(1 - x_i) \sum_{j \neq i} L_{ij}(1 - x_j) + \lambda_i - \eta = 0 , & \text{(7a)} \\[2ex]
\sum_{j \neq i} (r_i L_{ij} + r_j L_{ji})(1 - x_j) - c_i + \mu_i - \nu_i = 0 , & \text{(7b)} \\[2ex]
\lambda_i r_i = 0, \ \mu_i x_i = 0, \ \nu_i(1 - x_i) = 0 , & \text{(7c)} \\[2ex]
\eta \left(\sum_{i=1}^{N} r_i - r \right) = 0, \ \sum_{i=1}^{N} r_i \leq r , & \text{(7d)} \\[2ex]
r_i, \lambda_i, \mu_i, \nu_i, \eta \geq 0, \ 0 \leq x_i \leq 1 . & \text{(7e)}
\end{cases}
$$

A solution to the above system of equations indicates a Nash equilibrium for a given problem instance. Note that this problem has at least one NE, as the utilities are linear, and the action spaces are convex and compact [11]. To provide intuition on the properties of the equilibria of the game $\mathcal{G}^I$, we next propose a set of conditions on the problem parameters to simplify the KKT conditions in (7a)-(7e). We will then find the Nash equilibrium of the simplified game, and study its properties and dependence on the problem parameters.

Assumption 1. *For all i, and $j \neq i$, the loss L_{ij} can be written as $L_{ij} = \alpha_i \Lambda_j$, where Λ_j is a parameter that quantifies the size of the target j, while α_i models the importance of node i.*

Assumption 2. *For all i, the unit cost of security investment c_i is proportional to the size of the node Λ_i, i.e., c_i/Λ_i is a constant, $\forall i$.*

Without loss of generality, assume users are indexed such that α_i is a decreasing sequence, and that $\sum \Lambda_j = 1$. Then from Assumption 2, $\Lambda_i = c_i/\sum c_i$. We assume the number of nodes is large, so that we can approximate $\sum_{j \neq i} L_{ij} \approx \alpha_i \sum_j \Lambda_j = \alpha_i$, hence $\alpha_i \approx L_i$ from Section 2. Define $A := \sum \Lambda_j(1 - x_j)$, $B := \sum r_j \alpha_j (1 - x_j)$. Similarly for large networks, we can approximate $\sum_{j \neq i} \Lambda_j(1 - x_j) \approx A$ and $\sum_{j \neq i} r_j \alpha_j (1 - x_j) \approx B$, for all i. Using this approximation and Assumptions 1 and 2, we can characterize the Nash equilibrium of $\mathcal{G}^I$.

Theorem 2. *Assume $\alpha_i \neq \alpha_j$ for $i \neq j$. Let $\mathbf{r}^*$ and $\mathbf{x}^*$ be an equilibrium point for the game $\mathcal{G}^I$, and let $\boldsymbol{\lambda}^*$, $\boldsymbol{\mu}^*$, $\boldsymbol{\nu}^*$, η^* and A^*, B^* be the corresponding parameters. Then there exists some $1 \leq k \leq N$ such that $r_k^* \leq (c_k - B^*\Lambda_k)/A^*\alpha_k$, and,*

$$
(r_i^*, x_i^*) = \begin{cases} \left((c_i - B^*\Lambda_i)/A^*\alpha_i, \, 1 - \eta^*/A^*\alpha_i\right) & i < k , \\ (0,0) & i > k . \end{cases}
$$

If $r_k^* < (c_k - B^*\Lambda_k)/A^*\alpha_k$ then $x_k^* = 0$, and if $r_k^* - (c_k - B^*\Lambda_k)/A^*\alpha_k$, then any $0 \leq x_k^* \leq 1 - \alpha_{k+1}/\alpha_k$ constitutes an NE.

Consider an instance of $\mathcal{G}^I$ where $r_k^* = (c_k - B^*\Lambda_k)/A^*\alpha_k$ and $x_k^* = 0$. Using Theorem 2, we can find the equilibrium point for such a case, where k nodes have been completely saturated by the attacker, and the defender chooses not to secure the kth node. We can represent this equilibrium point as a function of k. Let $\mathbf{r}(k)$ and $\mathbf{x}(k)$ denote the NE, and $r^I(k) = \sum r_j(k)$ be the corresponding parameter. Defining $D(k) := \sum_{j=1}^{k} c_j \frac{\alpha_k}{\alpha_j}$ and $E(k) := \sum_{j=k+1}^{N} c_j$, we have:

$$
\begin{cases} r^I(k) = \sum_{j=1}^{k} r_j(k) = \dfrac{1}{\alpha_k} \dfrac{D(k)}{2D(k) + E(k)} \sum_{j=1}^{N} c_j , & (8a) \\[3mm] u_a^I(k) := u_a^I(\mathbf{r}(k), \mathbf{x}(k)) = D(k) \dfrac{D(k) + E(k)}{2D(k) + E(k)} , & (8b) \\[3mm] u_d^I(k) := u_d^I(\mathbf{r}(k), \mathbf{x}(k)) = \dfrac{D^2(k)}{2D(k) + E(k)} - \sum_{j=1}^{k} c_j . & (8c) \end{cases}
$$

4   Numerical Results

To illustrate the results of Section 2, we generate a network of N nodes by drawing V_i and L_i independently from a Rayleigh distribution, and plot the utilities of both parties at the NE, as a function of the attacker's resources.

As a reference point, in all the following simulations, we set $\mathbb{E}[V_i] = 1$. Also the scaling of L_i does not have an effect on the overall shape of the curve; it only affects the maximum capacity of the network. Therefore, we will let $\mathbb{E}[L_i] = 1$ throughout. Moreover, we will present the utilities of both parties as

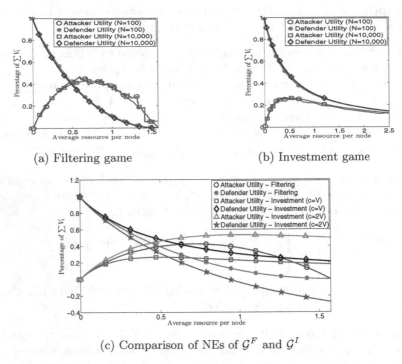

(a) Filtering game   (b) Investment game

(c) Comparison of NEs of $\mathcal{G}^F$ and $\mathcal{G}^I$

Fig. 1. Defender and attacker utilities at the NEs of $\mathcal{G}^F$ and $\mathcal{G}^I$

a percentage of the total value of the network, i.e., $\sum V_i$. The reason for this choice of scaling is that in the absence of any attack (i.e., $r = 0$), the defender obtains the entire value of the network at equilibrium. Therefore, the vertical axis depicts the fraction of network value lost as a result of the attacks. Finally, in order to obtain a better comparison among networks of different size, we scale the horizontal axis by the number of nodes N, resulting in plots that illustrate utilities as a function of the average attack resource per node.

Figure 1a plots the attacker and defender utilities under the filtering game $\mathcal{G}^F$, for two networks of size $N = 100$ and $N = 10,000$. An important aspect to this plot is that the utility of the attacker, $u_a(\boldsymbol{r}^*, \boldsymbol{f}^*)$ is not necessarily an increasing function of the total attack power r. In other words, the most successful attacker is not necessarily the one with the highest attack power. This observation can be intuitively explained as follows: assume an attacker with high r decides to spend only a smaller amount $r' < r$ of his attack resources. If the defender's response is such that the NE corresponding to r' is realized, a smaller number of nodes would be filtered, and both parties would receive a higher utility. Nevertheless, the attacker's action will no longer be a best response to the defender's strategy in this scenario, as the attacker has access to additional resources to further attack the unfiltered nodes. As the availability of these resources is common knowledge, and thus known to the defender, she will not under-filter the system against more powerful attackers. This increased filtering of nodes against a more powerful attacker in turn limits the attacker's ability to profit from the network, and ultimately, reduces his utility.

It is also interesting to note the sudden drops in the attacker's utility, which are more easily observable for $N = 100$. These drops correspond to points where the attacker's total power is such that exactly k nodes have been completely saturated ($r_i^* = V_i/L_i,\ 1 \le i \le k$), following which an attacker with more attack power would start putting his resources into the $k + 1^{\text{th}}$ node. As a result, the defender's filtering becomes more aggressive by limiting nodes under attack to an effective loss of L_{k+1} (i.e., $L_i(1 - f_i^*) = L_{k+1},\ 1 \le i \le k$), hence the drop in the attacker's utility. The defender's utility, however, is always decreasing in r.

Figure 1b illustrates the utilities at the NE of $\mathcal{G}^I$ as a function of r/N, by plotting $r^I(k)$, $u_a^I(k)$ and $u_d^I(k)$ from (8a)-(8c). The parameters of the game are generated similar to the filtering game simulations, i.e., α_i (which is parallel to L_i in $\mathcal{G}^F$) and c_i are drawn from a Rayleigh distribution with unit mean.

One important aspect of the investment game is that the x-axis extends further than the filtering game. In other words, the capacity of the network is larger in comparison to $\mathcal{G}^F$. An intuitive explanation for this phenomenon is the presence of internalities when nodes protect themselves via investment. When a node is blacklisted, the rest of the network is protected against attacks targeting that node, but this action does not protect the node itself. Therefore, filtering is an action that has externality, but not internality. This is not the case for investment, since investing in security protects oneself, as well as the rest of the network. When the attacker is powerful, a large portion of the network is investing in security, and the defender is well-protected by internalities. Thus the capacity of each node is relatively large.

To conclude this section, we look at the utilities of both parties under the investment and filtering games in Figure 1c. To this end, we set $N = 10,000$, and compare the two games under two different security cost vectors $c = V$ or $c = 2V$, the latter indicating relatively costly security measures.

First, we note that as expected, investing in network security is preferred by the defender when the cost of it is sufficiently low. The more surprising result is however in the trend of the attacker's utility under the different protection models. We see that with low attack power, both filtering and security yield similar utility to the attacker, as no considerable filtering or protection has yet been introduced by the defender. As the attack power grows, the attacker who is facing filtering gains a higher utility. Intuitively, this is also a consequence of the internality of investment actions. To further illustrate, note that an unfiltered attack on node i yields a payoff of $\sum_{j \ne i} L_{ij}$ per unit of effort. In contrast, an attack on an unprotected node i yields a payoff of $\sum_{j \ne i} L_{ij}(1 - x_j)$ per unit of effort. Lastly, for very powerful attackers, the attacker facing investment is more successful. This is due to the fact when the defender chooses to filter nodes, the network gets increasingly close to being fully saturated under high attack power. However, under investment, it takes more resources for the the attacker to saturate all nodes, leaving him more room to gain profit.

5   Conclusion

In this paper, we compare the efficacy of two security options, namely inbound/ outbound filtering based on RBLs and investing in self-protection methods, by a defender controlling a set of nodes facing a resource constrained strategic attacker. Specifically, our models take into account the indirect losses inflicted on neighboring nodes by a compromised node, loss of value due to the inevitable filtering of parts of the legitimate traffic, and the higher cost of self-protection as compared to filtering. Our analysis and simulation results show that the defender chooses to invest in security measures over filtering only when the cost of investing is sufficiently low. On the other hand, the attacker's potential to benefit in the face of each protection method is determined by his total attack power. Highly resourceful attackers are less successful when facing filtering actions rather than investment actions.

The current work can be continued in several directions. It would be interesting to study filtering and investment actions under less restrictive conditions, e.g. nonlinear risk functions with respect to the filtering policy, and taking diminishing returns into account when considering attacks originating from multiple sources (the latter can be modeled by setting the attacker's profit to a concave function of the sum in Equation (1)). The same game form can also be analyzed a dynamic framework, as both attacker and defender actions can be affected by the history of past events, including previous attack patterns, node takedowns, and the amount of time a node stays blacklisted. Modeling information asymmetries among the players, and strategic interactions among multiple non-cooperative attackers and/or defenders, are other possible extensions of the current model.

References

1. Spam data and trends, Q1 2009 (2009)
2. Spam volumes drop by two-thirds after firm goes offline (2008), http://voices. washingtonpost.com/securityfix/2008/11/spam_volumes_drop_by_23_after .html
3. Fultz, N., Grossklags, J.: Blue versus red: Towards a model of distributed security attacks. In: Dingledine, R., Golle, P. (eds.) FC 2009. LNCS, vol. 5628, pp. 167–183. Springer, Heidelberg (2009)
4. Hausken, K.: Income, interdependence, and substitution effects affecting incentives for security investment. Journal of Accounting and Public Policy 25(6) (2006)
5. Hausken, K.: Strategic defense and attack of complex networks. International Journal of Performability Engineering 5(1) (2009)
6. Hausken, K.: Strategic defense and attack of series systems when agents move sequentially. IIE Transactions 43(7) (2011)
7. Cisco System Inc.: SpamCop Blocking List - SCBL (May 2011), http://www.spamcop.net/

8. Laszka, A., Felegyhazi, M., Buttyán, L.: A survey of interdependent security games. CrySyS 2 (2012)
9. Barracuda Networks: Barracuda Reputation Blocklist (May 2011), http://www.barracudacentral.org/
10. Nochenson, A., Heimann, C.F.L.: Simulation and game-theoretic analysis of an attacker-defender game. In: Grossklags, J., Walrand, J. (eds.) GameSec 2012. LNCS, vol. 7638, pp. 138–151. Springer, Heidelberg (2012)
11. Osborne, M.J., Rubinstein, A.: A course in game theory. MIT Press (1994)

Online Learning Methods
for Border Patrol Resource Allocation

Richard Klíma[1,2,*], Christopher Kiekintveld[2], and Viliam Lisý[1]

[1] Department of Computer Science, FEE, Czech Technical University in Prague,
Prague, Czeck Republic
klimaric@fel.cvut.cz, lisy@agents.fel.cvut.cz
[2] Computer Science Department, University of Texas at El Paso,
EI Paso, TX, USA
cdkiekintveld@utep.edu

Abstract. We introduce a model for border security resource allocation with repeated interactions between attackers and defenders. The defender must learn the optimal resource allocation strategy based on historical apprehension data, balancing exploration and exploitation in the policy. We experiment with several solution methods for this online learning problem including UCB, sliding-window UCB, and EXP3. We test the learning methods against several different classes of attackers including attacker with randomly varying strategies and attackers who react adversarially to the defender's strategy. We present experimental data to identify the optimal parameter settings for these algorithms and compare the algorithms against the different types of attackers.

Keywords: security, online learning, multi-armed bandit problem, border patrol, resource allocation, UCB, EXP3.

1 Introduction

Border security is a major aspect of national security for many countries; in the United States alone billions of dollars are spent annually on securing the borders. However, the scale of the problem is massive, with thousands of miles of land and sea borders and thousands of airports to secure. Allocating limited resources to maximize effectiveness is a serious issue for the United States Customs and Border Protection agency (CBP). Indeed, the most recent strategic plan for the CBP places a great emphasis on mobilizing resources and using risk-based models to allocate limited resources. [1].

Game theory is an increasingly important paradigm for strategically allocating resources in security domains, and we argue that it can also be useful for border security. There are now many examples in which security games [6,11] have been used as a framework for randomizing security deployments and schedules in homeland security and infrastructure protection. This model has been successfully used to randomize the deployment of security resources in airports [7,8],

---

* Richard Klíma is affiliated with both CTU and UTEP; this research was conducted primarily while he was an exchange student at UTEP.

R. Poovendran and W. Saad (Eds.): GameSec 2014, LNCS 8840, pp. 340–349, 2014.

to create randomized schedules for the Federal Air Marshals [6,12], and to randomized patrolling strategies for the United States Coast Guard [10].

Existing models of security games rely on constructing a game-theoretic model of the security problem including the actions and payoffs of both the attacker and the defender. These models are based on whatever data is available combined with expert analysis and risk assessment to model the attacker preferences. One of the reasons for this style of modeling is that there is relatively little direct evidence about the attackers; we cannot directly elicit their preferences, and attack events are so rare that there is not enough data available to directly construct a model. This lack of data leads to a time intensive, expert-driven modeling process that still faces challenges in trying to validate the models and keep them up to date.

In border security the situation is quite different from many of the areas where security games have been applied. The CBP makes hundreds of thousands of apprehensions annually for illegal entry, smuggling, and other violations. This means that there is a large amount of data available for building and updating game models of the interactions between border patrol and the illegal entrants. The nature of the interaction is also different. A terrorist attack is a one-time, very high stake events. However, border security is more accurately characterized as a large number of repeated interactions with lower stakes. Similar situations with frequent incidents occurs also in cyber security domains, so we expect that our approachs is also applicable to these domains.

We propose to model border security using adversarial learning models that are related to both game theory and machine learning. These models are more dynamic, and account for the possibility of learning about the opponent during repeated interactions of a game. We introduce a basic model for a border security resource allocation task that is closely related to multi-armed bandit problems studied in the online learning literature. We then apply several different online learning algorithms to this model, including algorithms designed for adversarial bandit problems. We present an empirical evaluation of the performance of these algorithms and analyze the results to show the feasibility of modeling resource allocation for border patrol using this approach.

2   Model

We study a simplified model of the problem of resource allocation for border patrol. One of the main challenges that we try to capture in this model is the problem of *situational awareness*, which can also be thought of as a problem of balancing exploration and exploitation. There are limited resources available for patrolling different regions of the border. Ideally, these resources should be used in regions where there is a high level of illegal traffic. However, traffic patterns can change over time as the attackers (e.g., illegal entrants and criminal smuggling organizations) adapt to the border protection strategy. This means that it is necessary to maintain situational awareness even in areas that currently have low traffic so that any changes in the traffic patterns can be quickly detected.

We consider a model where a border region is divided into z distinct zones. The border patrol has only one resource available to patrol these zones, and must decide which zone to patrol[1]. The attackers try to cross the border without being detected. They must pick one of the z zones to attempt a crossing. The game is played in a series of n rounds representing discrete time periods (e.g., one day, or one hour). There are t attackers who attempt to cross during each round. Any attackers who chooses the same zone as the defender are apprehended, while attackers that chose different zones cross successfully.

We represent the defender and attacker strategies in a round using probability distributions over the zones. The defender strategy for round i is given by a vector $D^i = \langle d_1^i \ldots d_z^i \rangle$ where d_j^i represents the probability that the defender patrols zone j in period i. Similarly, the attacker strategy round i is given by a vector $A^i = \langle a_1^i \ldots a_z^i \rangle$ where a_j^i represents the probability that an attacker chooses zone j in period i. We assume that each of the t attackers chooses a zone independently according to the distribution A_i. This assumption is made by the idea that the attackers share common knowledge about the border; where it is more suitable to cross or there is high probability of being caught.

The goal for the defender is to maximize apprehensions. We assume that all zones are identical for the defender. The attacker has a penalty p for being caught as well as a base value that differs across the zones, denoted by c_j. For any zone j we calculate the attacker's expected value in round i as:

$$v_j^i = c_j - (d_j^i * p) \tag{1}$$

where d_j^i is the probability that the attacker will be caught in a given zone in this round, which comes from the defender's strategy. The values for the different zones can be interpreted as the value of successfully crossing in a given zone, less the costs associated with the crossing (e.g., payments to smuggling organizations, and the difficulty and time required to traverse the terrain). The asymmetry introduced by these values is also important, because if all zones are identical for both players there is a trivial equilibrium solution in which both players play the uniform random strategy (analogous to the symmetric game of Rock, Paper, Scissors).

3   Attacker Models

We introduce four different models of attacker behavior that represent a spectrum of levels of adaptation and intelligence. These are also designed to present different challenges for the online learning algorithms.

Random Fixed: This policy is a fixed attacker probability distribution over the zones. The strategy is generated randomly at the beginning of the scenario by drawing real random numbers from interval $(0, 1)$ for each zone and normalizing.

---

[1] We limit this to one resource to simplify the initial model, but plan to generalize to multiple resources in future work.

This is intended as a baseline that should be relatively easy for the online learning methods to learn.

Random Varying: In this models we generate a new random attacker strategy after a fixed number of rounds; the new strategy is unrelated to the previous one. This models an attacker that changes strategies, but not intelligently in response to the defender. We chose to generate large changes intermittently rather than making constant small changes because it allows us to average results over many runs and evaluate how quickly the learning methods are able to detect and respond to sudden changes in attacker behavior.

Adversarial Fixed: This model assumes that the attacker is intelligently adapting in response to the defender's strategy. We assume that the attacker knows the number of times the defender has visited each zone in the past.[2] The attacker adapts his strategy gradually to maximize the value given in Equation 1. Here, the attacker uses the observed frequencies of the defender patrols to estimate the probability of being caught in each zone. This is motivated by *fictitious play*, a well-known learning dynamic in which players play a best response to the history of actions played by the other players [4]. However, we parameterize this learning strategy so that we can control the rate at which the attacker moves towards a best response using the learning rate parameter α. The initial attack strategy is selected randomly, and it is updated on each iteration according to:

$$A^{i+1} = (1 - \alpha) * A^i + \alpha * M \tag{2}$$

where M represents a vector that has a 1 for the zone that gives the maximum value, and 0s for all other zones.

Adversarial Varying: This model is identical to the previous one, except that we randomly change the base values c_j for each zone after a fixed number of rounds, similar to the random varying policy. This model simulates an attacker that adapts intelligently, but also has preferences that can change over time.

4 Defender Strategies

Our model captures one of the central difficulties in accurately estimating traffic, which is the limited observations that the defender makes about the attacker's strategy. The defender only observes the level of traffic in the zone that is patrolled in each time period, and not in the other zones, just like a patrol in the real world. This means that a defender strategy that always tries to patrol the zones with the highest levels of activity to maximize apprehensions risks developing "blind spots" as the attacker strategy changes. What were previously low traffic zones may have increased traffic due to adaptations by the attacker, but the

---

[2] This assumes a very knowledgeable attacker, but is fairly realistic since major transnational smuggling organizations use sophisticated surveillance to track border patrol presence.

defender cannot observe this unless it allocates some resources to exploration–patrolling zones that are believed to have low traffic to detect possible changes in the traffic levels over time.

This becomes a problem of balancing exploration and exploitation when allocating the patrolling resources [9]. The online learning literature contains many examples of models that focus on this basic problem, including the well-known multi-armed bandit problem [2]. In a multi-armed bandit, a player must select from a set of possible arms to pull on each iteration. Each arm has a different sequence of possible rewards that is initially unknown to the player. The player selects arms with the goal of maximizing the cumulative reward received, and must balance between selecting arms that have a high estimated value based on the history of observations, and selecting arms to gain more information about the true expected value of the arm.

From the defender's perspective, our model very closely resembles the basic stochastic multi-armed bandit problem if we assume a Random Fixed attacker. The zones in our model map to arms in the bandit model, and the defender must select zones both to maximize apprehensions based on the current estimates of the attacker strategy, but also to explore other zones to improve the estimate of the strategy. Based on this mapping, we apply variations of some of the existing solution methods for multi-armed bandits to our border patrol scenario. For the other attacker models this is no longer a stochastic multi-armed bandit problem because the underlying distribution of rewards changes over time (in some cases based on an adversarial response). Therefore, we also consider solution algorithms that have been developed for other variations of the bandit problem that make different assumptions about how the underlying rewards can change. We now describe in more detail the specific algorithms we consider.

Uniform Random: A baseline in which the defender chooses a zone to patrol based on a uniform random distribution in every round.

Upper Confidence Bound (UCB): One of the standard policies used for multi-armed bandits is UCB [2]. This method follows a policy that selects the arm that maximizes the value of the following equation in each round:

$$x_j + \sqrt{\frac{2\ln(n)}{n_j}} \tag{3}$$

where x_j is the average reward obtained from arm j, n_j is the number of times arm j has been selected so far and n is the number of rounds completed.

Sliding-Window UCB: This algorithm is a variant of the standard UCB that is more suitable for non-stationary bandit problems [5]. This algorithm should do well in an abruptly changing environment which is suitable for our attacker models that can change strategies (or underlying preferences) quickly. The main difference from the standard UCB is that the algorithm uses a fixed window of data from the previous rounds to calculate the estimated average rewards. At time step t we get average of rewards not from the whole history but only the τ previous rounds. SW-UCB chooses a zone which maximize the sum of exploitation and exploration part. The exploitation part of the UCB formula is a local average reward:

$$\bar{X}_t(\tau, i) = \frac{1}{N_t(\tau, i)} \sum_{s=t-\tau+1}^{t} X_s(i) \mathbb{1}\{I_s = i\} \tag{4}$$

where N_i is the number of times arm i was played. $X_s(i)$ is a reward in time step s of ith zone and the indicator function returns a value of one if the chosen zone in the time step s is equal to ith zone, and zero otherwise.

The exploration part of the formula is defined by:

$$c_t = (\tau, i) = B\sqrt{\log(t \wedge \tau)/(N_t(\tau, i))} \tag{5}$$

where $(t \wedge \tau)$ denotes the minimum of two arguments and τ is a constant. B is a constant which should be tuned appropriately to the environment, which we address in out experiments.

EXP3: The Exponential-weight algorithm for Exploration and Exploitation (EXP3) [3] is designed for adversarial bandit problems in which an adversary can arbitrarily change the rewards returned by the arms. It is the most pessimistic algorithm due to the very weak assumptions about the structure of the rewards, but is still able to bound the total regret, similar to the guarantees provided by UCB for the standard model. It tends to result in greater rates of exploration than the UCB policies. The details of the algorithm are somewhat more complex, so due to space limitations we refer the reader to [3] for the full details.

5   Experiments

We now present the results of our initial empirical study of the performance of the different online learning strategies for the defender in the border patrol scenario. We test the algorithms against the four different attacker models that represent increasing levels of adaptation and intelligence. The performance of the learning strategies is evaluated based on the apprehension rate, which is the ratio between the number of apprehensions and total number of attackers that attempt to cross the border.

Unless otherwise specified, all of our experiments are conducted on a simulation with 8 zones. The simulation runs for 10000 rounds, and there are 10 attackers that attempt to cross each round according to the distribution specified by the attacker's strategy. Results are averaged over 50 runs of the simulation.

5.1   Parameter Selection

We begin by testing different parameter settings for the learning methods to find the best settings to compare the performance of the different methods (random and UCB do not have parameters). We are also interested in the sensitivity of the algorithm's performance to the parameter settings in our domain. Many of the parameters balance the tradeoff between exploration and exploitation, so we expect that different settings will perform well against relatively static opponents compared to more adaptive adversarial opponents. We choose the parameters which give the best result for adversarial attacker.

EXP3:   We first present parameter tuning results for the parameter γ of EXP3 that controls the level of exploration. The parameter has values in the interval $(0, 1]$, and for higher values the algorithm becomes similar to playing randomly. Table 1 shows the apprehension rates for different values against the four different attacker models. For values of γ close to 1 we get behavior identical to a random defender. The best value of γ is 0.7 against adversarial attacker, so we use this value in the next experiments.

Table 1. EXP3 parameter tuning

γ value	random	random with changes	adversarial	adversarial with changes
0.1	20.05%	15.33%	14.65%	15.03%
0.3	19.47%	15.11%	15.18%	15.84%
0.5	16.77%	14.54%	16.06%	16.69%
0.7	15.13%	13.74%	16.58%	15.90%
1	12.53%	12.51%	12.50%	12.52%

Sliding-Window UCB: There are no parameters in the basic version of UCB, but in sliding-window UCB there are several parameters specified in the original implementation [5]. We tune the parameter B which controls the exploration rate. The parameter τ controls the size of sliding window of history used in the calculations. For higher values of τ the algorithm of SW-UCB converges to standard UCB. We run several combinations of these parameters against two of the attacker models: the fixed random attacker and the adversarial attacker.

Table 2. SW-UCB τ tuning with different B parameter

	random fixed attacker					adversarial attacker				
τ value	0.5	1	5	10	20	0.5	1	5	10	20
50	21.11%	19.67%	14,25%	13.51%	12.92%	19.57%	18.21%	26.38%	26.52%	24.05%
100	21.74%	20.10%	14.93%	13.85%	13.15%	18.34%	19.03%	22.96%	24.26%	22.11%
500	21.64%	20.61%	17.08%	15.27%	13.82%	15.87%	16.39%	19.00%	21.91%	23.88%
1000	21.37%	23.08%	18.66%	15.87%	14.66%	15.09%	15.90%	18.72%	22.17%	26.53%
3000	21.58%	21.41%	19.75%	17.86%	15.08%	12.97%	13.45%	19.38%	23.02%	28.50%
5000	21.43%	21.51%	21.79%	18.81%	16.28%	12.68%	13.34%	19.84%	23.72%	28.98%

In Table 2 we present the performance for random fixed and adversarial attacker for different settings of the SW-UCB algorithm. As expected, against a fixed attacker the best results come from the longest sizes of sliding windows; if the environment is fixed, it does not make sense to throw out older data using a sliding window since this data is still informative. Also we can observe that lower values of B parameter give higher apprehension rates The value of 1 for the B parameter results in the best performance here.

On the right side of Table 2 we have SW-UCB algorithm against an adversarial attacker. We can see that the results are almost opposite than in random fixed attacker case. We get better results with higher values of B parameter and with longer sliding-windows. The best results here are with a *high* value of the exploration rate of 20, compared to the opposite result for the fixed attacker.

In remaining experiments we set the γ parameter for EXP3 equal to 0.7. For sliding-window UCB we will use parameter B equal to 20 and parameter of sliding window τ equal to 5000. For adversarial attacker we will use a learning rate of 0.5 learning rate. For attacker strategies with changes we select a new random strategy or set of preference parameters every 2000 rounds.

5.2   Comparing Apprehension Performance

We now present initial results directly comparing the performance of the different learning methods against the different attacker strategies. The figures show how the apprehension rates of the algorithms evolve over the course of the 10000 round simulation. Results are averaged over 50 runs, and the plots are also smoothed using a moving average over 500 round buckets.

The results in Figure 1a show the learning process of the defender strategies against the random fixed attacker. The x-axis shows the number of simulation rounds divided by 100, and the y-axis shows the apprehension rate. All of the learning methods show the ability to learn against the fixed attacker. The standard version of UCB learns the fastest and has the best total apprehension rate, while EXP3 has somewhat poorer performance due to a higher exploration rate. For the SW-UCB there is a drop in the apprehension rate after the size of the sliding window equal to 5000.

Figure 1b shows the performance for the defender strategies against the fixed random attacker strategy with a change in the probability distribution every 2000 rounds. The points when the attacker strategy changes are clear in the plot, since the performance for all of the learning strategies drops off abruptly. However, the strategies are able to quickly respond and re-learn the adversaries strategy. We note that UCB has a high variance here, but SW-UCB shows more stable behavior, since it is designed for environments with abrupt changes.

In Figure 1c we present the behavior of the defender strategies against the adversarial attacker. In this case SW-UCB performs the best out of all defender strategies but tend to decrease over time. EXP3 gives the second best result and is quite stable. For all of the algorithms the performance later on is poorer, which is due to the intelligent adversary adapting to the defender strategy over time. All of the algorithms appear to be converging to an equilibrium strategy with the attacker over time.

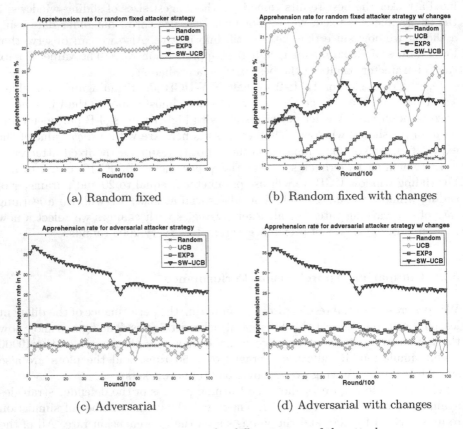

(a) Random fixed

(b) Random fixed with changes

(c) Adversarial

(d) Adversarial with changes

Fig. 1. Apprehension rates for different types of the attacker

Finally, Figure 1d shows the results against an adversarial attacker but with underlying zone preferences that change every 2000 rounds. The SW-UCB method again gives the best results, but the performance slightly decreases over time. We see here that the changes in the attacker preferences are not as dramatic as the direct changes in the attacker strategy, since they are muted and have an effect over time. The performance of the learning algorithms is somewhat degraded, but not dramatically worst than against the fixed adversarial attacker.

6   Conclusion

We have introduced a mathematical model for border patrol resource allocation that captures the important problem of allocating resources to maintain situational awareness via exploration. We have proposed several candidate solution methods drawn form the online learning literature that are suitable for making these decisions, including UCB, SW-UCB, and EXP3. They offer different levels of theoretical guarantees against changing and adversarial environments, with

EXP3 providing bounds on performance in even the most adversarial settings. Here, we have provided an initial empirical study comparing the performance of these algorithms in a simple border patrol scenario. We tested the parameters of the algorithms and determined the best settings, while also noting that the practical performance of the algorithms depends heavily on the parameter settings combined with how quickly the adversary changes. In comparison, SW-UCB often gives the best performance in the more adversarial cases, but all of the algorithms showed the ability to learn quickly and adapt even in the face of rapidly changing, adaptive adversaries. This demonstrates the potential for practical applications of these learning methods for resource allocation and situational awareness for border patrol.

Acknowledgements. This research was supported by the Office of Naval Research Global (grant no. N62909-13-1-N256).

References

1. 2012–2016 border patrol strategic plan. U.S. Customs and Border Protection (2012)
2. Auer, P.: Using confidence bounds for exploitation-exploration trade-offs. The Journal of Machine Learning Research 3, 397–422 (2003)
3. Auer, P., Cesa-Bianchi, N., Freund, Y., Schapire, R.E.: The non-stochastic multi-armed bandit problem. SIAM Journal on Computing 32(1) (2001)
4. Fudenberg, D., Levine, D.K.: The Theory of Learning in Games. The MIT Press (1998)
5. Garivier, A., Moulines, E.: On upper-confidence bound policies for non-stationary bandit problems. Technical report (2008)
6. Kiekintveld, C., Jain, M., Tsai, J., Pita, J., Ordonez, F., Tambe, M.: Computing optimal randomized resource allocations for massive security games. In: AAMAS 2009 (2009)
7. Pita, J., Jain, M., Western, C., Portway, C., Tambe, M., Ordonez, F., Kraus, S., Parachuri, P.: Depoloyed ARMOR protection: The application of a game-theoretic model for security at the Los Angeles International Airport. In: AAMAS 2008 (Industry Track) (2008)
8. Pita, J., Tambe, M., Kiekintveld, C., Cullen, S., Steigerwald, E.: GUARDS - game theoretic security allocation on a national scale. In: AAMAS 2011 (Industry Track) (2011)
9. Predd, J., Willis, H., Setodji, C., Stelzner, C.: Using pattern analysis and systematic randomness to allocate U.S. border security resources (2012)
10. Shieh, E., An, B., Yang, R., Tambe, M., Baldwin, C., Direnzo, J., Meyer, G., Baldwin, C.W., Maule, B.J., Meyer, G.R.: PROTECT: A Deployed Game Theoretic System to Protect the Ports of the United States. In: AAMAS (2012)
11. Tambe, M.: Security and Game Theory: Algorithms, Deployed Systems, Lessons Learned. Cambridge University Press (2011)
12. Tsai, J., Rathi, S., Kiekintveld, C., Ordóñez, F., Tambe, M.: IRIS - A tools for strategic security allocation in transportation networks. In: AAMAS 2009 (Industry Track) (2009)

A Distributed Optimization Algorithm for Attack-Resilient Wide-Area Monitoring of Power Systems: Theoretical and Experimental Methods

Jianhua Zhang[1], Prateek Jaipuria[2],
Aranya Chakrabortty[1,*], and Alefiya Hussain[2]
{jzhang25,achakra2}@ncsu.edu, {jaipuria,hussain}@isi.edu

[1] Electrical and Computer Engineering,
North Carolina State University, Raleigh, NC, USA
[2] Information Sciences Institute, University of Southern California,
Los Angeles, CA, USA

Abstract. In this paper we present a real-time distributed optimization algorithm based on Alternating Directions Method of Multipliers (ADMM) for resilient monitoring of power flow oscillation patterns in large power system networks. We pose the problem as a least squares (LS) estimation problem for the coefficients of the characteristic polynomial of the transfer function, and combine a centralized Prony algorithm with ADMM to execute this estimation via distributed consensus. We consider the network topology to be divided into multiple clusters, with each cluster equipped with a local estimator at the local control center. At any iteration, the local estimators receive Synchrophasor measurements from within their own respective areas, run a local consensus algorithm, and communicate their estimates to a central estimator. The central estimator averages all estimates, and broadcasts the average back to each local estimator as the consensus variable for their next iteration. By imposing a redundancy strategy between the local and the global estimators via mutual coordination, we show that the distributed algorithm is more resilient to communication failures as compared to alternative centralized methods. We illustrate our results using a hardware-in-loop power system testbed at NC State federated with a networking and cyber-security testbed at USC/ISI.

Keywords: Distributed estimation, wide-area monitoring, cyber-security, power systems.

1 Introduction

Following the Northeast blackout of 2003, Wide-Area Measurement System (WAMS) technology using Phasor Measurement Units (PMUs) has largely matured for the North American power grid [1]. However, as the number of PMUs scales up into the thousands in the next few years under the US Department

---

* The work of the third author was supported in part by the US National Science Foundation and the US Department of Energy.

R. Poovendran and W. Saad (Eds.): GameSec 2014, LNCS 8840, pp. 350–359, 2014.
© Springer International Publishing Switzerland 2014

of Energy's smart grid demonstration initiative, utility companies are struggling to understand how the resulting gigantic volumes of real-time data can be efficiently harvested, processed, and utilized to solve wide-area monitoring and control problems for any realistic power system interconnection. It is rather intuitive that the current state-of-the-art centralized communication and information processing architecture of WAMS will no longer be sustainable under such a data explosion, and a completely distributed, self-adaptive cyber-physical architecture will become imperative [2]. Motivated by this challenge, in this paper we address the problem of implementing wide-area monitoring algorithms over a distributed communication infrastructure using massive volumes of real-time PMU data. Our goal is to establish how distributing a monitoring functionality over multiple estimators can guarantee significantly more resiliency against extreme events. Such events may result from both malicious attacks on the cyber and physical assets as well as due to natural calamities such as storms and earthquakes. The specific monitoring algorithm that we study is the estimation of the frequency and damping of the electro-mechanical oscillations seen in the power flows in the grid after any disturbance. If the system size is small then it is straightforward to estimate these oscillation modes, or equivalently the eigenvalues and the eigenvectors of its state matrix, in a centralized way. Algorithms such as Eigenvalue Realization Algorithm (ERA), Prony analysis, and mode metering [3], for example, have been widely used by the WAMS community over the past decade for this purpose. However, as the system size and the number of PMUs scale up, the computational costs of these algorithms explode, and they completely fail to provide the required resiliency. As a solution, in this paper we present a distributed Prony-based algorithm combined with Alternating Direction Method of Multipliers [4] to estimate the frequency, damping and residue of each oscillation mode via distributed consensus. We partition the network topology into multiple clusters, with each cluster equipped with a local estimator at the local control center. At any iteration, the local estimators receive PMU measurements from within their own respective areas, run a local consensus algorithm, and communicate their estimates to a central estimator. The central estimator averages all estimates, and broadcasts the average back to each local estimator as the consensus variable for their next iteration. By imposing a redundancy strategy between the local and the global estimators via mutual coordination, we show that the distributed algorithm is highly resilient to communication failures. We also show that in case of an attack the local estimators only need to exchange a certain set of parameter estimates, and not actual PMU data, because of which the proposed algorithm is *privacy preserving*. We illustrate our results using a IEEE 39-bus system model emulated via a federated testbed between NC State University (Phasorlab) and the Information Sciences Institute (DETERLab) [5].

The remainder of the paper is organized as follows. Section 2 describes the dynamical model of a power system. Section 3 and Section IV present RLS and Prony/ADMM algorithms to estimate the the oscillation modes of this model.

Section 5 describes the attack resiliency of the proposed distributed method via simulations using the federated testbeds. Section 6 concludes the paper.

2   Power System Oscillation Model

Consider a power system network consisting of m synchronous generators and n_l loads connected by a given topology. Without loss of generality, we assume buses 1 through m to be the generator buses, and buses $m + 1$ through $m + n_l$ to be the load buses. Let P_i and Q_i denote the total active and reactive powers injected to the ith bus $(i = 1, \ldots, m + n_l)$ from the network, calculated as:

$$P_i = \sum_{k=1}^{m+n_l} {}' \left(\frac{V_i^2 r_{ik}}{z_{ik}^2} + \frac{V_i V_k}{z_{ik}} \sin(\theta_{ik} - \alpha_{ik}) \right), \quad Q_i = \sum_{k=1}^{m+n_l} {}' \left(\frac{V_i^2 x_{ik}}{z_{ik}^2} - \frac{V_i V_k}{z_{ik}} \cos(\theta_{ik} - \alpha_{ik}) \right), \quad (1a)$$

where $V_i \angle \theta_i$ is the voltage phasor at the ith bus, $\theta_{ik} = \theta_i - \theta_k$, r_{ik} and x_{ik} are the resistance and reactance of the transmission line joining buses i and k, $z_{ik} = \sqrt{r_{ik}^2 + x_{ik}^2}$, and $\alpha_{ik} = \tan^{-1}(r_{ik}/x_{ik})$. The electro-mechanical swing model of the ith generator is given as

$$\dot{\delta}_i(t) = \omega_s(\omega_i(t) - 1), \tag{2a}$$

$$M_i \dot{\omega}_i(t) = P_{m_i} - P_{e_i}(t) - D_i(\omega_i(t) - 1), \ i = 1, .., m \tag{2b}$$

with associated power balance equations given by

$$P_{e_i}(t) + P_i(t) - P_{L_i}(t) = 0, \quad Q_{e_i}(t) + Q_i(t) - Q_{L_i}(t) = 0,$$
$$P_k(t) - P_{L_k}(t) = 0, \quad Q_k(t) - Q_{L_k}(t) = 0, \tag{3}$$

for $i = 1, \ldots, m$ and $k = m + 1, \ldots, m + n_l$. Here, δ_i, ω_i, M_i, D_i, P_{m_i}, P_{e_i}, and Q_{e_i} denote the internal angle, speed, inertia, damping, mechanical power, active and reactive electrical powers produced by the ith generator, respectively, and P_{L_k} and Q_{L_k} denote the active and reactive powers of the loads at the kth bus. The Differential-Algebraic model (2)-(3) can be converted to a system of purely differential equations by relating the algebraic variables V_i and θ_i in (1) to the system state variables (δ, ω) and then substituting them back in (2) via Kron reduction. The resulting system is a fully connected network of m second-order oscillators with $l \leq m(m - 1)/2$ tie-lines. Let $\tilde{E}_i = E_i \angle \delta_i$ denote the internal voltage phasor of the ith machine. The electro-mechanical dynamics of the ith generator in *Kron*'s form can then be written as

$$\dot{\delta}_i(t) = \omega_s(\omega_i(t) - 1), \tag{4a}$$

$$M_i \dot{\omega}_i(t) = P_{mi} - D_i(\omega_i(t) - 1) - \sum_k E_i E_k \left(\frac{X_{ik}}{Z_{ik}^2} \sin(\delta_{ik}(t)) - \frac{R_{ik}}{Z_{ik}^2} \cos(\delta_{ik}(t)) \right), \tag{4b}$$

where, $i = 1, \ldots, m$, $Z_{ij}^2 = R_{ij}^2 + X_{ij}^2$, R_{ij} and X_{ij} denote the resistance and reactance of the line connecting the ith and j^{th} generator in the Kron's form, respectively, and $\delta_{ik}(t) = \delta_i(t) - \delta_k(t)$. Linearizing (4) about the equilibrium $(\delta_{i0}, 1)$ results in the small-signal state space model:

$$\begin{bmatrix} \Delta\dot{\delta}(t) \\ \Delta\dot{\omega}(t) \end{bmatrix} = \underbrace{\begin{bmatrix} 0_{m \times m} & \omega_s I_{m \times m} \\ \mathcal{M}^{-1}\mathcal{L} & \mathcal{M}^{-1}\mathcal{D} \end{bmatrix}}_{A} \begin{bmatrix} \Delta\delta(t) \\ \Delta\omega(t) \end{bmatrix} + \underbrace{\begin{bmatrix} 0 \\ \mathcal{M}^{-1}e_j \end{bmatrix}}_{B} u(t), \tag{5}$$

where, $\Delta\delta(t) = \left[\Delta\delta_1(t) \cdots \Delta\delta_m(t)\right]^T$, $\Delta\omega(t) = \left[\Delta\omega_1(t) \cdots \Delta\omega_m(t)\right]^T$, $I_{m \times m}$ denote the $(m \times m)$ identity matrix, $\mathcal{M} = \text{diag}(M_i)$ and $\mathcal{D} = \text{diag}(D_i)$ are the $(m \times m)$ diagonal matrices of the generator inertias and damping factors, $\mathbf{e}_j$ is the jth unit vector with all elements zero but the jth element that is 1, considering that the input is modeled as a change in the mechanical power in the jth machine. However, since we are interested only in the oscillatory modes or eigenvalues of $\mathcal{A}$, this assumption is not necessary and the input can be modeled in any other feasible way such as faults and excitation inputs. The matrix $\mathcal{L}$ in (5) is the $(m \times m)$ Laplacian matrix of the form

$$[\mathcal{L}]_{i,j} = \frac{E_i E_j}{Z_{ij}^2}\left(X_{ij}\cos(\delta_{i0} - \delta_{j0}) + R_{ij}\sin(\delta_{i0} - \delta_{j0})\right) \quad i \neq j,$$

$$[\mathcal{L}]_{i,i} = -\sum_{k=1}^{n}[\mathcal{L}]_{i,k}. \tag{6}$$

Let us denote the ith eigenvalue of the matrix $\mathcal{M}^{-1}\mathcal{L}$ by $\hat{\lambda}_i$. The largest eigenvalue of this matrix is equal to 0, and all other eigenvalues are negative, i.e. $\hat{\lambda}_m \leq \cdots \leq \hat{\lambda}_2 < \hat{\lambda}_1 = 0$. The eigenvalues of $\mathcal{A}$ are denoted by $\lambda_i = (-\sigma_i \pm j\Omega_i)$, $(j = \sqrt{-1})$. Our objective is to estimate $\lambda_i \; \forall i = 1,..,m$ using PMU measurements of voltage, phase angle, and frequency from multiple buses in the system in both centralized and distributed ways. We use Recursive Least Squares (RLS) for the centralized estimation, and propose an ADMM-based Prony algorithm for the distributed estimation. We illustrate that the distributed strategy is more resilient to communication failures than centralized.

3   Centralized Recursive Least-Squares

We open the problem by considering a fixed *input* bus, i.e., a node through which a disturbance input $u(t)$ enters the system, and two distinct output nodes, say bus p and bus q, which may or may not be the same as the input bus, where PMUs are installed. In reality, there may be many more than just two outputs. But for simplicity, we restrict our discussion to only two outputs, namely $y_p(t)$ and $y_q(t)$, measured by the two PMUs. Since there are m generators, each modeled by a second-order dynamic model, the total system order is $n = 2m$. The corresponding discrete-time transfer functions for the two outputs can be expressed as

$$\frac{Y_p(z)}{U(z)} = \frac{a_0 + a_1 z^{-1} + ... + a_{m_p} z^{-m_p}}{1 + b_1 z^{-1} + ... + b_n z^{-n}}, \frac{Y_q(z)}{U(z)} = \frac{c_0 + c_1 z^{-1} + ... + c_{m_q} z^{-m_q}}{1 + b_1 z^{-1} + ... + b_n z^{-n}} \tag{7}$$

where $m_p \leq n$ and $m_q \leq n$ are the orders of the respective zero polynomials. Taking inverse z-transform, (7) can be converted into the time-domain equations represented by the block-matrix at the sample index $k \in \{0, 1, ..., \infty\}$, as

$$y_p(k) = \left[\phi_p(k) \; U_p(k)\right]\begin{bmatrix} \gamma_3 \\ \gamma_1 \end{bmatrix}, \quad y_q(k) = \left[\phi_q(k) \; U_q(k)\right]\begin{bmatrix} \gamma_3 \\ \gamma_2 \end{bmatrix} \tag{8}$$

where, $\phi_p(k) = [y_p(k-1)\cdots y_p(k-n)]$, $\phi_q(k) = [y_q(k-1)\cdots y_q(k-n)]$, $U_p(k) = [u(k)\cdots u(k-m_p)]$, $U_q(k) = [u(k)\cdots u(k-m_q)]$, $\gamma_1 = [a_0\, a_1\cdots a_{m_p}]$, $\gamma_2 = [c_0\, c_1\cdots c_{m_q}]$, $\gamma_3 = [-b_1 - b_2\cdots - b_n]$. Our objective is to simply estimate the common characteristic polynomial of two transfer functions captured by the parameter vector γ_3 from the known input sequence $u(k)$ and the output sequences $y_p(k)$ and $y_q(k)$. Without any loss of generality, we assume the incoming disturbance $u(t)$ to be an impulsive input, and apply a real-time, centralized recursive least squares (RLS) approach to compute γ_3. From (8), we can write

$$\underbrace{\begin{bmatrix} y_p(k) \\ y_q(k) \end{bmatrix}}_{A(k)} = \underbrace{\begin{bmatrix} \phi_p(k)\ U_p(k) & 0 \\ \phi_q(k) & 0 & U_q(k) \end{bmatrix}}_{B(k)} \underbrace{\begin{bmatrix} \gamma_3 \\ \gamma_1 \\ \gamma_2 \end{bmatrix}}_{\Theta} \tag{9}$$

By assuming that any variable with a negative sample index is zero by default, we construct matrices A and B for $k \in \{0, 1, ..., M\}$ with $M > n$ being a sufficiently large integer, as below,

$$A = col(y_p(1), y_q(1), y_p(2), y_q(2), ..., y_p(M), y_q(M))$$

$$B = \begin{bmatrix} \phi_p(1) & U_p(1) & 0 \\ \phi_q(1) & 0 & U_q(1) \\ \vdots & \vdots & \vdots \\ \phi_p(M) & U_p(M) & 0 \\ \phi_q(M) & 0 & U_q(M) \end{bmatrix}_{2M\times(n+m_p+m_q+2)}. \tag{10}$$

The problem in the centralized case, therefore, is to generate the parameter vector Θ that solves $\Theta = B^+ A$, where $+$ denotes pseudoinverse, and then extract the first n entries of Θ, flip their sign to obtain the common parameter vector β. The entire operation is denoted as $\beta = -[B^{-1}A]^{+n}$. The computation of Θ can also be executed in a recursive fashion

$$\theta_{K+1} = \theta_K + P_{K+1}\phi_K(y_K - \phi_K^T\phi_K) \tag{11}$$

where K denotes an iteration index, Θ_0 is an initial guess for Θ, and the matrix P follows from the regressor equation using matrix inversion lemma (please see [6] for details). The stop condition of this recursive algorithm is when $\|\Theta_{K+1} - \Theta_K\| < \epsilon$ where ϵ is a chosen tolerance. The estimated common parameter vector is then denoted as $\beta = -[\Theta_K]^{+n}$. Once $\beta = \{b_1, b_2, \cdots, b_n\}$ is known, the eigenvalues λ_i of the matrix $\mathcal{A}$ in (5) can be computed simply by solving for the characteristic polynomial

$$1 + b_1 z^{-1} + b_2 z^{-2} + ... + b_n z^{-n} = 0, \tag{12}$$

and converting the roots to their continuous-time counterparts. The architecture for RLS is obvious from the foregoing analysis. Every PMU streams in its individual measurements in real-time to a central estimator; this estimator executes (11), and then (12).

4 Distributed Prony Algorithm with ADMM

To circumvent the single-point centralized architecture of RLS, we next propose the following distributed strategy combining the traditional centralized Prony's algorithm with Alternating Directions Multiplier Method (ADMM).

4.1 Prony's Algorithm

Consider that a set of N PMU measurements $\mathbf{y}(t) = col(y_1(t)\cdots y_N(t))$ are available over time t. From the structure of $\mathcal{A}$ in (5) we can write the output of the pth PMU to be of the form

$$y_p(t) = \sum_{i=1}^{n} r_{p,i} e^{(-\sigma_i+j\Omega_i)t} + r_{p,i}^* e^{(-\sigma_i-j\Omega_i)t}, \quad p = 1,\cdots,N \qquad (13)$$

The objective is to find the damping factors σ_i, the frequencies Ω_i, and the residues $\mathbf{r}_i = col(r_{1,i}\cdots r_{p,i}\cdots r_{N,i})$ for $i = 1,\ldots,n$. This can be done using the following three steps of the Prony's algorithm. For simplicity, we use only one measurement $y(t)$, i.e. $p = 1$.

Step 1. Let M be the total number of samples for y. The first step is to find the coefficients of the characteristic polynomial b_1 through b_n in (7) by solving

$$\underbrace{\begin{bmatrix} y(n) \\ y(n+1) \\ \vdots \\ y(n+\ell) \end{bmatrix}}_{\mathbf{c}} = \underbrace{\begin{bmatrix} y(n-1) & \cdots & y(0) \\ y(n) & \cdots & y(1) \\ \vdots & & \vdots \\ y(n+\ell-1) & \cdots & y(\ell) \end{bmatrix}}_{H} \underbrace{\begin{bmatrix} -b_1 \\ -b_2 \\ \vdots \\ -b_n \end{bmatrix}}_{\mathbf{b}} \qquad (14)$$

where ℓ is an integer satisfying $n + \ell \leq M - 1$, where $M - 1$ is the time index of the most recent measurement. One can find $\mathbf{b}$ by solving a least squares (LS) problem defined as

$$\min_{\mathbf{b}} \frac{1}{2}\|H\mathbf{b} - \mathbf{c}\|^2 \qquad (15)$$

Step 2. We next find the roots of the discrete-time characteristic polynomial, say denoted by z_i, $i = 1,\ldots,n$. Then, the eigenvalues of A in (5) are equal to $\ln(z_i)/T$, T being the sampling period.

Step 3. The final step is to find the residues $\mathbf{r}_i$ in (13). This can be done by forming the following so-called *Vandermonde* equation and solving it for $\mathbf{r}_1$ through $\mathbf{r}_n$:

$$\begin{bmatrix} y(0) \\ y(1) \\ \vdots \\ y(M) \end{bmatrix} = \begin{bmatrix} 1 & 1 & \cdots & 1 \\ (z_1)^{1/T} & (z_2)^{1/T} & \cdots & (z_n)^{1/T} \\ \vdots & \vdots & & \vdots \\ (z_1)^{M/T} & (z_2)^{M/T} & \cdots & (z_n)^{M/T} \end{bmatrix} \begin{bmatrix} r_1 \\ r_1^* \\ \vdots \\ r_n \\ r_n^* \end{bmatrix} \qquad (16)$$

The method can be easily generalized to the case of multiple output measurements $\mathbf{y}(t) = \mathrm{col}(y_1(t) \cdots y_N(t))$ by subscripting $\mathbf{c}$ and H in (14) as $\mathbf{c}_i$ and H_i for $i = 1, \ldots, N$, and concatenating them. Then, one can solve the following LS problem for $\mathbf{b}$:

$$\min_{\mathbf{b}} \frac{1}{2} \left\| \begin{bmatrix} H_1 \\ \vdots \\ H_N \end{bmatrix} \mathbf{b} - \begin{bmatrix} \mathbf{c}_1 \\ \vdots \\ \mathbf{c}_N \end{bmatrix} \right\|^2 \tag{17}$$

Step 3 then will be applied for each $y_i(t)$ individually.

4.2 Real-Time Distributed Prony Algorithm Using ADMM

The LS problem (17) is, in fact, a *global consensus problem* over a network of N regional utility companies, and, as shown in the following, can be estimated via distributed protocols with one central independent system operator (ISO) performing a *supervisory* step to guarantee convergence. Before we describe the mathematical formulation of the algorithm, we wish to briefly describe the way this strategy may be actually implemented by an ISO. We assume that each operating zone of the power system is equipped with its own regional PMUs, and estimators or *phasor data concentrators* (PDC). In reality, there may be a cluster of PDCs running in each area, but for convenience of analysis we only consider an aggregate regional PDC to be responsible for accepting its area-level PMU measurements, run the estimation using these measurements, and then share a set of estimated transfer function parameters with PDCs of other areas, albeit through one step of supervision through the central ISO-level PDC. This problem, in general, is described as,

$$\min_{\mathbf{b}_1, \ldots, \mathbf{b}_N, \mathbf{z}} \sum_{i=1}^{N} \frac{1}{2} \|H_i \mathbf{b}_i - \mathbf{c}_i\|^2 \tag{18}$$

subject to $\mathbf{b}_i - \mathbf{z} = 0$, for $i = 1, \ldots, N$,

where the global consensus solution, denoted by $\mathbf{z}$, is the solution of (17) that is obtained when the local estimates of the entire N regional PDCs, denoted by $\mathbf{b}_i, \forall\, i = 1, \ldots, N$, reaches the same value. We use ADMM to solve (18) using an *augmented Lagrangian*

$$L_\rho = \sum_{i=1}^{N} \left(\frac{1}{2} \|H_i \mathbf{b}_i - \mathbf{c}_i\|^2 + \mathbf{w}_i^T (\mathbf{b}_i - \mathbf{z}) + \frac{\rho}{2} \|\mathbf{b}_i - \mathbf{z}\|^2 \right),$$

where $\mathbf{w}_i$ is the vector of the *dual variables*, or the Lagrange multipliers associated with (18), and $\rho > 0$ denotes the *penalty factor*. Using this, the resulting ADMM problem, assuming $\mathbf{z} = \bar{\mathbf{b}} \triangleq \frac{1}{N} \sum_{i=1}^{N} \mathbf{b}_i$, can be defined by the following set of recursive optimization problems [4] to solve (18) in a distributed way:

$$\mathbf{b}_i^{(k+1)} = ((H_i^{(k)})^T H_i^{(k)} + \rho I)^{-1} ((H_i^{(k)})^T \mathbf{c}_i^{(k)} - \mathbf{w}_i^{(k)} + \rho \bar{\mathbf{b}}^{(k)}), \tag{19a}$$

$$\mathbf{w}_i^{(k+1)} = \mathbf{w}_i^{(k)} + \rho (\mathbf{b}_i^{(k+1)} - \bar{\mathbf{b}}^{(k+1)}). \tag{19b}$$

Each iteration of (19) consists of the following steps: 1) update the local variable $\mathbf{b}_i$ locally at PDC i (19a); 2) gather the values $\mathbf{b}_i$ at a central ISO-level PDC, and calculate their mean $\bar{\mathbf{b}}$; 3) broadcast $\bar{\mathbf{b}}$ to the other local PDCs; and 4) finally, update $\mathbf{w}_i$ at each PDC i (19b). It can be shown that $\bar{\mathbf{b}}$ as $k \to \infty$ converges to the global minimum of (18) [4].

5   Experimental Verification via Federated Testbeds

We next implement the RLS and Prony-ADMM algorithms using a federated testbed, which has been recently established via a collaboration between USC/ISI and NCSU. The testbed at ISI is a networking and cyber-security testbed called DETERLab [5], while the one at NCSU is a power system testbed called Pha-sorlab. We simulate a IEEE 39-bus power system model at Phasorlab using Real-time Digital Simulators (RTDS), and excite it with a three-phase fault last-ing for 0.3 mins. The system consists of 10 synchronous generators, partitioned into 4 coherent clusters with one PMU in each cluster. Traces of the frequency measurements from these four PMUs are shown in Figure 1. We represent the communication topology for RLS in DETER with four PMU nodes, generating the frequencies $y_1(k)$, $y_2(k)$, $y_3(k)$, and, $y_4(k)$, $k = 0, \ldots, 1800$, with a sampling rate of $T = 0.01$ seconds, and sending them to a central server node that ex-ecutes (11). To facilitate the estimation speed, for this case we down-sample the data at $T = 0.2$ second. The server node receives data packets with size of 250 bytes, which usually contains around 25 samples from total four PMU/PDC nodes in parallel with each other. For distributed estimation, each partitioned cluster has an additional Prony node representing a local PDC. The PMU nodes stream their local measurements $y_i(k)$, $i = 1, 2, 3, 4$ to the local Prony node. The Prony nodes run (19), and send their individual estimate $\mathbf{b}_i$ to the server node. The server nodes computes $\bar{\mathbf{b}}$, and broadcasts it back to the Prony nodes for the next iteration. Since every node exchanges only a parameter vector with the server, and vice versa, and not any actual PMU measurement, complete 'data privacy' is maintained between the clusters. The estimates for the three slow or *inter-area* modes for the 4-area system are shown in the second and third columns of Table 1. The actual values of the modes are shown in the first column. It can be seen that both RLS and distributed Prony yield reasonably accurate estimates. The slight mismatches in each from the actual values are mostly attributed to the sensitivity of the root finding step to small errors in b. The accuracy of RLS improves with more educated guesses for Θ_0.

Next, we consider three different types of attacks using DETERLab, namely (1) a malware attack that disrupts the operation of a PDC resulting in abnormal termination of an estimation algorithm, (2) flooding attack, where an attacker can flood a targeted network link with malicious traffic resulting in the reduc-tion of the estimation traffic due to overload, and (3) malfunctioning of physical infrastructure or hardware, eg. shutdowns due to power outages caused by earth-quakes and other natural disasters.

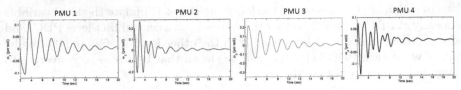

Fig. 1. PMU measurements of frequency from buses 1, 3, 4, 8 (in per unit)

We implement the flooding attack sequentially on each of the four communication links in the RLS topology, as shown in Figure 2. Once the algorithm detects an attack, it takes three immediate actions: 1) update the list of live PMUs, 2) resize the computation matrices A and B in (10), and 3) reset the initial guess for Θ to its value immediately before the attack. However, even with these accommodations, columns 2, 3 and 4 of Table 2 show that the accuracy of RLS estimation degrades significantly as more PMUs get disconnected by the attacks.

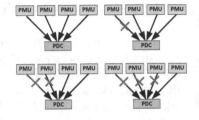

Fig. 2. Attacks on RLS links

In fact, as shown in the fourth column of Table 2, when three PMUs are disconnected RLS cannot identify the second slow mode at all. Additionally, the worse-case scenario is if the RLS server itself gets attacked by a malware or physical malfunction resulting in the server source code to terminate abnormally. In such a scenario, the RLS algorithm can not be resumed anymore.

In contrast, a unique feature of the proposed distributed Prony-ADMM algorithm is that each local PDC can adopt the dual roles of *local estimation* and *central averaging*. For example, as shown in Figure 3, if the central server is deactivated by a malware, then the local PDCs can halt their estimation updates immediately, and coordinate with each other to select a candidate PDC among themselves that can act as a *pseudo* central server. If, for instance, local PDC 1 is selected as the pseudo server, then one *module*, say $\mathcal{M}_1$, inside this PDC will continue to implement the local optimization (19a)-(19b), while another module $\mathcal{M}_2$ will implement the averaging of $\mathbf{b}_i$ communicated to it from $\mathcal{M}_1$ as well as every other local PDC, to generate $\bar{\mathbf{b}}$. Thus, the algorithm can continue uninterruptedly as soon as all the PDCs agree on the choice of the pseudo server.

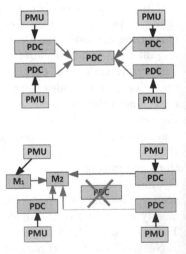

Fig. 3. Resiliency Strategy

For the next round, we assume that $\mathcal{M}_1$ has very high security firewalls and, therefore, is not *allowed* to be compromised, but the hacker has the choice of deactivating $\mathcal{M}_2$ similarly as the central server in the previous round. In that case, the PDCs will halt their estimation again, choose a different candidate for the pseudo-server among themselves, and reassign $\mathcal{M}_2$ to that chosen PDC

Table 1. Estimated Slow Eigenvalues for the 39-bus Power System

Actual Eigenvalues	Centralized RLS	Distributed Prony+ADMM
$-0.1993 \pm j3.1255$	$-0.197 \pm j3.125$	$-0.1966 \pm j3.1258$
$-0.6239 \pm j5.5644$	$-0.5113 \pm j5.5889$	$-0.5111 \pm j5.5773$
$-0.5112 \pm j6.1090$	$-0.3598 \pm j6.0868$	$-0.4932 \pm j6.0926$

Table 2. Accuracy Evaluation for Attack Scenarios

Actual Eigenvalues	Centralized RLS			Prony-ADMM
	1 PMU attacked	2 PMUs attacked	3 PMUs attacked	Server attacked
$-0.1993 \pm j3.1255$	$-0.1800 \pm j3.1258$	$-0.2615 \pm j3.1548$	$-0.3926 \pm j3.3219$	$-0.1963 + j3.1255$
$0.6239 \pm j5.5644$	$-0.6519 \pm j5.6453$	$-0.7269 \pm j5.5093$		$-0.5137 \pm j5.5872$
$-0.5112 \pm j6.1090$	$-0.2213 \pm j5.8828$	$-0.0682 \pm j6.4957$	$-0.6565 \pm j6.6205$	$-0.4944 \pm j6.0843$

to resume (19). Since none of the steps of the original proposed Prony-ADMM algorithm changes in these scenarios, the final estimate of β, and therefore of the modes, in this case is almost same as that for the unattacked case shown in the third column of Table 1. The fifth column of Table 2 testifies this fact.

6 Conclusion

In this paper we presented a distributed estimation algorithm for computing oscillation modes of power systems from Synchrophasor data, and illustrated its resiliency against component failures compared to conventional centralized techniques. Our future work will include the extension of these methods to closed-loop oscillation damping using delay-tolerant distributed Model Predictive Control, and to validate their resiliency properties using the two federated testbeds.

References

1. Chakrabortty, A., Khargonekar, P.: An Introduction to Wide-Area Control of Power Systems. In: American Control Conference, Washington, DC (2013)
2. Bakken, D., Bose, A., Hauser, C., Whitehead, D., Zweigle, G.: Smart Generation and Transmission with Coherent, Real-Time Data. Proc. of the IEEE (2011)
3. Zhou, N., Pierre, J.W., Hauer, J.F.: Initial Results in Power System Identification From Injected Probing Signals Using a Subspace Method. IEEE Transactions on Power Systems 21(3), 1296–1302 (2006)
4. Boyd, S., Parikh, N., Chu, E., Peleato, B., Eckstein, J.: Distributed Optimization and Statistical Learning via the Alternating Direction Method of Multipliers. Foundations and Trends in Machine Learning 3(1), 1–122 (2011)
5. The DETERLab, http://www.deter-project.org
6. Ljung, L.: System Identification: Theory for the User. Prentice Hall, NJ (1999)

Testing for Hardware Trojans: A Game-Theoretic Approach[*][**]

Charles A. Kamhoua, Manuel Rodriguez, and Kevin A. Kwiat

Air Force Research Laboratory, Information Directorate,
Cyber Assurance Branch, Rome, NY, USA
{charles.kamhoua.1,manuel.rodriguez-moreno.1.ctr,kevin.kwiat}@us.af.mil

Abstract. The microcircuit industry is witnessing a massive outsourcing of the fabrication of ICs (Integrated Circuit), as well as the use of third party IP (Intellectual Property) and COTS (Commercial Off-The-Shelf) tools during IC design. These issues raise new security challenges and threats. In particular, it brings up multiple opportunities for the insertion of malicious logic, commonly referred to as a *hardware Trojan*, in the IC. Testing is typically used along the IC development lifecycle to verify the functional correctness of a given chip. However, the complexity of modern ICs, together with resource and time limitations, makes exhaustive testing commonly unfeasible. In this paper, we propose a game-theoretic approach for testing digital circuits that takes into account the decision-making process of intelligent attackers responsible for the infection of ICs with hardware Trojans. Testing for hardware Trojans is modeled as a zero-sum game between malicious manufacturers or designers (*i.e.*, the attacker) who want to insert Trojans, and testers (*i.e.*, the defender) whose goal is to detect the Trojans. The game results in multiple possible mixed strategy Nash equilibria that allow to identify optimum test sets that increase the probability of detecting and defeating hardware Trojans in digital logic.

Keywords: Hardware Trojan, Cyber security, Game Theory, Functional Testing, Integrated Circuit.

1   Introduction

A hardware Trojan is a malicious modification of the circuitry of an Integrated Circuit (IC). A hardware Trojan is inserted into a main circuit at manufacturing or during design, and is mostly inactive until it is activated by a rare condition. When activated, it produces an error in the circuit, potentially leading to catastrophic consequences. The threat of serious, malicious IC alterations is of special concern to government agencies, military, finance, energy and political sectors.

The threat of hardware Trojans has become more pronounced due to today's massive outsourcing of the IC manufacturing processes (*fab-less model*), as well as from the increased reliance on hardware COTS (Commercial Off-The-Shelf) components. A good example of the latter are legacy military systems including aerospace and defense platforms, which are facing obsolescence due to their extended lifetime (e.g., often be-

---

[*] The rights of this work are transferred to the extent transferable according to title 17 § 105 U.S.C.
[**] Approved for Public Release; Distribution Unlimited: 88ABW-2014-2398, 19 MAY 2014.

R. Poovendran and W. Saad (Eds.): GameSec 2014, LNCS 8840, pp. 360–369, 2014.
© Springer International Publishing Switzerland 2014 (outside the US)

cause of budgetary decisions) and which rely on the use of COTS for the maintenance and replacement of their electronics. Most chip designers have now gone *fab-less*, outsourcing their manufacturing to offshore foundries. In doing so, they avoid the huge expense of building a state-of-the-art fab. Trust for Trojan-free fabrication of the chip is often placed upon foundries overseas. This gives many possibilities for potential attackers to maliciously alter the IC circuitry and insert hardware Trojans.

Functional testing of a digital system aims at validating the correct operation of the system with respect to its functional specification [1]. It commonly consists of generating inputs to the system and comparing the obtained outputs against a so-called *golden reference*. Testing of Integrated Circuits (IC) is usually carried out during the IC development cycle via *Automatic Test Pattern Generation* (ATPG). However, because of the stealthy nature of hardware Trojans, standard functional or ATPG testing is usually rendered as insufficient for detecting hardware Trojans. This is underscored in [2] by showing a test set for a simple digital circuit that detects all stuck-at-zero and stuck-at-one faults yet fails to detect the hardware Trojan. The specific application of functional testing to the detection of hardware Trojans has led to a number of approaches in the literature [2, 3, 4]. In addition, *Design for Testability* (DFT) techniques have been used by a number of authors to specifically support the testing of hardware Trojans [5, 6].

To efficiently detect hardware Trojans, testing techniques need to be specifically designed to target intelligent faults. While a number of testing approaches exist for hardware Trojans, they do not explicitly address the decision-making process of intelligent attackers responsible for infecting circuits with Trojans. Compared to the approaches cited above, our proposed testing methodology is built on a game-theoretic technique. To the best of our knowledge, this is the first work that adopts a game-theoretic approach to detect hardware Trojans. Our approach, based on a mixed strategy Nash equilibrium, captures the defender's and attacker's possible best responses that are used to enhance testing for hardware Trojans. The use of game theory in our approach allows for selecting optimum sets of verification tests that maximize the chances of detecting a hardware Trojan inserted in a circuit by an intelligent attacker. Unlike many other Trojan testing approaches, our methodology is independent of the circuit type (either combinational or sequential) and the lifecycle development phase in which the testing happens (either IC design or manufacturing).

From our study of the literature, only the work by Kukreja *et al.* [7] applies game theory to the domain of *testing*. The authors model software test scheduling as a Stackelberg game between testers and developers. The testers act as defenders commit to a testing strategy. Developers play the role of attackers who may check-in insufficiently tested code to complete application functionality sooner. They compute a strong Stackelberg equilibrium that provides the optimal probability distribution of the test cases to be selected for a randomized schedule.

However, none of the approaches described above on game theory have addressed the problem of hardware Trojans. As far as we know, our paper is the first application of game theory to the detection of hardware Trojans in digital circuits.

The main contribution of this paper is to propose the use of *game theory* to explicitly model the interactions between intelligent attackers and testers of digital circuits.

Our work represents the first application of a game-theoretic approach to the field of hardware Trojan detection. The proposed game-theoretic methodology helps identify optimum test sets that increase the probability of uncovering hardware Trojans. It also allows finding the optimum test conditions that can be used to force an attacker not to insert any Trojan so as to avoid severe penalties. In this respect, we formulate a non-cooperative game between malicious designers or manufacturers who seek to insert Trojans, and testers (who act as defenders) whose goal is to detect the Trojans. To solve this game, we analyze the Nash equilibrium point – representing the solution of the game – in which neither attacker nor defender has an incentive to change their strategy. The results show that the attacker is less likely to insert a high value (damage) Trojan because such high values Trojans are frequently tested by the defender. Moreover, a rational attacker is better off not to insert any hardware Trojans when the defender follows our testing procedure.

The paper is organized as follows. The game theoretic model is introduced in Section 2. Section 3 illustrates the methodology via numerical analysis of the game theoretic model. Finally, Section 4 concludes the paper.

2   Game Model

We consider an *attacker* who inserts a single hardware Trojan in a digital IC. To minimize detection, the attacker does not insert multiple Trojans. The attacker's strategy consists of inserting a Trojan from one of N possible classes. Thus, the attacker has N strategies that we denote $S_1, S_2, ..., S_N$.

We consider that Trojans from different classes have different impacts in the system and thus different values for the attacker. The attacker's values for the Trojan classes are denoted by $V_1, V_2, ..., V_N$ ($V_i \geq 0$). We consider that the game is zero-sum. Any win for the attacker is a loss to the defender and vice-versa.

We consider a defender who has limited resources to test and detect hardware Trojans. In other words, the defender can only perform a partial test on each circuit. The defender tests the hardware for a limited number k of Trojan classes, $k < N$. Therefore, the defender has $\binom{N}{k}$ possible strategies. We assume that a Trojan is detected if the corresponding class is included in the subset tested; otherwise the Trojan is not detected.

Further, we consider that there is a fine F ($F > 0$) that the attacker pays to the defender when a Trojan is detected. Unlike software attacks where attribution is difficult, in an IC attack, the hardware manufacturer is known and can be made responsible for Trojan infection by paying a fine. The new requirements from the Defense Logistics Agency (DLA) mandate that hardware manufacturers with DoD contracts must use a botanic DNA to mark their chips, which will increase the attribution reliability in hardware [8].

We consider the attacker and the defender to be both rational. The attacker inserts a Trojan that minimizes the likelihood of detection while the defender looks for a test set that maximizes the probability of detecting the Trojan. We assume that the test resources are common knowledge between the attacker and the defender. More precisely, both the attacker and the defender share the following knowledge:

- The defender's test size k
- The attacker's values of the Trojan classes $V_1, V_2, ..., V_N$
- The attacker's fine F

In this model, the defender's tests cannot be deterministic because a rational attacker would simply avoid inserting a Trojan whose class is part of the test procedure. We propose to find the distribution that is most likely to detect a Trojan and yields the maximum payoff to the defender. Given this model, we have a strategic non-cooperative game having two players $i.e.$, an attacker and a defender. The attacker's strategy is to choose a Trojan to insert in an IC while the defender selects a subset of Trojans classes to be tested. Both players' payoffs will depend on the value of the Trojan inserted and the detection fines imposed on the attacker. We will investigate the mixed-strategy Nash equilibrium as a solution to this game. As previously mentioned, within the IC testing game, the use of mixed-strategies is suitable since the defender has an incentive to randomize over the test cases. Moreover, the mixed-strategy Nash solution will allow us to find the frequency with which the defender and attacker will choose certain strategies at the equilibrium. In a nutshell, in this robust testing game approach, a mixed strategy Nash equilibrium is still well-founded to fight irrational attackers or those with limited knowledge because any sub-optimum action they take yields a benefit to the defender, $i.e.$, zero-sum.

3   Numerical Results

Without loss of generality, we consider a digital circuit with 4 input partitions ($N = 4$). This leads to four classes of Trojans, thus the attacker has 4 different strategies. We denote the attacker's strategies by A, B, C, D. For this illustration, the values of the attacker's strategies are: $V_A = 1, V_B = 2, V_C = 4, V_D = 12$. However, the results in this section can be generalized for different choice of values V. The current values are chosen with different orders of magnitude of Trojans' impact on a system.

The defender tests 2 of the 4 possible Trojan classes, $i.e.$, $k = 2$. Therefore, the defender has 6 possible strategies ($\binom{4}{2} = 6$). The defender's strategies are: AB, AC, AD, BC, BD, and CD, as represented in the normal form game in Table 1.

Table 1. Hardware Trojan detection game in normal form

		Defender					
		AB	AC	AD	BC	BD	CD
Attacker	A	-F, F	-F, F	-F, F	1, -1	1, -1	1, -1
	B	-F, F	2, -2	2, -2	-F, F	-F, F	2, -2
	C	4, -4	-F, F	4, -4	-F, F	4, -4	-F, F
	D	12, -12	12, -12	-F, F	12, -12	-F, F	-F, F

Note that our model is still valid and will work when considering any other set of values for N, V_i and k. The Nash equilibria are calculated using the game solver in [9]. Next, we will discuss the properties of the Nash equilibria and their implications on hardware testing. Subsection 3.1 elaborates on the mixed strategy Nash equilibria of the game in Table 1. Those mixed strategies will be used as the basis for testing hardware Trojans.

3.1 Mixed Strategy Nash Equilibrium

For this section, we set the fine $F = 8$. Therefore, $V_A < V_B < V_C < F < V_D$. We can verify that the game in Table 1 does not admit a pure strategy Nash equilibrium. If the defender announces that he is testing only Trojan classes C and D (playing pure strategy CD may be because C and D are the two most dangerous Trojan classes), an intelligent attacker will insert a Trojan from class B which will go undetected giving the best payoff, i.e., a payoff of 2 for the attacker (thus -2 for the defender). The defender's best response to an attacker inserting a Trojan of class B is to play AB, BC or BD so that the Trojan can be detected and the attacker pays fine F, i.e., a payoff of -8 for the attacker and 8 for the defender. If the defender plays BC, the attacker's best response is to play D and go undetected, i.e., a payoff of 12 for the attacker and -12 for the defender. This circular reasoning shows that the game in Table 1 does not admit a pure strategy Nash equilibrium.

However, the game admits three mixed strategy Nash equilibria (as calculated by the game solver from [9]) that we denote (1), (2) and (3).

$$\{0.323A + 0.29B + 0.242C + 0.145D; 0.355AD + 0.29BC + 0.129BD + 0.226CD\} \tag{1}$$

$$\{0.323A + 0.29B + 0.242C + 0.145D; 0.29AC + 0.065AD + 0.419BD + 0.226CD\} \tag{2}$$

$$\{0.323A + 0.29B + 0.242C + 0.145D; 0.29AB + 0.065AD + 0.129BD + 0.516CD\} \tag{3}$$

The attacker's strategy is the same for the three mixed strategy Nash equilibria. The mixed-strategy Nash solution for the attacker is shown in Fig. 1. On the other hand, the defender has three possible mixed strategy Nash equilibria represented in the clustered column of Fig. 2. In all the three Nash equilibria, the attacker's payoff is -2.193 while the defender's payoff is 2.193. The three Nash equilibria are payoff equivalent to each other. Both players are indifferent to which mixed strategy is used for testing hardware Trojans.

Fig. 1 shows a surprising result, i.e., the attacker's probability to insert a Trojan decreases with the value of the Trojan class. In other words, the attacker is less likely to insert a high value Trojan. This is because high values Trojans are heavily protected by the defender as shown in Fig. 2 and Fig. 3. A high value Trojan has more chances to get detected by a defender who adopts the strategic, game-theoretic testing procedure proposed here. Thus the attacker is better off inserting low value Trojans. This will result in the maximum possible benefit to the attacker who is taking into account the defender's best strategy.

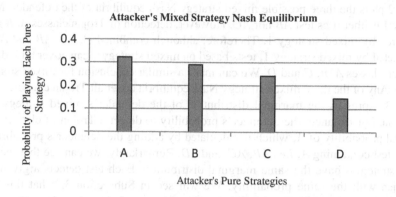

Fig. 1. Attacker's mixed strategy Nash equilibrium

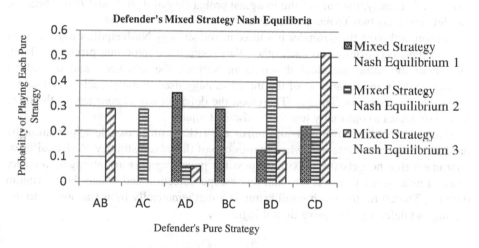

Fig. 2. Defender's mixed strategy Nash equilibria

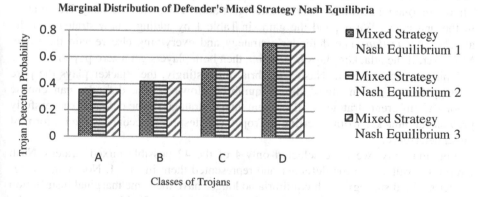

Fig. 3. Marginal distribution of the defender's mixed strategy Nash equilibria

Fig. 2 plots the three possible mixed strategy Nash equilibria of the defender. Mixed strategy 1 neither uses test *AB* nor *AC*. However, detection of Trojan classes *A*, *B* and *C* is covered by mixed strategy 1. Therefore, although combined classes *AB* and *AC* are not targeted by mixed strategy 1, tests based on mixed strategy 1 can cover all individual Trojan classes *A*, *B*, *C* and *D*. We can make a similar conclusion for mixed strategies 2 and 3. Any of the three mixed strategy Nash equilibria cover all Trojan classes.

Fig. 3 expresses the marginal distribution of the defender's mixed strategy Nash equilibria. For instance, the defender's probability to detect a Trojan of class *A* is the marginal probability of *A*, which is calculated by adding the defender's probability of using a test containing *A*, *i.e.*, *AB*, *AC*, and *AD*. Remarkably, we can see that the three mixed strategies have the same marginal distribution. Each test detects a given class of Trojan with the same probability. We will see in Subsection 3.2 that this result cannot be generalized. A Trojan with low impact, say of class *A*, is less often tested (35% of the times), while the high impact class-*D* Trojan is tested with a high probability, 71%. Finally, the sum of the marginal probability of *A, B, C* and *D* is 2 because the defender tests two Trojan classes at a time.

In short, although the defender has three mixed strategy Nash equilibria that appear to be completely different to each other, they share three important properties. First, they have the same marginal distribution. Second, the attacker's and defender's payoff are the same regardless of the mixed strategy used. Finally, each mixed strategy covers all classes of Trojans. Therefore, the defender can choose any of the three mixed strategies to optimally test for hardware Trojans.

Using the Nash equilibrium minimizes the risk of any possible exploitation by hardware Trojans. A defender who chooses one of the mixed strategy Nash equilibria, guarantees that he gets the maximum possible payoff against an intelligent attacker who, in turn, seeks to insert Trojans that escape detection and create the maximum damage. Therefore, the Nash equilibrium is a mathematically robust approach to detecting and defeating deceptive digital logic.

3.2   Considering *No Trojan* as an Attacker's Additional Strategy

We saw that the game in Table 1 yields a negative payoff to the attacker, *i.e.*, -2.193. Therefore, one may wonder why an intelligent attacker should insert any Trojan at all in the first place. We expand the game in Table 1 by adding a new strategy for the attacker, *No Trojan*. The defender's strategy and everything else remain the same. Moreover, if the attacker plays *No Trojan*, then both players get a zero payoff.

This new game has 42 Nash equilibria. Interestingly, the attacker plays the pure strategy *No Trojan* in all of those equilibria. However, the defender can choose among 42 different strategies, all of them mixed strategies. The defender successfully prevents an attacker from inserting a Trojan. Our testing procedure is then a form of cyber deterrence.

For simplicity, we have selected only 4 of the 42 possible mixed strategy Nash equilibria available to the defender and represented them in Fig. 4. Note that the defender's mixed strategy Nash equilibria no longer have the same marginal distribution as opposed to the result in Subsection 3.1 (Fig. 2 and Fig. 3). Moreover, none of the three mixed strategy Nash equilibria in Fig. 2 remain an equilibrium for the new game. Therefore, a defender who considers the *No Trojan* strategy as a plausible attacker's strategy, must completely change his test procedure.

However, the set of 42 possible tests cover all Trojan classes *A, B, C,* and *D*. Finally, both players get a payoff of zero in all those 42 equilibria. This is because the attacker always plays *No Trojan*. If the fine *F* was such that the attacker had a positive payoff in the game of Table 1, then the attacker could adopt a mixed strategy for this new game in which he would sometimes insert a Trojan in the hardware.

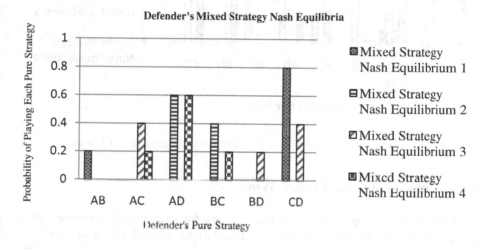

Fig. 4. Defender's mixed strategy Nash equilibria considering the *No Trojan* strategy

3.3   Considering *No Test* among the Defender's Strategy

The defender's goal for performing testing is to deter intelligent attackers from inserting hardware Trojans by making the attacks unprofitable. Therefore, a defender may want to conserve testing resources by letting some of the hardware go untested. This is the defender's *No Test* strategy. The attacker's strategy *No Trojan* remains an option in this subsection.

This new game has 62 Nash equilibria. As in Subsection 3.2, the attacker plays the pure strategy *No Trojan* in all of those equilibria. Moreover, the defender can choose among 62 different strategies, all of them mixed strategies. Notably, all of the 42 mixed strategy Nash equilibria in Subsection 3.2 remain as Nash equilibria for this new game. However, the defender has 20 new mixed strategies based on *No Test*. Fig. 5 shows 4 of the 20 new strategies. Both players still get a payoff of zero in all of those 62 equilibria. All Trojan classes are covered by the 62 equilibria.

Although the 62 tests have the same performance, the mixed strategy 1 of Fig. 5 will lead to the fastest test because it has the highest proportion of *No Test*. Using mixed strategy 1 as the test baseline, 37 % of the hardware does not need to be tested. This results in a test procedure that is 37 % faster than the 42 tests in Subsection 3.2.

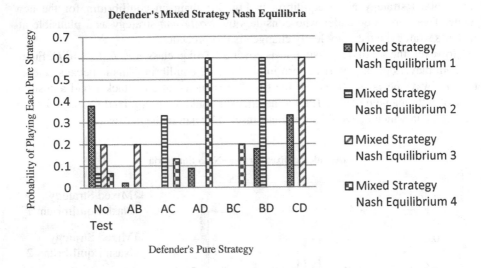

Fig. 5. Defender's mixed strategy Nash equilibria considering the *No Test* strategy

4   Conclusion and Future Work

We have proposed a game-theoretic methodology for analyzing the decision making processes involved in testing digital logic that may be infected with a hardware Trojan. The methodology allows for identifying optimum sets of test cases that take into account the attacker's decision-making, and that helps increase Trojan detection coverage. This is done via the calculation of the mixed strategy Nash equilibria that allow the tester of an IC to optimally combat Trojan attacks deployed by an intelligent attacker. The studied model also included the cases of: attackers who may decide not to insert a Trojan in the circuit to avoid the risk of being caught and paying a penalty; and testers who may decide not to test a subset of chips to conserve testing resources. The proposed game-theoretic approach leads to three important properties that improve testing: (i) it allows to save testing resources by identifying test sets with the minimum number of test cases; (ii) when manufacturing large volumes of the same IC design, it allows to find the minimal subset of produced chips to be tested; (iii) it provides means to find the conditions that deter the attacker from infecting circuits with Trojans.

An important result of our methodology is that it helps the defender find the conditions that make the attacker's payoff becomes negative and thus it discourages the attacker from inserting any Trojan at all, i.e., the *No Trojan* strategy (which is a pure strategy Nash equilibrium) becomes the best one for the attacker.

Future work will look into extensions for both the game theory model and the test procedure. For example, the game theoretic model can be extended to consider incomplete information games or lack of common knowledge between the attacker and the tester. We will also analyze the case in which a circuit infected with a Trojan does not provide reliable information about its manufacturer/designer (e.g., it has no botanic DNA) and thus the attacker cannot easily be identified. On the other hand, the test procedure can be extended with means for considering false positives/negatives and for estimating performance. Computational complexity of Nash equilibrium algorithms will be taken into consideration in regard of the size and complexity of the IC under test.

Acknowledgments. This research was performed while Dr. Manuel Rodriguez held a National Research Council (NRC) Research Associateship Award at the Air Force Research Laboratory (AFRL). This research was supported by the Air Force Office of Scientific Research (AFOSR).

References

[1] Lai, K.-W., Siewiorek, D.P.: Functional Testing of Digital Systems. In: Proc. of the 20th Design Automation Conference, DAC 1983 (1983)

[2] Wolff, F., Papachristou, C., Bhunia, S., Chakraborty, R.S.: Towards Trojan-Free Trusted ICs: Problem Analysis and Detection Scheme. In: Proc. of the 2008 Design, Automation and Test in Europe, DATE 2008 (2008)

[3] Jha, S., Jha, S.: Randomization Based Probabilistic Approach To Detect Trojan Circuits. In: Proc. of the 11th IEEE High Assurance Systems Engineering Symposium, HASE 2008 (2008)

[4] Chakraborty, R.S., Wolff, F., Paul, S., Papachristou, C., Bhunia, S.: MERO: A Statistical Approach for Hardware Trojan Detection. In: Clavier, C., Gaj, K. (eds.) CHES 2009. LNCS, vol. 5747, pp. 396–410. Springer, Heidelberg (2009)

[5] Salmani, H., Tehranipoor, M., Plusquellic, J.: New Design Strategy for Improving Hardware Trojan Detection and Reducing Trojan Activation Time. In: Proc. of the 2009 IEEE International Workshop on Hardware-Oriented Security and Trust, HOST 2009 (2009)

[6] Chakraborty, R.S., Paul, S., Bhunia, S.: On-demand transparency for improving hardware trojan detectability. In: Proc. of the 2008 IEEE International Workshop on Hardware-Oriented Security and Trust, HOST 2008 (2008)

[7] Kukreja, N., Halfond, W., Tambe, M.: Randomizing Regression Tests Using Game Theory. In: Proc. of the 2013 IEEE/ACM 28th Int. Conf. on Automated Software Engineering, ASE (2013)

[8] Howard, C.: Counterfeit Component Chaos. Military and Aerospace Electronics (12) (December 24, 2013)

[9] Savani, R.: Solve a Bimatrix Game, http://banach.lse.ac.uk

Surveillance for Security
as a Pursuit-Evasion Game

Sourabh Bhattacharya[1], Tamer Başar[2], and Maurizio Falcone[3]

[1] Department of Mechanical Engineering, Iowa State University, Ames, IA 50011, USA
sbhattac@iastate.edu
[2] Department of Electrical and Computer Engineering, University of Illinois at Urbana
Champaign, Urbana, IL 61801, USA
basar1@illinois.edu
[3] Dipartimento di Matematica, Universita di Roma "La Sapienza",
p. Aldo Moro 2, 00185 Roma, Italy
falcone@mat.uniroma1.it

Abstract. This work addresses a visibility-based target tracking problem that arises in autonomous surveillance for covert security applications. Consider a mobile observer, equipped with a camera, tracking a target in an environment containing obstacles. The interaction between the target and the observer is assumed to be adversarial in order to obtain control strategies for the observer that guarantee some tracking performance. Due to the presence of obstacles, this problem is formulated as a game with state constraints. Based on our previous work in [6] which shows the existence of a value function, we present an off-line solution to the problem of computing the value function using a Fast Marching Semi-Lagrangian numerical scheme, originally presented in [15]. Then we obtain the optimal trajectories for both players, and compare the performance of the current scheme with the Fully Discrete Semi-Lagrangian Scheme presented in [6] based on simulation results.

Keywords: pursuit-evasion games, semi-Lagrangian schemes, fast marching.

1 Introduction

Security is an important concern in infrastructure systems. Although advanced electronic and biometric techniques can be used to secure facilities reserved for military activities, vision-based monitoring is primarily used for persistent surveillance in buildings accessible to civilians. The idea is to cover the environment with cameras in order to obtain sufficient visual information so that appropriate measures can be taken to secure the area in case of any suspicious activity. However, the number of static cameras needed to cover and monitor activities in a moderately sized building is substantial, and this leads to fatigue in security personnel. In this work, we explore a scenario in which mobile agents that can visually track entities in the environment are deployed in a surreptitious manner for surveillance applications. This gives rise to a problem that is often called the *target tracking* problem.

R. Poovendran and W. Saad (Eds.): GameSec 2014, LNCS 8840, pp. 370–379, 2014.

Target tracking refers to the problem of tracking a mobile object, called a *target*. Based on the sensing modality and sensing constraints, there is a range of problems that can be addressed under this category. In this work, we assume that the autonomous observer is equipped with a vision sensor for tracking the target. The environment contains obstacles that occlude the view of the target from the observer. The goal of the observer is to maintain a persistent line-of-sight with the target. Therefore, the mobile observer has to control its motion, keeping in mind the sensing constraints and the motion constraints posed by the obstacles. In order to compute motion strategies for the observer that can provide some performance guarantees, the target is assumed to be an adversary. Several variants of the target-tracking problem have been considered in the past that consider constraints in motion as well as sensing constraints for both agents. For an extensive discussion regarding the previous work and its applications, we refer to [13,12]. In this work, we consider the target tracking problem without any constraints in sensing or motion models for both agents except for those posed by the obstacles present in the environment.

Past efforts to provide a solution to the aforementioned problem can be primarily divided into two categories: (1) Formulating the problem as a game of kind, and providing necessary conditions for pursuit and evasion in the presence of polygonal obstacles [13,10,11]; (2) Formulating the problem as a game of degree, and using the theory of differential games to provide necessary and sufficient conditions for pursuit [12,14,7,9]. Although, the structure of optimal solutions has been characterized extensively in previous works, a complete construction of the solution in a general environment containing polygonal obstacles is still open. In [8], the authors analyze the problem in a simple environment containing a circular obstacle, and characterize the optimal trajectories near termination using differential game theory. In [6], we use a semi-Lagrangian iterative numerical scheme to provide a solution to the aforementioned problem. In this work, we use another numerical technique, *Fast Marching Semi-Lagrangian scheme*, based on the ideas of front propagation to provide an off-line solution to the problem. The numerical techniques introduced in this work can be used for any 2-player generalized pursuit-evasion game with state constraints.

Numerical techniques for games are primarily based on the principles of Dynamic Programming (DP). Finite differences approximation schemes based on generalized gradients were proposed by Tarasyev[20] who also considered the problem of the synthesis of optimal controls using approximate values on the finite grid. Convergence results to the value function of the generalized pursuit-evasion games for the approximation scheme based on Discrete Dynamic Programming (also called semi-Lagrangian scheme) were first presented in [4], under either continuity assumptions on the value function or for problems with a single player (i.e. control problems). The extension of the scheme and of the convergence theorem to the discontinuous case was obtained in [2]. Later these results have been extended to pursuit-evasion games with state constraints in [5,16]. Our work is in a similar vein, and uses the fully discrete scheme proposed in the aforementioned works to address the target tracking game. For a general introduction to semi-Lagrangian schemes and their applications in control and game problems, we refer to [18].

The paper is organized as follows. In Section 2, we present the problem formulation, and address the issue of existence of the value function for our problem setting. In Section 3, we reduce the dimensionality of the problem by reformulating it in relative coordinates. In Section 4, we present the numerical scheme. In Section 5, a comparison of the different schemes is presented based on simulation results. Finally, Section 6 includes some concluding remarks.

2   Problem Statement

In this section, we present the problem formulation (see Figure 1(a)). Consider a circular obstacle in the shape of a disc of radius a_1 in the plane enclosed inside a concentric circular boundary of radius a_2. The centers of both circles are assumed to be at the origin of the reference frame. Consider a mobile observer and a target in the plane. Each agent is assumed to be a point in the plane. Let $\mathbf{y} \in \mathbb{R}^2$ and $\mathbf{z} \in \mathbb{R}^2$ denote the coordinates of, respectively, the observer and the target in the plane. Both agents are assumed to be

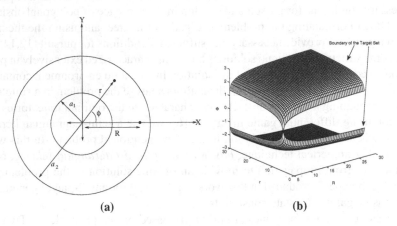

(a)   (b)

Fig. 1. Figure (a) shows the geometry around a circular obstacle with a circular boundary. Figure (b) shows the boundary of the terminal manifold of the game in relative coordinates.

simple kinematic agents, and their motions are governed by the following equations

$$\dot{\mathbf{y}} = \mathbf{u}_1, \quad \dot{\mathbf{z}} = \mathbf{u}_2$$

subject to the constraints $\mathbf{y} \in K_U$, $\mathbf{z} \in K_V$ where

$$K_U \equiv \{\mathbf{y} \in \mathbb{R}^2 : (\|\mathbf{y}\|_2^2 - a_1^2)(\|\mathbf{y}\|_2^2 - a_2^2) \le 0\}, \; K_V \equiv \{\mathbf{z} \in \mathbb{R}^2 : (\|\mathbf{z}\|_2^2 - a_1^2)(\|\mathbf{z}\|_2^2 - a_2^2) \le 0\}$$

Let $\mathbf{x} = (\mathbf{y}, \mathbf{z})^T$ and $f(\mathbf{x}, \mathbf{u}_1, \mathbf{u}_2) = (\mathbf{u}_1, \mathbf{u}_2)^T$. The controls $\mathbf{u}_1(\cdot)$ and $\mathbf{u}_2(\cdot)$ belong to the following sets

$$\mathbf{u}_1(\cdot) : \mathbb{R} \to U, \quad U = B_1(0,0), \quad \mathbf{u}_2(\cdot) : \mathbb{R} \to V, \quad V = B_\mu(0,0)$$

where $B_r(a)$ is a ball of radius r with center a, and μ is a parameter which represents the maximum speed of the target. We will see later that we have to pick $\mu \leq 1$ to make the problem meaningful. The line-of-sight between the pursuer and the evader is defined as the line joining the two players on the plane. The line-of-sight is considered to be broken if it intersects with the circular obstacle. In order to account for the worst-case scenario, the target is assumed to be adversarial in nature. Therefore, the interaction between the observer and the target is modeled as a game. The observer is assumed to be the pursuer, and the target is assumed to be the evader. The objective of the pursuer is to maximize the time for which it can continuously maintain a line of sight to the evader. The objective of the evader is to break the line-of-sight in the minimum amount of time. The game terminates when the line-of-sight between the pursuer and the evader is broken. The problem is to compute the strategies of the players as a function of their positions. Since this is a 2-player zero-sum game [1], we use the concept of *saddle-point equilibrium* [12] to define the optimal strategy for each player.

Let $T(\mathbf{x_0})$ denote the optimal time of termination of the game when the players start from the initial position $\mathbf{x_0}$. A strategy for a player will be defined as a map from the control set of the opponent to its own control set, with some informational constraints imposed, as appropriate. Let α and β denote the strategies of the pursuer and the evader, respectively. A pair of strategies (α^*, β^*) for the two players is said to be in saddle-point equilibrium if the following pair of inequalities is satisfied

$$T(\mathbf{x_0}; \alpha^*, \beta) \geq T(\mathbf{x_0}; \alpha^*, \beta^*) \geq T(\mathbf{x_0}; \alpha, \beta^*) \qquad \forall \alpha, \beta \text{ admissible}$$

(here we write explicitly the dependence of T on the strategies). If the pair (α^*, β^*) exists, then the function $T^*(\mathbf{x_0}) = T(\mathbf{x_0}; \alpha^*, \beta^*)$ is called the value of the game and T^* is called the *value function*. The existence of the value function depends on the class of strategies under consideration for both the players. In this work, the notion of *non-anticipating strategies* [17] will be used to define the information pattern between the players.

Definition: A strategy α for player P is non-anticipating if $\alpha \in \Gamma$, where

$$\Gamma = \{\alpha : V \to U \mid b(t) = \tilde{b}(t), \forall t \leq t' \text{ and } b(t), \tilde{b}(t) \in V \Rightarrow \alpha[b](t) = \alpha[\tilde{b}](t), \forall t \leq t'\}$$

Similarly, we can define a non-anticipating strategy $\beta \in \Delta$ for E, where

$$\Delta = \{\beta : U \to V \mid a(t) = \tilde{a}(t), \forall t \leq t' \text{ and } a(t), \tilde{a}(t) \in V \Rightarrow \beta[a](t) = \beta[\tilde{a}](t), \forall t \leq t'\}$$

Frequently, in problems involving games and optimal control, it is the case that the value function ceases to exist in the class of strategies used by the players. In [6], we show that the value of the game exists. Since the existence of the value function is established from the above transversality conditions, we can address the problem of computing it.

3 Dimensionality Reduction

In this section, we present a formulation of the problem in reduced coordinates where we exploit the symmetry of the problem in order to reduce dimensionality. To this end, we formulate the problem in polar coordinates. We express the position of the players in relative coordinates. Let the polar coordinates of the pursuer and the evader be denoted as (r_p, θ_p) and (r_e, θ_e), respectively. Instead, we can use the relative coordinates $(R = r_p, r = r_e, \phi = (\theta_p - \theta_e))$ to define the state of the game. The equations of motion of the two players in relative coordinates are given by the following

$$f_R = \dot{R} = u_{r_p}; \quad f_r = \dot{r} = u_{r_e}; \quad f_\phi = \dot{\phi} = \frac{u_{\theta_e}}{r} - \frac{u_{\theta_p}}{R}, \tag{1}$$

where (u_{r_p}, u_{θ_p}) and (u_{r_e}, u_{θ_e}) are the radial and tangential components of the velocities of the pursuer and the evader, respectively, and satisfy the following constraints

$$u_{r_p}^2 + u_{\theta_p}^2 \leq 1; \quad u_{r_e}^2 + u_{\theta_e}^2 \leq \mu^2 \tag{2}$$

The problem statement dictates that $a_1 \leq R, r \leq a_2$ and $-\pi \leq \phi \leq \pi$. The problem is to determine the time of termination of the game, and the optimal strategies of the individual players given the initial position $\mathbf{x} = (r, R, \phi)$ of the pursuer and the evader:

$$(u_{r_p}^*, u_{\theta_p}^*, u_{r_e}^*, u_{\theta_e}^*) = \arg \max_{u_{r_p}, u_{\theta_p}} \min_{u_{r_e}, u_{\theta_e}} T(\mathbf{x}; u_{r_p}, u_{\theta_p}, u_{r_e}, u_{\theta_e}) \tag{3}$$

The existence of the value function was established in [6], as indicated in the previous section, and hence the max and min operations commute in the above equation. Since the evader always wins from any given initial position of the players for $\mu > 1$, we only consider the case $\mu \leq 1$. The winning strategy of the evader for $\mu > 1$ is to move along the boundary of the obstacle with its maximum speed in a fixed direction. Based on the problem formulation, the game terminates when the line-of-sight between the pursuer and the evader intersects with the circular obstacle. Therefore, the boundary of the terminal manifold is given by the set of states for which the line-of-sight between the pursuer and the evader is tangent to the circular obstacle.

Figure 1(b) shows the boundary of the terminal manifold in relative coordinates for $a_1 = 5$ and $a_2 = 30$. The line-of-sight is in the free space only if the state of the players lies between the two symmetric surfaces. Otherwise, the game has terminated. The set of states for which the line-of-sight intersects the obstacles is also the target set, denoted as $\mathcal{T}$. The objective of the evader is to drive the state of the system to the target set. The objective of the pursuer is to prevent the state from reaching it. Let $\mathcal{R}$ denote the reachable set, i.e., the set of initial points from which it is possible for the evader to drive the state of the system to the target set in finite time irrespective of the pursuer's control action. One can clearly see that $\mathcal{R}$ depends on $\mathcal{T}$ and the dynamics of the players.

We have the following result from [4].

Theorem 1. *If $\mathcal{R} \setminus \mathcal{T}$ is open, and $T \in C^0(\mathcal{R} \setminus \mathcal{T})$, then $T(\cdot)$ is a viscosity solution of the following equation:*

$$\min_{a \in U} \max_{b \in V} \{-f(\mathbf{x}, \mathbf{a}, \mathbf{b}) \cdot \nabla T(\mathbf{x})\} - 1 = 0, \quad \mathbf{x} \in \mathcal{R} \setminus \mathcal{T} \tag{4}$$

Let $v(\mathbf{x})$ denote the *Kružkov transform* [3] of $T(\mathbf{x})$

$$v(\mathbf{x}) = \begin{cases} 1 - e^{-T(\mathbf{x})} & \text{if } T(\mathbf{x}) < +\infty \quad (\mathbf{x} \in \mathcal{R}) \\ 1 & \text{if } T(\mathbf{x}) = +\infty \quad (\mathbf{x} \notin \mathcal{R}) \end{cases} \tag{5}$$

Since $T(\mathbf{x})$ takes values in the interval $[0, \infty)$, $v(\mathbf{x})$ takes values in the interval $[0, 1]$. Using $v(\mathbf{x})$ instead of $T(\mathbf{x})$ leads to better numerical schemes due to the bounded values of $v(\mathbf{x})$. Moreover, there is a bijective map between $v(\mathbf{x})$ and $T(\mathbf{x})$ given by the following:

$$T(\mathbf{x}) = -ln(1 - v(\mathbf{x}))$$

In terms of $v(\mathbf{x})$, the reachable set is given by the following expression

$$\mathscr{R} = \{\mathbf{x} | v(\mathbf{x}) < 1\}$$

Therefore, we address the problem of computing $v(\mathbf{x})$ numerically in the following sections. If $v(\mathbf{x})$ is continuous, then it is the unique viscosity solution of the following Dirichlet problem [4]

$$\begin{cases} v(\mathbf{x}) + \min_{\mathbf{a} \in U} \max_{\mathbf{b} \in V} \{-f(\mathbf{x}, \mathbf{a}, \mathbf{b}) \cdot \nabla v(\mathbf{x})\} - 1 = 0, & \text{for } \mathbf{x} \in \mathbb{R}^n \setminus \mathscr{T} \\ v(\mathbf{x}) = 0 & \text{for } \mathbf{x} \in \partial \mathscr{T} \end{cases}$$

4 Numerical Scheme

First, we describe the discretization of the state space. The entire state space $\mathbf{X}(\mathbb{R}^3)$ is discretized by constructing a three dimensional lattice of cubes with edge lengths k. The lattice points are placed at the corners of cubes with the origin as one of the lattice points. The numerical scheme computes the approximation of $v(\mathbf{x})$ at the lattice points. Let Q denote a closed and bounded subset of $\mathbf{X}$ containing the entire free space including the obstacles. Once the state space is discretized, we are only concerned with values of v at those lattice points which belong to Q. We will call these lattice points as *nodes*. Let the nodes be ordered as $\{1, \ldots, N\}$, where N is the number of nodes in Q. Let $(\mathbf{x}_1, \ldots, \mathbf{x}_N)$ denote the state of the nodes in Q. Let $I_\mathscr{T}$ denote the set of nodes in Q that belong to the target set. The values of these nodes are set to zero since the game would already have terminated if it started from any of these nodes. Therefore, if $x_i \in I_\mathscr{T}$, $T_{\mathbf{x}_i} = 0$, which implies $v(\mathbf{x}_i) = 0$. We arrange the values of v at all the nodes in the form of a vector $V = (V_1, \ldots, V_N)$. The solution is usually obtained via a fixed point iteration $V^{n+1} = SV^n$ starting from a given V^0 [18].

In the Fast Marching Method (FMM), the state space is initially discretized in a manner described in the previous paragraph. At every instant of time, the nodes are divided into the following three groups. The *accepted nodes* are those where the solution has already been computed, and it cannot change in the subsequent iterations. The *narrow band nodes* are those where the computation actually takes place, and their values can change in the subsequent iterations. The *far nodes* are those in the space where an approximate solution has never been computed. The front in our problem represents the surface that updates the initial value of $v(\mathbf{x}_i)$ at node i to its approximate value as it propagates in the state space. The accepted region represents the nodes in the state space through which the front has already passed. The narrow band represents the nodes in the region around the current position of the front where the values are being updated. The far region represents the nodes where the front has not yet passed.

The algorithm initializes by labeling all the nodes in the target set as accepted nodes. In order to compute the narrow band nodes, we need to first define the concept of *reachable sets*. The reachable set at any iteration is defined as the set of nodes from which the pursuer can drive the state of the system to a node that belongs to the accepted set irrespective of the controls of the evader. A sketch of the algorithm is given below:

1. The nodes belonging to the target set $\mathcal{T}$ are located and labeled as *accepted*, setting their values to $v(x) = 0$. All other nodes are set to $v(x) = 1$ and labeled as *far*.
2. The initial *narrow band* is defined as the set of all the neighbors of the accepted nodes. Their values are valid only if they are in the reachable set.
3. The node in the narrow band with the minimal valid value is accepted, and it is removed from the narrow band.
4. Neighbors of the last accepted node that are not yet accepted are computed and inserted in the narrow band. Their values are valid only if they are in the reachable set.
5. If the narrow band is not empty, the next iteration starts at step 3.

The complete algorithm is given in the table below as Algorithm 1.

Algorithm 1. FMSL

1: **declare** $\mathcal{D}_{Accepted}$, $\mathcal{D}_{NarrowBand}$, $\mathcal{D}_{Far}$ be the sets of accepted nodes, narrow band nodes and far nodes
2: **for each** $x_i \in \mathcal{D}$ **do**
3: **if** $x_i \in \mathcal{T}$ **then**
4:   $V_{x_i} = 0$ and $x_i \in \mathcal{D}_{Accepted}$
5: **else**
6:   $V_{x_i} = 1$ and $x_i \in \mathcal{D}_{Far}$
7: **end if**
8: **end for**
9: $\mathcal{D}_{NarrowBand} = \{x_j | x_j \in \bigcup_{x_i \in \mathcal{D}_{Accepted}} N(x_i) \cap \mathcal{R}^h\}$
10: **while** $\mathcal{T}_{NarrowBand} \neq \emptyset$ **do**
11: **if** $x_k = \arg\min V(x_j)$ **then**
12:   Remove x_k from $\mathcal{D}_{NarrowBand}$ and add it to $\mathcal{D}_{Accepted}$
13:   Add $N(x_k) \cap \mathcal{R}^h$ to $\mathcal{D}_{NarrowBand}$
14: **end if**
15: **end while**

From [19], it is well known that the performance of FM deteriorates rapidly when the characteristic and the gradient lines do not coincide. In order to overcome this limitation, the Buffered Fast Marching Method (BFMM) was introduced in [15]. BFMM is an amalgamation of SL and FM methods that retains the advantages of both techniques. In BFMM, in addition to the accepted nodes, narrow band and far nodes, we have a *buffer* zone. Every iteration of BFMM starts with the implementation of the FM scheme. Once the nodes having the least value in narrow band are computed, they are moved to buffer. All the nodes in the buffer are recomputed using the Fully Discrete Semi-Lagrangian scheme for two different initial boundary conditions of the nodes. In the first step, the values of all the nodes in the narrow band are set to 1. In the second step, the values of all the nodes in the narrow band are set to 0. If there is any node in the buffer for which the value remains unchanged with two different boundary conditions, then the node is considered to be accepted.

In the next section, we present some numerical results obtained from the FMM and BFMM.

5   Results

In this section, we present simulation results, and compare it with our previous results in [6]. All the simulations were performed on a Core 2 Duo P7450 processor. The radii of the inner and outer obstacles are $a_1 = 1$ and $a_2 = 10$, respectively. The speed of the evader is set at 0.8 for all simulations. Figure 2 depicts the value function for all the three numerical schemes, and trajectories of the players for a specific initial position. Figure 2(d) shows the trajectories of the players computed from the Fully Discrete Semi-Lagrangian technique presented in [6]. Figures 2(e) and 2(f) show the trajectories of the players from the Fast Marching techniques proposed in this work. Figure 3 illustrates the variation of the performance of the three techniques on the basis of the computational time and capture time with respect to the grid size. Figure 3(a) shows the time expended to compute the value functions for the three different techniques as the grid size increases. We can see that for a fixed grid size the iterative scheme takes more

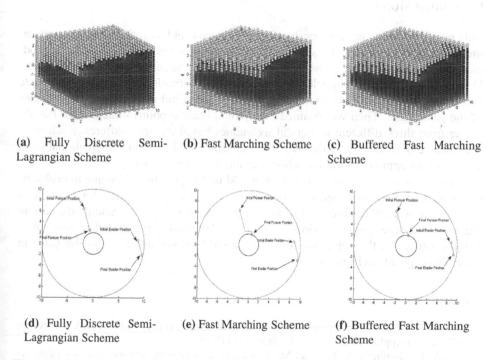

(a) Fully Discrete Semi-Lagrangian Scheme

(b) Fast Marching Scheme

(c) Buffered Fast Marching Scheme

(d) Fully Discrete Semi-Lagrangian Scheme

(e) Fast Marching Scheme

(f) Buffered Fast Marching Scheme

Fig. 2. The figure shows variation of the value function computed at the nodes, and the trajectories of the players for the three techniques

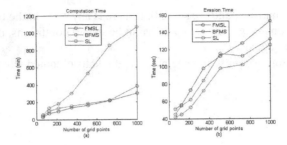

Fig. 3. Figure (a) shows a plot of the computational time required to compute the $v(x_i)$ using the three different techniques. Figure (b) shows the variation of the time required for the target to escape with increasing number of grid points used for computation.

time to compute the value function as compared to the FM schemes. Moreover, the results clearly show that the time required for computation of the value function increases as the grid resolution becomes finer. Figure 3(b) shows the variation of the termination time for the game for a fixed trajectory of the target using the three techniques. One can clearly see that the SL scheme is expensive in terms of computational time compared to the other two techniques.

6   Conclusions

This work has addressed a vision-based surveillance problem for securing an environment. The task of keeping a suspicious target in the observer's field-of-view was modeled as a pursuit-evasion game by assuming that the target is adversarial in nature. Due to the presence of obstacles, this problem was formulated as a game with state constraints. We first showed that the value of the game and the saddle-point strategies of the game exist. Then we obtained the optimal (saddle-point) strategies for the observer from three different numerical techniques based on finite-difference schemes. The relative performance of the three different schemes based on computational time, and degree of approximation was illustrated through simulations.

An immediate extension of this work would be to apply the technique to problems that have non-holonomic agents having more complicated dynamics, for example, a Dubin's car or a differential drive robot. We are also working on extending the current technique to more general environments. A fundamental question that remains open is the existence of the value function and the saddle-point strategies for the game in general polygonal environments.

References

1. Başar, T., Olsder, G.J.: Dynamic Noncooperative Game Theory, 2nd edn. SIAM Series in Classics in Applied Mathematics, Philadelphia (1999)
2. Bardi, M., Bottacin, S., Falcone, M.: Convergence of discrete schemes for discontinuous value functions of pursuit-evasion games. In: New Trends in Dynamic Games and Applications, pp. 273–304. Springer (1995)

3. Bardi, M., Capuzzo-Dolcetta, I.: Optimal control and viscosity solutions of Hamilton-Jacobi-Bellman equations. Springer (2008)
4. Bardi, M., Falcone, M., Soravia, P.: Fully discrete schemes for the value function of pursuit-evasion games. Advances in Dynamic Games and Applications 1, 89–105 (1994)
5. Bardi, M., Koike, S., Soravia, P.: Pursuit-evasion games with state constraints: Dynamic programming and discrete-time approximations. Discrete and Continuous Dynamical Systems 6(2), 361–380 (2000)
6. Bhattacharya, S., Başar, T., Falcone, M.: IEEE Conference on Intelligent Robots and Systems (to appear, 2014)
7. Bhattacharya, S., Başar, T., Hovakimyan, N.: Singular surfaces in multi-agent connectivity maintenance games. In: 50th IEEE Conference on Decision and Control and European Control Conference (CDC-ECC), pp. 261–266 (2011)
8. Bhattacharya, S., Başar, T., Hovakimyan, N.: Game-theoretic analysis of a visibility based pursuit-evasion game in the presence of a circular obstacle. In: AIP Conference Proceedings, vol. 1479, p. 1222 (2012)
9. Bhattacharya, S., Basar, T., Hovakimyan, N.: On the construction of barrier in a visibility based pursuit evasion game. In: European Control Conference (ECC), pp. 1894–1901. IEEE (2014)
10. Bhattacharya, S., Candido, S., Hutchinson, S.: Motion strategies for surveillance. In: Robotics: Science and Systems (2007)
11. Bhattacharya, S., Hutchinson, S.: Approximation schemes for two-player pursuit evasion games with visibility constraints. In: Robotics: Science and Systems (2008)
12. Bhattacharya, S., Hutchinson, S.: On the existence of Nash equilibrium for a two player pursuit-evasion game with visibility constraints. International Journal of Robotics Research 29(7), 831–839 (2010)
13. Bhattacharya, S., Hutchinson, S.: A cell decomposition approach to visibility-based pursuit evasion among obstacles. International Journal of Robotics Research 30(14), 1709–1727 (2011)
14. Bhattacharya, S., Hutchinson, S., Başar, T.: Game-theoretic analysis of a visibility based pursuit-evasion game in the presence of obstacles. In: Proceedings of American Control Conference, St. Louis, MO, pp. 373–378 (June 2009)
15. Cristiani, E.: A fast marching method for Hamilton-Jacobi equations modeling monotone front propagations. Journal of Scientific Computing 39(2), 189–205 (2009)
16. Cristiani, E., Falcone, M.: Numerical solution of the Isaacs equation for differential games with state constraints. In: 17th IFAC World Congress, vol. 17, pp. 11352–11356 (2008)
17. Elliott, R.J., Kalton, N.J.: The existence of value in differential games Number 1-126. AMS Bookstore (1972)
18. Falcone, M., Ferretti, R.: Semi-Lagrangian Approximation Schemes for Linear and Hamilton Jacobi Equations. SIAM (2014)
19. Sethian, J.A., Vladimirsky, A.: Ordered upwind methods for static Hamilton–Jacobi equations: Theory and algorithms. SIAM Journal on Numerical Analysis 41(1), 325–363 (2003)
20. Taras'yev, A.M.: Approximation schemes for constructing minimax solutions of Hamilton-Jacobi equations. Journal of Applied Mathematics and Mechanics 58(2), 207–221 (1994)

Author Index